Nanotechnology

Nanotechnology
Principles and Applications

edited by

Rakesh K. Sindhu
Mansi Chitkara
Inderjeet Singh Sandhu

JENNY STANFORD
PUBLISHING

Published by

Jenny Stanford Publishing Pte. Ltd.
Level 34, Centennial Tower
3 Temasek Avenue
Singapore 039190

Email: editorial@jennystanford.com
Web: www.jennystanford.com

British Library Cataloguing-in-Publication Data
A catalogue record for this book is available from the British Library.

Nanotechnology: Principles and Applications

ISBN 978-981-4877-43-5 (Hardcover)
ISBN 978-1-003-12026-1 (eBook)

Contents

Preface

The 21st century has witnessed an enormous increase in interest in the field of nanoscience and nanotechnology. The community of nanoscientists has been focusing more and more on nanotechnology-based applications. The study of nanoscience as an interdisciplinary subject in academic institutions also holds the key to its success. Because of its multidisciplinary nature, nanotechnology has brought together scientists and technologists from all over the world to find solutions to many global problems to benefit humankind.

This book takes a comprehensive approach toward the understanding of nanotechnology's principles and applications. Experts and researchers from all over the globe working in this field have contributed their findings in this book.

We hope that this book will provide useful scientific insights to students and will prove to be a suitable companion to them as they take their first steps into the marvellous world of nanotechnology.

Finally, we would like to express our sincere thanks to all the contributing authors and colleagues without whose active participation it would not have been possible to bring out this book.

We also thank the team at Jenny Stanford Publishing for sharing their expertise to bring out this book.

Rakesh K. Sindhu
Mansi Chitkara
Inderjeet Singh Sandhu
January 2021

Chapter 1

Introduction and Basics of Nanotechnology

Anjali Saharan,[a] Pooja Mittal,[a] Kashish Wilson,[a] and Inderjeet Verma[b]
[a]MM School of Pharmacy, Maharishi Markandeshwar University, Sadopur-Ambala, India
[b]MM College of Pharmacy, MM(DU), Mullana-Ambala, India
inderjeet.verma@mmumullana.org

Nanotechnology is a broad term that includes a variety of technologies and materials involving a vast variation of particle size helpful in different aspects of technology and research. In terms of technology and science, nanotechnology deals with smaller (nano) sized particles. The size of the created nanomaterials ranges from 1 to 500 nm, which allows the exploration of materials at the nanoscale. Nanomaterials have gained eminence in the field of advanced technologies because of their alterable physical, biological, and chemical properties. Their uniqueness is due to their higher surface-area-to-volume ratio, which is responsible for their numerous advantages over conventional materials, such as higher optical clarity, higher entrapment of the material enclosed, and

Nanotechnology: Principles and Applications
Edited by Rakesh K. Sindhu, Mansi Chitkara, and Inderjeet Singh Sandhu
Copyright © 2021 Jenny Stanford Publishing Pte. Ltd.
ISBN 978-981-4877-43-5 (Hardcover), 978-1-003-12026-1 (eBook)
www.jennystanford.com

hence faster action, which make them suitable for various industrial applications. All these exceptional and unique features of these particles lead to their demand in various commercial and research facilities. Nanotechnology has grabbed its roots in various fields such as in the food industry, fertilizers, biosensors, nanoelectronics, biopharmaceuticals, and biopolymers. This chapter will explore the concept of nanotechnology with a brief overview of the method of production and applications of nanomaterials in various fields.

1.1   Introduction

Nanotechnology and nanoscience have spread roots in various product domains during the past few years. Nanotechnology helps in the development of materials along with their clinical applications where traditional techniques are limited in the desired criteria. The word "nano" has been used as a prefix in a variety of reports for the last few decades with vast applications in technology [1]. The word nanotechnology was formed using two terms, the first being "nano," which comes from the Greek term "nanos" meaning dwarf, and the second being "technology." Nanotechnology involves the use and knowledge of tools, machines, and techniques that promote research along with problem solving or specific function performance. Nanotechnology is defined as the study of controlling or manipulating matter on an atomic or molecular scale [2]. The structure range of nanoparticles varies between 1 and 100 nm in one dimension [3]. As from all such specifications of this technology, humans were the first to have their role in all fields of nanotechnology [4]. In research also, it gets expanded to two prominent areas— nanoparticles (NPs) and nanostructured materials (NSMs). These factors lead to greater contribution in the fields of research and technoeconomic world with a great distinction in technology due to their variant physicochemical characteristics such as melting point, wettability, thermal and electrical conductivity, light absorption, catalytic activity, and scattering, resulting in enhanced performance over their bulk counterparts [5]. Different evaluation parameters of nanotechnology are provided in different standards of measurement bureaus, such as nanometer (nm), which is an SI (International System of Units) unit (10^{-9} m in length). Different legislations such as the European Union (EU) and the United States postulated different definitions. Thus, there is diverse variation till date and no single internationally accepted definition for nanomaterials (NMs) exists.

A few definitions are as follows:

1. According to the **Environmental Protection Agency** (EPA), "Nano materials can exhibit unique properties dissimilar than the equivalent chemical compound in a larger dimension" [6].

2. The **US Food and Drug Administration** (USFDA) also refers to NMs as "materials that have at least one dimension in the range of approximately 1 to 100 nm and exhibit dimension-dependent phenomena." The International Organization for Standardization (ISO) has described NMs as a "material with any external nanoscale dimension or having internal nanoscale surface structure" [6].

3. The **European Commission** (EU) describes the term nanomaterial as "a manufactured or natural material that possesses unbound, aggregated or agglomerated particles where external dimensions are between 1–100 nm size range."

As mentioned, diverse definitions have led to a major hurdle for nanomaterials in the field of regulatory efforts to lead and implement a legal hesitation in applying regulatory approaches to all. To satisfy the diverging needs, an international definition should be coined to mimic the differences and for smooth implementation of regulatory provisions.

The term nanoparticle is coined from the relative bonding of two words: nanospheres and nanocapsules. Nanospheres are defined as matrix-like cage, which encapsulates the drug uniformly, while in nanocapsules, drug is surrounded by a unique polymeric membrane. Nanotechnology and nanoscience studies have emerged rapidly during the past years in a broad range of product domains [7]. These approaches have led to their vast applications in the fields of conventional medicine. Nanotechnology provides opportunities for the development of materials through medical and conventional approaches. Due to these unconditional benefits in every field, it cannot be standardized as a single unit. Nanotechnology is involved in the production of designs and applications of materials at atomic, molecular, and macromolecular scales used to produce novel nanosized structures [8]. Pharmaceutical nanoparticles are defined as solid, submicron-sized (less than 100 nm in diameter) drug carriers that may or may not be biodegradable [9].

1.2 Historical Background

The term "nano" is not new in this era. Nanotechnology has been a well-known field in era of research for the last few decades. The concept of nanotechnology is not new to nature or humankind [10]. A prominent example of human-made nanomaterial is stained glass. As discussed above, the role of nanotechnology has been well reported in the field of medicine and in conventional approaches. Nanoscale features are often consolidated into bulk materials and large surfaces [8, 11]. Nobel laureate *Richard Feynman* was the first person who explored and presented the conceptual foundations of nanotechnology in 1959. During their evolution, various materials have been explored and formulated at the nanoscale level. Nanoparticles include a variety of materials that include particulate substances, which have dimension less than 100 nm [12]. Depending on the shape, materials are classified as zero dimensional (0D), one dimensional (1D), two dimensional (2D), or three dimensional (3D) [13].

The diversity in the sizes leads to differences in the physio-chemical properties of a substance, such as optical properties. The influence of these optical properties shows characteristic colors and properties with the variation of size and shape, which can be utilized in bioimaging applications [14]. The alteration of the above-mentioned factor influences the absorption properties of the nano-particles, and hence different absorption colors are observed. NPs are mainly composed of three layers; the first layer is the surface layer, which is functionalized within a variety of small molecules, metal ions, polymers, and surfactants. The second layer is a shell layer, which is a chemically different material from the core, and the third layer is the core, which is mainly considered the central position of the nanoparticles [15]. Owing to such exceptional char-acteristics, these materials have got immense interest of scientific researchers in multidisciplinary fields. To conclude, pharmaceutical nanoparticles are defined as solid, submicron-sized (less than 100 nm in diameter) drug carriers that may be biodegradable.

1.3 Classification of Nanoparticles

Classification of nanoparticles is done on various basis, including functional features such as structural configuration; origin, such as natural or synthetic engineered nanoparticles; dimensional structures, such as one, two, and three dimensional; morphology, size, and chemical properties, such as carbon-based nanoparticles, metallic nanoparticles, ceramic nanoparticles, semiconductor nanoparticles, and polymeric nanoparticles; organic profiles, such as carbon-based nanomaterials, inorganic nanomaterials, organic nanomaterials, and composite-based nanomaterials. Their variable characteristic steric effects make them distinguishing and productive for various profiles [16]. An elaborate description of these nanoparticles is provided in the following sections.

1.3.1 Nanoparticles Based on Origin

Based on origin, nanoparticles are broadly classified into two categories, natural and synthetic nanomaterials.

1.3.1.1 Natural nanomaterials

Natural nanomaterials are present in the hydrosphere, atmosphere, lithosphere, and biosphere. The earth is surrounded by various nanoparticles in the form of rocks, soils, magma, or lava in the biosphere. Various microorganisms and humans are also included in the same. The naturally occurring nanomaterials and nanoparticles are mainly produced by two aspects, through biological species or via anthropogenic activities. Naturally occurring NPs are immensely found in the form of volcanic ash, fine sand and dust, ocean spray, and even biological matter (e.g., viruses).

1.3.1.2 Synthetic (engineered) nanomaterials

Synthetic (engineered) nanomaterials are synthesized primarily by four methods: physical, chemical, biological, or hybrid methods [17]. Due to diverse configuration methods, risk strategies have arisen on a wider scale due to increased production and customer expectations. Some of the cited examples of synthetically engineered nanoparticles with metal oxides such as zinc oxide and titanium dioxide are used in paints, varnish, and cosmetics. Zircon and aluminum oxide nanopowders are used as components in ceramics to improve both hardness and breaking strength.

The only arising challenge regarding engineered nanoparticles is whether the existing knowledge is enough for their behavior or whether they exhibit a distinct environmental behavior that makes them distinguishing from naturally occurring nanoparticles. In this new era, various sources related to novel applications are used in the production of engineered nanoparticles.

1.3.2 Nanoparticles Based on Dimensional Structures

Moving ahead, the second type of classification is based on their dimensional structures as one, two, and three dimensional [18].

1.3.2.1 One-dimensional nanoparticles

During the last few decades, one-dimensional nanoparticles have been widely used in electronics, chemistry, and engineering as they have a thin film or manufactured surfaces. The thin films have size between 1 and 100 nm and are used as a monolayer in the production of solar cells or in catalysis. These films have vast application in the field of chemical sensors, biological biosensors, memory and storage, magneto-optics, optical devices, and fiberoptic systems.

1.3.2.2 Two-dimensional nanoparticles

Two-dimensional nanoparticles are well known as carbon nanotubes (CNTs). They are composed of a hexagonal network of carbon atoms, 100 nm in length and 1 nm in diameter. CNTs are wrapped in the form of a cylinder surrounded by a layer of graphite. CNTs are of two types: single-walled CNTs and multi-walled CNTs. The small dimensions and the significant physical, mechanical, and electrical properties of CNTs make them unique. Metallic and semi-conductive properties are depicted based on carbon atoms surrounded by them. These nanotubes possess the ability to carry high current density, which can reach up to 1 billion m^2 and makes them superconductors. The mechanical strength of CNT is 60 times greater than that of steel, due to which they have greater molecular absorption capabilities and chemical stability.

1.3.2.3 Three-dimensional nanoparticles

Three-dimensional nanoparticles can carry more than two-dimensional structures, based on which they have been classified into three types: fullerenes (carbon 60), dendrimers, and quantum dots.

Fullerenes: Fullerenes (C60) are spherical cage-like structures with the core central atom C60 surrounded by 28–100 carbon atoms. The whole configuration looks like a hollow ball composed of interconnected carbon pentagons and hexagons. They resemble a soccer ball or hollow ball composed of interconnected carbon pentagons and hexagons. Fullerenes are composed of a special class of materials having unique physical properties [19]. They retain their original shape even in the case of extreme pressure. The molecules present in them are not combined with each other and give them major support with a wider usage as lubricants. They have distinguishing electrical properties that make them productive in electronics and fit for a range of applications from data storage to production of biologically active molecules, medical applications, and solar cells [20].

Dendrimers: Dendrimers are a novel form of polymers with nanometric dimensions, which facilitate in drug delivery and imaging. Dendrimers with an approximate size of 10 to 100 nm have been used for determining functional groups, which is used to determine their ideal behavior. The structure and function of dendrimers are well elaborated as they have specialized encapsulating functional molecules in their surface, which have been used for diagnostics on a wider scale. Dendrimers are considered fundamental elements for large-scale production of organic and inorganic nanostructures with varying dimensions of 1 to 100 nm. They carry different reactive surface groupings called nanostructure, which makes them compatible with organic structures such as DNA and applicable in various fields such as medical and biomedical pharmaceuticals. Dendrimers have vast application in the production of pharmaceutical compounds such as nonsteroidal anti-inflammatory formulations, antiviral drugs, antimicrobial formulations, anticancer drugs, pro-drugs, and screening agents for drug discovery [21].

Quantum dots: Quantum dots (QDs) are described as small devices having a set of electrons with colloidal semiconductor ranging from 2 to 10 nm. Quantum is produced mainly by two methods: colloidal synthesis and electrochemistry. Quantum dots have variation in electrons from 1 to 1000, and the size, shape, and number of electrons should be adequately monitored. They are being used in the production of semiconductors, metals, insulators, magnetic materials, metallic oxides, optoelectronic devices, quantum computing, and information storage. Some special color-coded quantum dots are used for fast DNA testing. They also provide a wider surface area for the attachment of therapeutic agents, which are used for targeted drug delivery, bioimaging, and tissue engineering. Easily available quantum dots are indium arsenide, cadmium telluride, indium phosphide, and cadmium selenide.

1.3.3   Nanoparticles Based on Material

Nanoparticles are classified based on their composition of material into four types: carbon-based nanomaterials, inorganic nanomaterials, organic nanomaterials, and composite-based nanomaterials.

1.3.3.1   Carbon-based nanomaterials

Carbon-based nanomaterials are mainly found in structures such as hollow tubes, spheres, or ellipsoids. Some of the well-known examples are fullerenes (C60), CNTs, carbon nanofibers, carbon black, and graphene (Gr). These nanomaterials are formed using primarily two different methods: arc discharge and chemical vapor deposition. Chemical vapor deposition is used to formulate all types of carbon derivatives, except carbon 8.

1.3.3.2   Inorganic nanomaterials

These types of nanomaterials include metals and metal oxides. These NMs can be used to synthesize silver or gold NPs, metal oxides such as titanium oxide and zinc oxide, and semiconductor nanoparticles such as silicon and ceramics.

Ceramics NPs: Inorganic nonmetallic solids are used for the composition of ceramics. Ceramics are synthesized using heat and successive cooling. Their appearance has been found to be amorphous, polycrystalline, dense, porous, or hollow. Due to which these NPs have a wider range of applications such as in catalysis, photocatalysis, photodegradation of dyes, and imaging [22].

1.3.3.3 Organic nanomaterials

These types of nanomaterials are composed of organic materials only. The noncovalent bond and interactions of the molecules assist in self-assembling and design of molecules, which helps to transform them into desired structures of organic nanomaterials such as micelles, liposomes, and polymeric NPs.

Polymeric NPs: These are normally organic NPs in which a special polymer is used to encapsulate the nanoparticle, leading to the formation of polymeric nanoparticles. The newly formed particles are in the form of nanospheres and nanocapsules. The nanospheres are depicted as particles in which the major mass is present at the center and surrounded by molecules adsorbed on the outer surface of the sphere. But in the case of nanocapsules, the solid mass is completely encapsulated within the particle. These polymeric nanoparticles are easily configured and have lipid moieties, leading to their use in biomedical applications. Surfactants or emulsifiers are used to stabilize the external core of nano lipid polymers, which helps in designing and synthesizing these nanoparticles. Their prime application has been in the field of drug delivery and RNA synthesis in cancer therapy.

1.3.3.4 Composite-based nanomaterials

Composite-based nanomaterials are monophasic in nature having multiphase ends which are conjugately joined with the other bulk materials such as hybrid nanofibers. The resulting framework is known as metal organic framework. The framework composites consist of combinations of carbon-based, metal-based, or organic-based, which exist in the form of ceramics.

Semiconductor NPs: Semiconductor materials are composed of both metallic and nonmetallic materials carrying all properties with a vast range of applications. Semiconductors have wider band gaps due to which they show significant alteration in their properties.

Major applications are seen in photocatalysis, photo-optics, and electronic devices [23].

1.4 Various Approaches for Synthesis of Nanoparticles

For the formulation of nanoparticles, primarily two approaches of synthesis of nanoparticle are used: bottom-up approach and top-down approach.

The **bottom-up approach** is a process that leads to the building of larger and more complex systems by starting from the molecular level. In this method, once the molecular structure is maintained precisely and accurately, then the primary focus is put on the formation of nuclei, which ultimately forms nanoparticles.

In the **top-down approach**, various methods and assemblies are used in which large particles or bulk material is converted into fine particles [24].

To conclude, both approaches have a positive effect and significant influence on the synthesis of particles. Based on this, broadly two methods have been developed for the synthesis of nanoparticles: physical method and chemical method.

1.4.1 Physical Methods

1.4.1.1 Sol-gel technique

In the sol-gel technique, discrete particles are adhered to an integrated network precursor placed in the chemical solution that aids in the formation of metal oxides. The formulated precursor sol is used to be deposited on the substrate to form a film that was further used to synthesize powders.

1.4.1.2 Solvothermal synthesis

In solvothermal synthesis, the polar solvents are exposed to dissimilar condition such as tremendous rise in temperature as rising above their boiling point and pressures above the 1 bar. By using this process, the solubility of the reaction can be increased by reducing the temperature [24].

1.4.1.3   Emulsion-solvent evaporation method

It is a very common and versatile method of preparation of nanoparticles. In this method, there is first emulsification of the polymer solution into an aqueous phase. Then the evaporation of polymer solvent is being carried out, which forms a layer of precipitated polymers, which is known as nanospheres. The newly formulated nanoparticles are collected using the ultracentrifugation process and washed using distilled water to remove stabilizer residue or any free drug and lyophilized for storage. High-pressure emulsification and solvent evaporation method are modifications of this method. In the method of high-pressure emulsification, the emulsion is prepared, which is processed for homogenization under high pressure with constant stirring to remove organic solvent [25]. The process variables for size maintenance include adjustment of stirring rate, type and amount of dispersing agent, viscosity of organic and aqueous phases, and temperature. This method is applicable to liposoluble drugs, and scaleup issue is a major limitation in this case. The major polymers used in this method are polylactic acid or polylactide (PLGA is generally an acronym for poly D,L-lactic-co-glycolic acid where D- and L-lactic acid forms are in equal ratio) and cellulose acetate phthalate.

1.4.1.4   Double emulsion and evaporation method

This method was developed to remove the flaws of the poor entrapment of hydrophilic drugs. By using this method, the entrapment efficiency of hydrophilic drugs can be enhanced. In this method, the w/o emulsion is formed using the addition of aqueous drug solutions to organic polymer solution with the use of vigorous stirring. The formulated emulsion is incorporated into the second aqueous phase with continuous stirring to form the w/o/w emulsion. The resulting emulsion is subjected to evaporation for the formation of nanoparticles. The formulated nanoparticles are isolated using high-speed centrifugation and lyophilized for storage. The characterization of these formulated nanoparticles is done on the basis of these process variables such as amount of hydrophilic drug incorporated in formulation, the concentration of stabilizer used, the polymer concentration, and the volume of aqueous phase [26].

1.4.1.5 Emulsion–diffusion method

In this method, the encapsulating polymer is dissolved in a partially water-miscible solvent (such as propylene carbonate, benzyl alcohol), and saturation is done with water that maintains the initial thermodynamic equilibrium of both liquids. The resulting solution is now a polymer–water saturated solvent phase, which is further subjected to emulsification in an aqueous solution containing stabilizer. Then the formulated solution is subjected for diffusion. Diffusion is done on the basis of the oil-to-polymer ratio present in the outer surface of the layer, and the difference in the boiling point helps in the removal of solvent by evaporation and filtration, which ultimately aids in the formulation of nanospheres and nanocapsules. The benefits of this method are found in high encapsulation efficiencies (generally 70%), homogenization free, high batch-to-batch reproducibility, ease of scaleup probability, simplicity, and narrow size distribution. Apart from various benefits, some of the disadvantages of this method include high water volumes required for elimination from the suspension, along with the leakage of water-soluble drugs into the saturated aqueous external phase during emulsification, and reduction in encapsulation efficiency. Examples of this method include drug-loaded nanoparticles such as mesotetra (hydroxyphenyl), porphyrin-loaded PLGA (p-THPP) nanoparticles, doxorubicin-loaded PLGA nanoparticles, and cyclosporine (cy-A-)-loaded sodium glycolate nanoparticles [27].

1.4.1.6 Solvent displacement/precipitation method

In the solvent displacement method, a preformed polymer layer is precipitated from an organic solution by the diffusion of the organic solvent in the aqueous medium in the presence or absence of a surfactant. A semi-polar water-miscible solvent such as acetone or ethanol is used for dissolving polymers, drugs, and lipophilic surfactant. The resulting solution is poured or injected into an aqueous solution that contains a stabilizer by using magnetic stirring process. By using this process, the solvents are rapidly diffused, which helps in the quick formation of nanoparticles. The size of the nanoparticles is influenced by the rate of addition of the organic phase into the aqueous phase. However, it has also been found that decrease in both particle size and drug entrapment occurs with the rate of mixing of the two phases. The above technique has great impact on poorly soluble drugs [28].

1.4.1.7 Solvent evaporation technique

In this method, the emulsification of emulsion is done to prepare polymeric solutions using a large amount of volatile solvents. The solvents used for the preparation of polymers are chloroform or dichloromethane, but nowadays ethyl acetate is used due to its better toxicological profile. In solvent evaporation, polymer diffusion takes place from the nanoparticulates, which are in the continuous phase of emulsions, which are processed under high-speed homogenization followed by constant magnetic stirring or under decreased pressure. The formulated nanoparticles are collected using the ultracentrifugation method, washed with distilled water, and lyophilized for storage [29]. The formulated nanoparticles can be used for targeted drug delivery for the management of ophthalmic bacterial diseases. The novel formulated polymers exhibit improved stability and corneal mucoadhesion of nanoformulation.

1.4.1.8 Solvent displacement technique

This technique is used for the synthesis of nanospheres and nanocapsules in which the drug is emulsified into aqueous phases, which are further subjected to organic phase with constant mechanical stirring. With the help of stirring, solvent displacement takes place, which helps in further formation of nanoparticles. The newly formed nanoparticles are characterized using particles size determination, percentage entrapment efficiency, and pattern of drug release. These nanoparticles vary in the size range of 100–240 nm having greater than 99% of percentage entrapment efficiency. The benefits of this method include controlled and sustained drug delivery [30].

1.4.1.9 Emulsification/solvent diffusion technique

This method involves the preparation of solid lipid nanoparticles with high recovery. In this method, for the emulsification, the drug and lipophilic surfactant are dissolved in a water-miscible solvent, which is further incorporated into an aqueous surfactant with continuous stirring. The resultant solution is subjected to high-pressure homogenization for the diffusion of solvent. The solvent diffuses completely from the solution and leads to the formation of nanoparticles [31].

1.4.2   Chemical Methods

1.4.2.1   Supercritical antisolvent

In this method, a solid sample is dissolved in a common (organic or inorganic) solvent and then injected into a supercritical fluid, which is held under pressure, resulting in a large decrease in solution density. This effect leads to the reduction in solubility of the solid and precipitation. The commonly used methods are rapid expansion of supercritical solution (RESS) and precipitation with compressed antisolvent process in which supercritical fluids are used. This technique involves two completely miscible solvents in which one is a supercritical liquid and the other is a fluid solvent. If the solutes are to be insoluble in the supercritical liquid, then nanoparticulate formation occurs because immediate precipitation of solutes happens via the extraction of fluid solvent due to the supercritical fluid. In the RESS technique, the solutes loss the solvent power, which leads to their dissolution into the supercritical liquid. Thus, pressure is decreased in the small nozzle area and solutes get precipitated. Examples of various techniques are prepared nanoparticles of poly(heptadecafluorodecyl acrylate) via the supercritical fluid technique whose size is greater than 50 nm. Although the above-stated method has the advantage that no organic solvent is used, the main disadvantage is that micro-scaled products are formed in the primary stage instead of nanoscale product. To overcome this drawback, a new supercritical fluid technology named RESOLV has been launched, in which a fluid solvent reduces the growth of particulates in the expansion jet nozzle, which ultimately forms nanoscaled particulates in the primary stage [32] using the milling methods. By using forces of attrition, the size of particles has been reduced; microsized drug materials are reduced to nanosized materials. The evaluation parameters include structural, optical, and surface properties, which are studied using XRD, TEM, UV spectroscopy, and zeta potential analyzer [30].

1.4.2.2 Chemical reduction

The commonly used reducing agents are sodium borohydride, hydrazine hydrate, and sodium citrate. In this method, the ionic salts are involved in the reduction process in an appropriate medium containing the surfactant along with the above-mentioned reducing agents.

1.4.2.3 Laser ablation

In this technique, laser energy is used for the removal of materials from the solid surface. When the material is heated at low laser flux, laser energy is absorbed and the material is evaporated. If the material is heated at a higher flux, the material is converted to plasma. For example, CNTs can be produced by this method.

1.4.2.4 Inert gas condensation

In this method, an ultra-high vacuum chamber is filled with helium or argon gas at a pressure of a few 100 pascals in which separate crucibles are used to evaporate different metals. Thus, due to interatomic collisions with gas atoms in the chamber, the evaporated metal condenses to form small crystals as they lose their kinetic energy extensively, and ultimately these particles are collected and accumulated on liquid-nitrogen-filled cold finger.

The cited examples of this method are gold nanoparticles synthesized from gold wires.

1.4.2.5 Salting out method

The salting out method is used for the separation of the water-miscible solvent from aqueous solution using the salting out effect. It is based on the separation of a water-miscible solvent from the aqueous solution via the salting out effect. Further, the polymer and the drug are initially dissolved in a solvent, which is subsequently emulsified into an aqueous gel containing the salting out agent, such as electrolytes magnesium chloride and calcium chloride, or non-electrolytes such as sucrose, and a colloidal stabilizer such as polyvinyl pyrrolidone or hydroxyethyl cellulose. This oil/water emulsion is diluted with a sufficient volume of water or aqueous solution to enhance the diffusion of solvent into the aqueous phase, which helps in the formation of nanospheres. Various manufacturing parameters can be varied, including stirring rate, internal/external phase ratio, concentration of polymers in the organic phase, type of electrolyte concentration, and type of stabilizer in the aqueous phase. Polymers drafted using this method, such as poly(methacrylic) acids and ethyl cellulose nanospheres, have very high efficiency. A major advantage of this method is that it does not require heat for manufacturing and can be highly effective for manufacturing thermolabile products easily. The major flaws in this method are application of lyophilic drugs and extensive steps of washing nanoparticles.

1.4.2.6 Salting out technique

In this method, the water-miscible solvent system takes place from hydrophilic solution. The resulting solution has been found to contain active pharmaceutical ingredients and polymer deposited in the organic solution, which is further emulsified in an aqueous solution containing salting out agents such as calcium chloride and magnesium chloride, along with some stabilizers such as hydroxyl ethyl cellulose and polyvinyl pyrrolidone. Finally, the diffusion of organic solvent can be enhanced by the addition of a standard amount of aqueous solution in the emulsion, which would form nanospheres. The evaluation parameters such as encapsulation efficiency utterly depend on the salting out agent, so special care has to be taken to choose the suitable salting out agent [29]. Some of the examples of this methods are synthesis of polyanions loaded gelatin nanoparticulates in ocular surface epithelial cells [33].

1.4.2.7 Dialysis method

In this method, dialysis tubes are used, which consist of a polymer with an appropriate molecular weight and an organic solvent, which aids in the preparation of narrowly distributed and smaller sized nanoparticles. The solvent inside the tube is replaced by the agglomeration of polymers, due to which solubility is lost leading to the formation of a nanosuspension. The physiological barriers act as a semipermeable membrane, which helps in the passage of solvent through passive diffusion that allows the mixing of nonsolvent with polymer solutions. By using this method, various synthetic and natural polymeric nanoparticles are formed [34].

1.4.2.8 Emulsification/solvent diffusion

This method is an elaborated form of the evaporation technique. In this method polymers are mixed with the hydrophilic solvents such as propylene carbonate in order to maintain the thermodynamic equilibrium of the solution the water was added gradually. The nanoparticles are formulated by using polymer precipitation by dispersing in the dispersed phase by diluting with an extra amount of solvent. The aqueous polymer solution formed is further emulsified with a hydrophilic solution, which consists of stabilizers that cause diffusion of solvent to the external phase, leading to the formation of nanomaterials as per the oil–polymer ratio. At the last stage of the process, evaporation of solvents takes place according to their boiling points. This process consists of advantages such as high encapsulation efficiencies, easy method of scaleup, and negligible homogenization requirements [29].

1.4.2.9 Nanoprecipitation method

In this method, the polymer is precipitated in the organic solvent, and the resultant organic solvent is diffused in the hydrophilic medium with the help of a surfactant. The nanosized polymers are formulated using precipitation in a water-miscible solvent. Now the resulting solution is introduced into the hydrophilic medium having a stabilizer, i.e., surfactant with stirring. By using quick diffusion process, the polymer is developed on the interface between organic and aqueous solvent, which leads to the formation of nanosized materials. Nanocapsules can also be prepared by using this method with a little amount of nontoxic oil in the organic phase. This technique is used to formulate water-miscible solvents in which the diffusion rate is enough to produce spontaneous emulsification.

1.4.2.10 Polymerization method

This method involves polymerization of monomers in an aqueous solution. For the preparation of aqueous solution, two different techniques are used: (a) emulsion polymerization—emulsification of monomer in nonsolvent phase and (b) dispersion polymerization—dispersion of monomer in nonsolvent phase. The drug is incorporated in the nanoparticle either by dissolution of drug in polymerization medium or by adsorption on the nanoparticle. This suspension of nanoparticles contains excipients like surfactants and stabilizers, which are finally removed by the ultracentrifugation method and the resulting suspension is suspended in an isotonic medium that is surfactant free. The nanoparticles produced by this method are poly butyl cyanoacrylate or poly(alkylcyanoacrylate). The process variables of particle size are concentration of stabilizer and surfactant involved in preparation.

1.4.2.11 Coacervation or ionic gelation method

This method involves the use of hydrophilic biodegradable polymers such as chitosan, sodium alginate, and gelatin for the preparation of nanoparticles. The hydrophilic chitosan nanoparticles are prepared by the ionic gelation method. This method involves the following procedure. Two aqueous phases are prepared in which one is the polymer chitosan, a di-block co-polymer ethylene oxide or propylene oxide (PEO-PPO), and the other is a polyanion sodium tripolyphosphate. Both these solutions are mixed. This mixing leads to the reaction of positively charged amino group with the negatively charged tripolyphosphate, which ultimately forms coacervates in the nanometer size range. The coacervates were formed due to electrostatic interaction between the two aqueous phases.

1.4.2.12 Ionic gelation or coacervation of hydrophilic polymers

The synthesis of nanoparticles using this method will be in totally aqueous media. Ionic nanogels are produced at low gel point concentrations from dilute aqueous solutions of charged polysaccharides. Small clusters are formed from chains of polymers in the pre-gel state, which are stabilized by complex formation with oppositely charged polyelectrolytes [29].

1.5 Characterization Parameters of Nanoparticles

Two main parameters are needed to be studied for the characterization of nanoparticles: particle size and distribution. However, we can also determine their surface charge or zeta potential, surface area, crystallographic structure, and surface morphology.

1.5.1 Particle Size and Surface Morphology Characterization

The determination of particle size is of prime importance when we deal with nanoparticles. The particle size, polydispersity index can be determined by using various techniques such as transmission electron microscopy (TEM) and scanning electron microscopy (SEM) for the determination of particle shape. A particle size analyzer can be used to study the particle size as well as polydispersity index (uniformity of nanoparticles), and X-ray crystallographic techniques can be utilized for determining the crystalline nature of nanoparticles. To study the surface morphology, atomic force microscopy (AFM) can be used.

1.5.2 Zeta Potential

Zeta potential is a measure of the magnitude of the electrostatic or charge repulsion/attraction between particles and is one of the fundamental parameters known to affect stability. Its measurement brings detailed insight into the causes of dispersion, aggregation, or flocculation and can be applied to improve the formulation of dispersions, emulsions, and suspensions. Zeta potential can also be determined by using a particle size analyzer.

Other parameters for the characterization of nanoparticles in the case of health care can be encapsulation efficiency determination, which will determine the amount of drug encapsulated inside the nanoparticles, drug release studies to determine the release pattern of drugs, loading efficiency to determine the amount of drug loaded inside the nanoparticle shell, etc. Also to determine the chemical nature of nanoparticles, high-resolution SEM (HR-SEM) can be used.

Further, nuclear magnetic resonance (NMR) and mass spectroscopy can also be utilized for the same purpose [35, 36].

1.6 Application of Nanotechnology in Various Fields

Advancement in the field of nanotechnology in various fields has proven its significance for the betterment of life. Advanced nanotechnology techniques have transformed a vast number of scientific and industrial areas, including drug delivery and food technology. It has shown its potential in fields such as drug delivery, surgery, diagnosis of diseases, food processing and packaging, batteries, solar cells, fuel, air quality, water quality, chemical sensors, and fabric improvement.

1.6.1 Applications in Medical Technology, Drug Delivery, and Diagnosis of Diseases

Drug delivery using nanotechnology has emerged with the idea of formulating nanoparticles that have the ability to get delivered to particular diseased tissues or organs by applying advanced technologies such as active targeting. This concept can be advantageous since pathophysiological characteristics of diseased tissue/organs can be utilized for active targeting, and the nano size of nanoparticles helps them escape to tissues and blood capillaries, which is responsible for their higher accumulation in diseased tissues/organs. Nanoparticles also carry the advantage of crossing the blood–brain barrier, so they can be effectively used to deliver drugs to the brain for the treatment of various neurological disorders. Selective drug delivery to tissues or organs offers the advantage of loading lesser dose as compared to the conventional dosage, which enhances the safety profile of the drug and reduces toxicity as well as cost of the drug.

Nanotechnology is currently employed for the delivery of drugs to cancerous tissue as conventional drug delivery suffers with the disadvantage of causing toxicity to healthy cells, which leads to their permanent impairment and even causes their death. By virtue of their small size, they escape from the leaky vasculature around the cancerous tissue, which leads to their accumulation at the cancerous site. This effect is known as the enhanced permeation and retention (EPR) effect. Also, they can be engineered in such a way that they directly get targeted to their respective diseased sites, which is achieved by the modification of their properties by some physical and chemical methods and also by the attachment of specific receptors or antibodies onto their surfaces so that they can be attracted to their specific sites only. This process of architecting nanoparticles for their specificity is called active targeting, while targeting due to their inherent properties is called passive targeting [37, 38].

Nanotechnology is also getting utilized for the detection of various deadly diseases. Some scientists are using CNTs attached to various antibodies for the detection of cancer. Similarly, for the early detection of kidney damage, a test has been developed, which utilizes gold nanorods attached with a protein being generated by damaged kidneys. If the kidneys are damaged, more protein will get attached to the nanorods, which will shift the color of the reagent and the kidney damage can be predicted, and from the intensity of the color, the extent of damage can also be predicted. Some researchers have combined nanosensors with the techniques of artificial intelligence and claimed that a single virus can be detected by using this technology. Biomarker molecules released by cancer cells can also be utilized in the detection of cancer. The concentration of these specific biomarkers in blood can be tested at early stages, and if the concentration exceeds the normal concentration, then cancer can be detected at early stages. Another method for diagnosing brain cancer is utilizing the potential of NMR technology. Magnetic nanoparticles are prepared that possess the property to get attached to the particles in the blood stream and formulate microvesicles, which can be detected by NMR techniques. Nanoflares have also been prepared, which have the ability to bind genetic targets present in cancer cells. They are designed in such a way that a light is illuminated when they come across the cancer target, which can be detected by using MRI or other techniques.

Carbon nanotubes and gold nanoparticles can detect proteins released in oral cancer. So they can be used in biomarker tests, and the results can be produced within 1 h of testing. Nanofibers coated with antibodies can also be used for the detection of cancer, and they will capture individual cancer cells. Nanoparticles attached with protein, antibodies, and other molecules allow the detection of cancer cells at very early stages. Gold nanoparticles attached with antibodies have proven to be successful in the detection of flu virus. The phenomenon is based on the intensity of light given and received. Nanoparticles are added to the sample to be tested, and if the virus is present, the intensity of light will get increased as the nanoparticles will form a cluster around the viruses [39–41].

1.6.2 Nanotechnology in Electronics/Nanoelectronics

Utilizing the potential of nanosystems has significantly reduced the weight and power consumption of electronic devices. Huge TVs, computers, etc. have been converted into thin LED screens. Power consumption by air conditioners has been reduced to a larger extent. Now the focus of researchers is on creating a microchip with the density of 1 terabyte memory per square inch. Touchscreen mobile phones have made a boom in the field of electronics. Researchers at the Melbourne Institute of Technology have concluded that thin indium tin oxide sheets containing touch are less costly, more flexible, and also less power consuming.

Researchers are also focused on developing flexible electronics by depositing cadmium selenide nanocrystals on plastic sheets. Fabrication of these screens will be simpler as well. Data transmission at higher speed can be made possible by integrating silicon nanophotonics into CMOS-integrated circuits. Researchers have also demonstrated the use of nanomagnets as switches in transistors, electrical circuits, etc. It will help in reducing power consumption. Silver nanoparticle ink has been used to print prototype circuit boards where standard inkjet printers were involved. The conductive lines of circuit boards have also been formulated by the help of silver nanoparticle ink. Nano-patterned silicon surface laser light has been utilized for the production of light of tighter frequency, which can be responsible for the higher rates of data transmission over fiber optics [37, 42–44].

Carbon nanotubes can be utilized for the production of transistors and integrated circuits. To let them work, scientists have developed methods to remove metallic nanotubes and leave semiconducting tubes. They have developed an algorithm to deal with them, and the fabricated circuits contained 178 transistors. Flat panel displays are made with the help of electrodes consisting of nanowires. Similarly, semiconductor nanowires are utilized to prepare transistors.

Researchers have also developed new methods for the formulation of PN junctions in graphene. They have prepared the P and N junctions on the substrate and applied the graphene film on them, which will reduce the disruption in the graphene lattice. The nanoparticle organic memory field-effect (NOMFET) transistors have been developed by combining gold nanoparticles with organic molecules.

Integrated circuits, which can measure in nanometers, can be developed by using nanotechnology. Magnetic random-access memory (MeRAM) can be developed using nanoscale magnets.

Nanotechnology has also been applied in the development of molecule-sized transistors, which allow us to shrink the width of transistor gates to 1 nm, which can significantly increase the density of the transistors. Nanowires have been utilized in the production of transistors without N junctions.

Magnetic quantum dots are another nanotechnology-based devices that can be significantly used in the production of spintronic semiconductor devices. Silver nanowires embedded in polymer can be used to make conductive layers.

1.6.3   Nanotechnology in Food Science

Nanotechnology is impacting all aspects of food, for example how food can be sown and grown in fields, how it can be processed, and how it can be packed. Even the nature of packaging material is getting impacted by the help of nanotechnology. Nanomaterials are getting constructed, which will maintain not only the taste of food, but also the stability, safety, and health benefits of food.

Clay nanocomposites as well as silicate nanoparticles can be utilized to create an impermeable barrier to pass gases such as oxygen and CO_2 in bottles, cartons, and packets. We know about the antibacterial properties of silver, so silver nanoparticles can be utilized to create a layer in plastic bins, which will kill all the bacteria present and will not allow other harmful bacteria to enter the bins and destroy the food.

Nowadays, research is being carried out to develop low-cost foils that can detect spoiled food. CNTs are being utilized for this work.

Spraying CNTs on plastic surfaces will produce some sensors that can detect spoiled food present in them.

Zinc oxide nanoparticles can be added to the plastic packaging material that can block UV rays and also provide antibacterial action. This property can be utilized for the safety of packed food [40, 41].

The development of nanosensors is under way, which will sense the presence of bacteria at the packaging plant itself. This point of package testing will cut the cost of the product as the costlier end point testing can be avoided and the rejection will occur at the first point itself, which will help to reduce the wastage of raw materials. Also nanocapsules containing sensors are being developed to sense the deficiency of micro- and macronutrients in the body and then release only the required constituent at the required concentration only. It will create a super storehouse of vitamins in the body, which will be able to supply the product on demand.

Similarly, to protect food in the fields from pests, nanoparticles containing pesticides are being developed, which will provide site-specific delivery of pesticides, i.e., in insect's stomach only, and thereby they will minimize the contamination of plants itself.

Production of nanosensors and dispensers is under way, which will sense the deficiency in plants and the dispenser will dispense out the required nutrients immediately to combat the deficiency [45].

1.6.4   Nanotechnology in Fuel Cells and Solar Cells

The most expensive material, platinum, is utilized in companies as a catalyst. To reduce cost, companies are now using nanoparticles of platinum so that we can carry out our work with lesser amount of platinum. Also fuel cells contain a membrane that allow only hydrogen ions to pass through the cells but prevent the passage of other ions and atoms such as oxygen to the cells. In this regard, nanoparticles are being utilized to produce efficient membranes.

Direct methanol fuel cells (DMFCs) are the recent upgradation of the conventional rechargeable battery systems works by methanol insertion. When the whole battery is discharged, you need not to go for charging the battery, in spite of it, directly we can insert the cartridge of methanol into it [46].

Cars running with fuel cells instead of batteries are under development. Hydrogen is the most proposed fuel by researchers. Despite catalyst improvement and formulation of safe biomembranes, it is also important to have safe hydrogen fuel tank systems in cars.

Researchers have developed a model to predict the optimum size for platinum nanoparticle catalysis. They verified that particles 1 nm in diameter and containing approximately 40 platinum atoms showed increased catalytic effectiveness [47–51].

Nanotechnology can also play an important role in the manufacturing of solar cells. It has various advantages: It can reduce the manufacturing cost by the use of low-temperature processes instead of high-temperature deposition processes that are typically used in the production of conventional cells.

The involvement of nanotechnology in the production of solar cells has also reduced the installation cost by the production of flexible rolls instead of rigid crystalline panels.

Some researchers are using phosphorene nanosheets to formulate low-temperature perovskite solar cells, which are more efficient and cheaper than silicon cells. Another group of scientists has developed a honeycomb structure of graphene, which are kept separated by lithium carbonate. They utilized the property of 3D graphene structure to replace platinum in a dye-sensitized solar cell and achieved a good percentage of conversion of sunlight into energy.

1.6.5 Nanotechnology for Better Air Quality

Nanotechnology has also paved its way in the betterment of air quality by reducing air pollution. There are many ways by which nanotechnology can reduce air pollution. Catalysts are agents that help in the conversion of one agent into another at a lower temperature. So we can utilize nanotechnology and improve the efficiency of catalysts, which can change the harmful pollutant gases into harmless ones. Nanoparticles containing catalyst molecules possess higher surface area to react with pollutants, which makes the reaction more effective.

Nanostructured biomembranes can prove to be effective in separating carbon dioxide from industrial plant exhaust. Gold nanoparticles embedded in porous manganese oxide can be utilized as a room-temperature catalyst, which can breakdown the harmful compounds in air [42, 46, 47, 52].

1.6.6 Nanotechnology for Improvement of Fuel Availability

Nanotechnological principles can combat the short availability of fossil fuels such as diesel, petrol, and gasoline by producing fuels from low-grade materials and also making procedures more efficient. The increased efficiency of processes will make the production of fuels from raw materials more economical and also permit the production of fossil fuels by utilizing unstable raw materials.

In biotechnological products as well as in enzymology, nanotechnology has led the way. The performance of enzymes involved in the conversion of cellulose into ethanol has been improved with the help of nanotechnology. By this technique, the waste cellulose obtained from wood blocks and grass is converted into sugar, which produces ethanol after fermentation [41–43].

1.6.7 Nanotechnology in Reduction of Water Pollution

As nanotechnology is being utilized in every aspect in all fields, it can also be used for the betterment of water quality. Nanotechnology can enhance the water quality in three different ways: The first and foremost challenge is to remove industrial waste from groundwater, so by using nanoparticles, we can convert harmful contaminants into harmless ones. Now the second challenge is to remove salt and metals from water, i.e., purification of water. It can also be achieved by using electrodes with nanocomposite fibers. These nanocomposite fibers are cheap and have shown promising results in reducing the cost of power used. The third challenge is removing virus cells from water, which is not possible by using standard filters. So filters with nanopores are being developed, which are expected to show good results. Researchers have also demonstrated that nanofilters containing thin sheets of aluminum oxide can filter both heavy metals and oils from groundwater. Another application of nanotechnology is that chlorinated contaminants can be separated from groundwater by using nanoparticles of gold and palladium, but the treatment is costly enough. So to make the treatment cheaper, pellets of nanoparticles can be formulated where every molecule will react fully and thus reduce the cost of treatment. Dissolved carbon dioxide can be removed by using enzyme-functionalized micromotors. Vesicle nanoparticles can be made to remove antibiotics from water. Research has also been carried out to remove radioactive material from groundwater, and it has been found that flakes of graphene oxide can absorb radioactive material, which will form clumps after absorption and can be separated easily. Also membranes containing graphene with nanodiameter can be used to remove salt and ions from water. The cost of desalination will be lower than the present cost of reverse osmosis. Also CNTs can be used in the pores, which will decrease power consumption because the water molecules can pass through these pores by utilizing this concept. The level of mercury pollution in water can be determined by using hair-like nanoparticles [44, 47, 48].

1.6.8 Nanotechnology for Improvement of Fabrics

Fiber properties can also be improved by producing composite fabrics with nanosized pores, which occurs without significantly increasing the weight, thickness, or stiffness of the fabric. Nanowhiskers can be used to produce water- and stain-resistant fabric. Silver nanoparticles can also be utilized in the production of odor-resistant fabric because of their antibacterial properties.

1.6.9 Nanotechnology in Chemical and Biological Sensors

Sensors can be enabled to detect very small chemical vapors by nanotechnology. Materials such as CNTs and zinc oxide are used in sensors for detecting elements. Sensors have the capability to change their characteristics such as resistance whenever they absorb a molecule of gas. Nanotubes and nanoparticles, because of their small size, can enforce a few gas molecules to alter the electrical properties of sensors. The aim is to have inexpensive sensors that can identify chemicals similarly to dogs that are used to sniff chemicals in airports for identifying explosives or drugs. The sensors will not require sleep or exercise and have the ability to identify chemical vapor. They can be used in a number of ways since they are small and inexpensive. They can be straight away installed at airports or be part of any security measure to identify vapors coming from explosives.

These sensors can be used in industrial plants to detect the release of vapors from chemicals. A sensor that detects escaped hydrogen from hydrogen fuel cells in cars or other devices is useful in case of leaks. Air quality monitoring can be made inexpensive by making use of nanosensors that may track air pollution sources [44, 47, 48].

Researchers are in the process of developing sensors that may bind polymers on nanotubes to sense different explosives or chemical vapors coming from them.

Carbon nanotube sensors embedded in gel have been developed by researchers, which can be used to monitor the levels of nitric oxide in blood stream. They can be injected under the skin. Nitric oxide concentration in blood is useful to determine inflammation and, therefore, helpful in monitoring inflammatory diseases. The sensor has been functional in tests with laboratory mice for over a year. Researchers have put in place a method that demonstrates how CNTs can be sprayed on the plastic surface to generate sensors. The plastic film used to wrap food can be useful for such sensors to detect spoiled food.

Researchers have also developed hydrogen sensors by using palladium nanoparticles. When they absorb hydrogen, palladium nanoparticles swell, which causes shorts between the nanoparticles and the resistance of nanoparticles gets lowered. To sense the chemicals present in nerve gas, sensors with a single layer of molybdenum can be utilized, which shows changes in the resistance as they sense the gas. To sense the vapors of chemicals such as ammonia, graphene nanoparticle sheets can be utilized to formulate the sensors as they also show changes in resistance when they sense gases. The same application can be obtained by using CNT detection elements.

Gold nanoparticles deposited on a polymer film can make a sensor for the detection of volatile organic compounds. The polymeric film swells when it absorbs volatile organic compounds, and the spaces between gold nanoparticles change, which results in changes in resistance values [37, 39, 46, 52, 53].

1.7   Future Prospects of Nanotechnology

Much debates and discussions have been carried out on the future prospects of nanotechnology. As we have seen, nanotechnology can formulate various new materials as well and procedures in a vast number of fields such as medicine, health care, electronics, fabrics, and sensors. Moreover, the recent approach has been surrounded by various issues such as cost effectiveness, safety, health hazards, environmental and climate safety, and toxicity, and their effects on the global economics have been speculated. Despite the existing disputes, nanotechnology has a massive hope for the betterment of future. Innovations in all aspects and all fields can be made with its principles. Nowadays, it has been utilized in various fields from the delivery of genes for the treatment of genetic diseases to molecular level and real-time imaging to safety enhancement of foods and food foil sensors. The prime topic of research nowadays is combating deadly diseases such as cancer. For which, active targeting and site-specific delivery of drugs and genes in the human body have played an important role. We have walked miles in this field and must go more miles. Till date, site-specific delivery is not active practically.

The technology has shown hopes for the future even after some disputes. Innovations may be made by making use of this technology in various sectors such as gene therapy, drug delivery, and molecular imaging. The target-specific drug therapy can be a source of study in this technology. Specifically, target-specific drug delivery can lead to early detection and treatment of diseases by efficient diagnosis. Applications are being developed for clinical diagnosis and R&D. Quantum dot technology has already been licensed, and monitoring of cellular activities in tissue applications is in the process of being licensed. The other application that is of high importance for this technology is the development of means for detection of nucleic acids and proteins. Soon these products will be part of commercialization and of industrial, chemical, and medical sector. Devices for molecular filtration and cell isolation will then make it to the market. Many drug-delivery systems will be commercialized or will be in advanced clinical trials. One such drug-delivery system for the encapsulation of taxol has been developed by researchers as the parent drug Paclitaxel is useful in cancer. Infrastructure for the development of nanotechnology is needed to develop and commercialize devices since this technology will play an essential part in the future. Nanotechnology is useful in drug target manipulation as well as device implantation. Advancements at a rapid rate in nanomedicine have provided opportunities for a variety of medical disciplines. Investigations are being carried out to make nanotechnology a part of diagnostic as well as regenerative medicine. Detection of diseased cells in diagnosis up to the level of a single sick cell and avoiding infection to other cells would be faster as only diseased cell can be cured by making use of nanotechnology. Nanomedicine can also benefit individuals suffering from traumatic injuries or impaired organ functions.

1.8 Conclusion

Since the last decade, applications of nanotechnology in all sectors have risen. The use of nanotechnology has significantly risen in the field of medicines, biotechnology and agriculture, and food technology. New techniques have been designed based on the nanotechnological principles for farming, food preparations, safety, and stability. Similarly, treatment of a vast number of diseases can become possible with the help of nanotechnological principles. Even the prevalence of most dangerous diseases such as cancer and AIDS has been reduced to a larger extent. Day by day, new techniques are getting evolved for the betterment of life. Detection and surgical procedures have become less panicky and safer due to the evolution of nanotechnology. The most advanced technology under process is the detection of spoiled food in the tiffin box/cover itself, which is based on nanoprinciples. A vast number of formulations or types of nanosystems have been researched till date, and to prepare them, a huge number of techniques have been evolved. But the safety of nanomaterials is the biggest challenge to be sorted out. Along with the large number of advantages, there are some disadvantages such as toxicity due to the nature of nanomaterials, toxicity due to their reduced size, and their removal/elimination from the body specially after the detection of diseases and cancer treatment, which utilizes some radioactive materials and synthetic polymers that are not biocompatible and biodegradable. These issues have to be resolved completely by the end of this decade.

References

1. Bowman, D. M. and Hodge, G. A. (2007). A small matter of regulation: An international review of nanotechnology regulation, *Columbia Science and Technology. Law Review*, **8**(1): 1–36.

2. Teo, W. E. and Ramakrishna, S. (2006). A review on electrospinning design and nanofibre assemblies, *Nanotechnology*, **17**(14): R89.

3. Kavitha, K., Baker, S., Rakshith, D., Kavitha, H., Yashwantha Rao, H., Harini, B., et al., (2013). Plants as green source towards synthesis of nanoparticles, *International Research Journal of Biological Sciences*, **2**(6): 66–76.

4. Firdhouse, M. J., Lalitha, P., and Sripathi, S. K. (2012). Novel synthesis of silver nanoparticles using leaf ethanol extract of *Pisonia grandis* (R. Br), *Der Pharma Chemica*, **4**(6): 2320–2326.

5. Firdhouse, M. J. and Lalitha, P. (2012). Green synthesis of silver nanoparticles using the aqueous extract of *Portulaca oleracea* (L.), *Asian Journal of Pharmaceutical and Clinical Research*, **6**(1): 92–94.

6. Linsinger, T., Chaudhry, Q., Dehalu, V., Delahaut, P., Dudkiewicz, A., Grombe, R., et al. (2013). Validation of methods for the detection and quantification of engineered nanoparticles in food, *Food Chemistry*, **138**(2–3): 1959–1966.

7. Addo Ntim, S., Thomas, T. A., Begley, T. H., and Noonan, G. O., (2015). Characterisation and potential migration of silver nanoparticles from commercially available polymeric food contact materials, *Food Additives and Contaminants. Part A*, **32**(6): 1003–1011.

8. Cho, M., Cho, W-S., Choi, M., Kim, S. J., Han, B. S., Kim, S. H., et al. (2009). The impact of size on tissue distribution and elimination by single intravenous injection of silica nanoparticles, *Toxicology Letters*, **189**(3): 177–183.

9. Kreuter, J. (1996). Nanoparticles and microparticles for drug and vaccine delivery, *Journal of Anatomy*, **189**(Pt 3): 503.

10. Biswas, P. and Wu, C.-Y. (2005). Nanoparticles and the environment, *Journal of the Air and Waste Management Association*, **55**(6): 708–746.

11. Schubert, S., Delaney Jr., J. T., and Schubert, U. S. (2011). Nanoprecipitation and nanoformulation of polymers: From history to powerful possibilities beyond poly (lactic acid), *Soft Matter*, **7**(5): 1581–1588.

12. Villaraza, A. J., Bumb, A., and Brechbiel, M. W. (2010). Macromolecules, dendrimers, and nanomaterials in magnetic resonance imaging: The

interplay between size, function, and pharmacokinetics, *Chemical Reviews*, **110**(5): 2921–2959.

13. Vardharajula, S., Ali, S. Z., Tiwari, P. M., Eroğlu, E., Vig, K., Dennis, V. A., et al. (2012). Functionalized carbon nanotubes: Biomedical applications, *International Journal of Nanomedicine*, **7**: 5361.

14. Dreaden, E. C., Alkilany, A. M., Huang, X., Murphy, C. J., and El-Sayed, M. A. (2012). The golden age: Gold nanoparticles for biomedicine, *Chemical Society Reviews*, **41**(7): 2740–2779.

15. Wang, X., Yang, L., Chen, Z., and Shin, D. M. (2008). Application of nanotechnology in cancer therapy and imaging, *CA: A Cancer Journal for Clinicians*, **58**(2): 97–110.

16. Park, C., Huang, J. Z., Ji, J. X., and Ding, Y. (2012). Segmentation, inference and classification of partially overlapping nanoparticles, *IEEE Transactions on Pattern Analysis and Machine Intelligence*, **35**(3).

17. Gjerde, D. T., Burge, A., Jones, R., and Abel, M. (2010). Method and device for sample preparation. Google Patents.

18. Köhler, A. R. and Som, C. (2008). Environmental and health implications of nanotechnology: Have innovators learned the lessons from past experiences? *Human and Ecological Risk Assessment*, **14**(3): 512–531.

19. Mohanraj, V. and Chen, Y. (2006). Nanoparticles: A review, *Tropical Journal of Pharmaceutical Research*, **5**(1): 561–573.

20. Tomalia, D. and Huang, B. (2006). Stabilized and chemically functionalized nanoparticles. Google Patents.

21. Cheng, Y., Xu, Z., Ma, M., and Xu, T. (2008). Dendrimers as drug carriers: Applications in different routes of drug administration, *Journal of Pharmaceutical Sciences*, **97**(1): 123–143.

22. Lucarelli. M., Gatti, A. M., Savarino, G., Quattroni, P., Martinelli, L., Monari, E., et al. (2004). Innate defence functions of macrophages can be biased by nano-sized ceramic and metallic particles, *European Cytokine Network*, **15**(4): 339–346.

23. Spanhel, L., Weller, H., and Henglein, A. (1987). Photochemistry of semiconductor colloids. 22. Electron ejection from illuminated cadmium sulfide into attached titanium and zinc oxide particles, *Journal of the American Chemical Society*, **109**(22): 6632–6635.

24. Prathna, T., Mathew, L., Chandrasekaran, N., Raichur, A. M., and Mukherjee, A. (2010). Biomimetic synthesis of nanoparticles: Science, technology and applicability. *Biomimetics Learning from Nature*, Intechopen.

25. Voura, E. B., Jaiswal, J. K., Mattoussi, H., and Simon, S. M. (2004). Tracking metastatic tumor cell extravasation with quantum dot nanocrystals and fluorescence emission-scanning microscopy, *Nature Medicine*, **10**(9): 993–998.

26. Weyenberg, W., Filev, P., Van den Plas, D., Vandervoort, J., De Smet, K., Sollie, P., et al. (2007). Cytotoxicity of submicron emulsions and solid lipid nanoparticles for dermal application, *International Journal of Pharmaceutics*, **337**(1–2): 291–298.

27. Vargas, A., Pegaz, B., Debefve, E., Konan-Kouakou, Y., Lange, N., Ballini, J.-P., et al. (2004). Improved photodynamic activity of porphyrin loaded into nanoparticles: An in vivo evaluation using chick embryos, *International Journal of Pharmaceutics*, **286**(1–2): 131–145.

28. Pal, S. L., Jana, U., Manna, P. K., Mohanta, G. P., and Manavalan, R. (2011). Nanoparticle: An overview of preparation and characterization, *Journal of Applied Pharmaceutical Science*, **1**(6): 228–234.

29. Nagavarma, B., Yadav, H. K., Ayaz, A., Vasudha, L., and Shivakumar, H. (2012). Different techniques for preparation of polymeric nanoparticles: A review, *Asian Journal of Pharmaceutical and Clinical Research*, **5**(3): 16–23.

30. Brahamdutt, V. K. K., Kumar, A., Hooda, M. S., and Sangwan, P. (2018). Nanotechnology: Various methods used for preparation of Nanomaterials, *Asian Journal of Pharmacy and Pharmacology*, **4**(4): 386–393.

31. Kumari, A., Yadav, S. K., and Yadav, S. C. (2010). Biodegradable polymeric nanoparticles based drug delivery systems, *Colloids and Surfaces B: Biointerfaces*, **75**(1): 1–18.

32. Meziani, M. J., Pathak, P., and Sun, Y.-P. (2009). Supercritical fluid technology for nanotechnology in drug delivery. *Nanotechnology in Drug Delivery*, Springer, pp. 69–104.

33. Kaptay, G. (2012). On the size and shape dependence of the solubility of nano-particles in solutions, *International Journal of Pharmaceutics*, **430**(1–2): 253–257.

34. Chronopoulou, L., Sparago, C., and Palocci, C. (2014). A modular microfluidic platform for the synthesis of biopolymeric nanoparticles entrapping organic actives, *Journal of Nanoparticle Research*, **16**(11): 2703.

35. Li, Z., Wei, L., Gao, M., and Lei, H. (2005). One-pot reaction to synthesize biocompatible magnetite nanoparticles, *Advanced Materials*, **17**(8): 1001–1005.

36. Roco, M. (1999). Nanoparticles and nanotechnology research, *Journal of Nanoparticle Research*, **1**(1): 1.

37. Drobne, D. (2007). Nanotoxicology for safe and sustainable nanotechnology, *Arhiv za Higijenu Rada i Toksikologiju*, **58**(4): 471–478.

38. Wen, M. M., El-Salamouni, N. S., El-Refaie, W. M., Hazzah, H. A., Ali, M. M., Tosi, G., et al. (2017). Nanotechnology-based drug delivery systems for Alzheimer's disease management: Technical, industrial, and clinical challenges, *Journal of Controlled Release*, **245**: 95–107.

39. Binulal, N., Deepthy, M., Selvamurugan, N., Shalumon, K., Suja, S., Mony, U., et al. (2010). Role of nanofibrous poly (caprolactone) scaffolds in human mesenchymal stem cell attachment and spreading for in vitro bone tissue engineering: Response to osteogenic regulators, *Tissue Engineering Part A*, **16**(2): 393–404.

40. Hill, E. K. and Li, J. (2017). Current and future prospects for nanotechnology in animal production, *Journal of Animal Science and Biotechnology*, **8**(1): 26.

41. Mihranyan, A., Ferraz, N., and Strømme, M. (2012). Current status and future prospects of nanotechnology in cosmetics, *Progress in Materials Science*, **57**(5): 875–910.

42. Benfu, Y. Q. C. H. H. (2002). Review of recent advances and future prospects for nanotechnology [J], *Materials Review*, **8**.

43. Petersen, J. L. and Egan, D. M. (2002). Small security: Nanotechnology and future defense: Center for Technology and National Security Policy, National Defense University.

44. Ramgir, N. S., Yang, Y., and Zacharias, M. (2010). Nanowire-based sensors, *Small*, **6**(16): 1705–1722.

45. Moraru, C., Huang, Q., Takhistov, P., Dogan, H., and Kokini, J. (2009). Food nanotechnology: Current developments and future prospects. *Global Issues in Food Science and Technology*, Elsevier, pp. 369–399.

46. Tambe, N. S. and Bhushan, B. (2005). Identifying materials with low friction and adhesion for nanotechnology applications, *Applied Physics Letters*, **86**(6): 061906.

47. Quadros, M. E., Pierson IV, R., Tulve, N. S., Willis, R., Rogers, K., Thomas, T. A., et al. (2013). Release of silver from nanotechnology-based consumer products for children, *Environmental Science & Technology*, **47**(15): 8894–8901.

48. Roco, M. C. (2001). International strategy for nanotechnology research, *Journal of Nanoparticle Research*, **3**(5–6): 353–360.

49. Santos-Magalhães, N. S. and Mosqueira, V. C. F. (2010). Nanotechnology applied to the treatment of malaria, *Advanced Drug Delivery Reviews*, **62**(4–5): 560–575.

50. Stephanopoulos, G. and Reklaitis, G. V. (2011). Process systems engineering: From Solvay to modern bio-and nanotechnology: A history of development, successes and prospects for the future, *Chemical Engineering Science*, **66**(19): 4272–4306.

51. Tambe, N. S. and Bhushan, B. (2004). Scale dependence of micro/nano-friction and adhesion of MEMS/NEMS materials, coatings and lubricants, *Nanotechnology*, **15**(11): 1561.

52. Yinghuai, Z., Cheng Yan, K., Maguire, J. A., and Hosmane, N. S. (2007). Recent developments in boron neutron capture therapy (BNCT) driven by nanotechnology, *Current Chemical Biology*, **1**(2): 141–149.

53. Zuo, W., Li, R., Zhou, C., Li, Y., Xia, J., and Liu, J. (2017). Battery-supercapacitor hybrid devices: Recent progress and future prospects, *Advanced Science*, **4**(7): 1600539.

Multiple Choice Questions

1. Nanoparticles target the rare _____ causing cells and remove them from blood.

   (a) Tumor

   (b) Fever

   (c) Infection

   (d) Cold

2. _____is the field in which nanoparticles are used with silica-coated iron oxide.

   (a) Magnetic applications

   (b) Electronics

   (c) Medical diagnosis

   (d) Structural and mechanical materials

3. DNA detection through the _____ by using oligonucleotide-functionalized gold nanocrystals has been developed.

   (a) Colorimetric

   (b) Diathermy

   (c) Electrotherapy

   (d) Treatment tables

4. Industrial catalysts should have _____ surface area.

   (a) High

   (b) Low

   (c) Moderate

   (d) No

5. The nanoparticles extensively used as catalyst are_____
   - (a) Silver
   - (b) Copper
   - (c) Gold
   - (d) Cerium

6. Which of the following is the application of nanotechnology in food science and technology?
   - (a) Agriculture
   - (b) Food safety and biosecurity
   - (c) Product development
   - (d) All of the mentioned

7. The efficiency of today's best solar cell is about _____
   - (a) 15–20%
   - (b) 40%
   - (c) 50%
   - (d) 75%

8. 1 nanometer = _____ cm.
   - (a) 10^{-9}
   - (b) 10^{-8}
   - (c) 10^{-7}
   - (d) 10^{-6}

9. The size of *E. coli* bacteria is _____ nm.
   - (a) 75,000
   - (b) 2000
   - (c) 200
   - (d) 5

10. The most important property of nanomaterials is
    - (a) Force
    - (b) Friction
    - (c) Pressure
    - (d) Temperature

11. 1 meter = _____ nm
    - (a) 10^9
    - (b) 10^{-9}
    - (c) 10^{10}
    - (d) 10^{-10}

12. Nanotechnology, in other words, is
    - (a) Carbon engineering
    - (b) Atomic engineering
    - (c) Small technology
    - (d) Microphysics

13. The width of carbon nanotube is _____nm.
    - (a) 1
    - (b) 1.3
    - (c) 1.55
    - (d) 10

14. The size of red and white blood cells is in the range of _____ μm.
    - (a) 2–5
    - (b) 5–7
    - (c) 7–10
    - (d) 10–15

15. The ratio of thermal conductivity of silver to that of carbon nanotube is _____.
    - (a) 100 : 1
    - (b) 1 : 100
    - (c) 10 : 1
    - (d) 1 : 10

16. The tensile strength of a carbon nanotube is ____ times that of steel.
    (a) 10   (b) 25
    (c) 100   (d) 1000

17. The width of carbon nanotube is ____nm.
    (a) 1   (b) 1.3
    (c) 1.55   (d) 10

18. Nanoparticles are used for the treatment of cancer.
    (a) True   (b) False

19. Biological applications of nanotechnology include
    (a) Cancer treatment
    (b) Cancer detection
    (c) Treatment of dental problems
    (d) All of the above

20. Quantum dots are the type of nanoformulations that can be used for the treatment of various diseases.
    (a) True   (b) False

21. Methods to formulate nanoparticles include
    (a) Salting out
    (b) Emulsification solvent evaporation
    (c) Nanoprecipitation
    (d) All

22. Carbon nanotubes can be formulated by
    (a) Laser ablation technique   (b) Nanoprecipitation method
    (c) Both   (d) None

23. Method to formulate nanofibers includes
    (a) Electrospinning technique   (b) Coacervation phase separation
    (c) Salting out   (d) Ethanol injection

24. Nanofibers can be used for the delivery of drugs.
    (a) True   (b) False

25. Lipid-based nanoparticles are more beneficial and less toxic to human body.
    (a) True   (b) False

26. In food sciences, nanotechnology can be applied to
    (a) Food packaging   (b) Food processing
    (c) Nanosensors   (d) All

27. The nanoformulation that can be used to target brain cancer is
    - (a) Nanoflares
    - (b) Carbon nanotubes
    - (c) Gold nanoparticles
    - (d) All

28. Nanoparticles escape through reticuloendothelial systems during cancer treatment.
    - (a) True
    - (b) False

29. Better air quality can be achieved by nanotechnology.
    - (a) True
    - (b) False

30. Polymers that can be used for the formulation of nanoparticles by the ionic gelation method include
    - (a) Chitosan
    - (b) Sodium alginate
    - (c) Gelatin
    - (d) All

31. Nanoparticles show the effect due to
    - (a) Loading dose
    - (b) Entrapped dose
    - (c) Both

32. The solvent used in the solvent evaporation technique includes
    - (a) Ethyl acetate
    - (b) Dichloromethane
    - (c) Chloroform
    - (d) All

33. The types of nanoparticles based upon the method of drug entrapment include
    - (a) Reservoir type
    - (b) Matrix type
    - (c) Both
    - (d) None

34. The diversity in size observed in nanoparticles is responsible for the change in which of the following properties:
    - (a) Physical
    - (b) Chemical
    - (c) Optical
    - (d) All

35. The nanoprecipitation method can be used to prepare
    - (a) Nanoparticles
    - (b) Nanocapsules
    - (c) Nanofibers
    - (d) Both A and B

36. The risk of drug spillage is higher in which type of nanoparticles?
    - (a) Reservoir type
    - (b) Matrix type
    - (c) Both

37. To separate chlorinated contaminants from water, which type of nanoparticles are used?
    - (a) Gold nanoparticles
    - (b) Palladium nanoparticles
    - (c) Both
    - (d) None

38. Nanostructured lipid carriers are more stable and can carry more amount of drug than solid lipid nanoparticles.

    (a) True   (b) False

39. Nanostructured lipid carriers contain both solid and liquid lipids as constituents.

    (a) True   (b) False

40. GRAS certification is required for the biomedical applications of nanoparticles.

    (a) True   (b) False

Answer Key

1. (a)	2. (c)	3. (a)	4. (a)	5. (c)	6. (d)	7. (b)
8. (c)	9. (b)	10. (b)	11. (a)	12. (b)	13. (b)	14. (a)
15. (d)	16. (c)	17. (b)	18. (a)	19. (d)	20. (a)	21. (d)
22. (a)	23. (a)	24. (a)	25. (a)	26. (d)	27. (d)	28. (a)
29. (a)	30. (d)	31. (c)	32. (d)	33. (c)	34. (d)	35. (d)
36. (a)	37. (c)	38. (a)	39. (a)	40. (a)		

Long Answer Questions

1. Classify the methods for the preparation of nanoparticles.
2. Explain various physical methods for the preparation of nanoparticles.
3. Explain in detail the biomedical applications of nanoparticles.
4. What are the different types of nanoformulations. Explain their construction and applications in specific fields.
5. Explain the applications of nanoparticles in electronics.

Short Answer Questions

1. How nanoparticles are beneficial in cancer detection and treatment? Explain.
2. Can we change the quality of air by using nanotechnology?
3. Differentiate between emulsification solvent evaporation and salting out techniques for the preparation of nanoparticles.
4. Name various physical methods for the preparation of nanoparticles and explain the emulsification solvent evaporation method in detail.

5. Explain the role of nanosensors for the determination of food quality.

6. How nanotechnology can be beneficial in the food and agricultural sciences? Explain.

7. Explain the supercritical antisolvent technique for the preparation of nanoparticles.

Chapter 2

Application of Nanotechnology in Pharmaceutical Sciences

Rakesh K. Sindhu,[a] Gagandeep Kaur,[a] Arashmeet Kaur,[a] Shivam Garg,[a] Shantanu K. Yadav,[a] and Sumitra Singh[b]
[a]*Chitkara College of Pharmacy, Chitkara University, Punjab-140401, India*
[b]*Department of Pharmaceutical Sciences, Guru Jambheshwar University of Science and Technology, Hisar, Haryana-125001, India*
rakeshsindhu16@gmail.com

2.1 Introduction

The field of nanotechnology is defined in terms of manufacturing and manipulation of atoms at the nanoscale level [1]. Nanoparticles can be defined as particles that have size less than 100 nm [2]. Generally, these nanoscale particles constitute three layers: surface, shell, and core. These nanoscale particles have an advantage of biodegradability as well as nontoxicity in the human body. These downregulate various side effects. The desired drugs can be incorporated into the nanoparticles, resulting in the release of the drug into the desired target. Thereby, a low-dose drug can be easily administered. These nanoparticles can be in natural origin such as phospholipids, lactic

Nanotechnology: Principles and Applications
Edited by Rakesh K. Sindhu, Mansi Chitkara, and Inderjeet Singh Sandhu
Copyright © 2021 Jenny Stanford Publishing Pte. Ltd.
ISBN 978-981-4877-43-5 (Hardcover), 978-1-003-12026-1 (eBook)
www.jennystanford.com

acid or may be chemical in origin such as metals, polymers, and carbon [2]. Targeted drug delivery enables the specific targeted delivery of drugs to the targeted site in addition to the delivery of drugs to non-targeted sites [4]. Over the recent decade, the use of biodegradable polymers has elevated the drug-delivery system. In addition, continuous research for the development of polymeric nanoparticles has been in demand, mainly for the targeted delivery of drugs, sustained release as well as stability of labile compounds such DNA, peptides, and proteins from the process of degradation [5].

2.2   Historical Background

In general, the nineteenth century marks the presence of nanoparticles, primarily Mesopotamia, which usually produce glittering effect on the surface of the pot. The glittering effects on the surface are generally in the form of glitter from copper or gold. About 4000 years ago, the utilization of nanoparticles was wholly human made along with the chemical process of PBS, which is almost similar to the size of 5 nm. The above-mentioned instance is regarded as the oldest human-made utilization of nanoparticles. In the beginning of the twentieth century, studies were carried out on the analysis of amorphous and nanocrystalline substances. Therefore, there has been rapid advancement in the field of research and development [3]. However, the mechanism of nanoparticles is still a major challenge in the field of nanotechnology [4].

2.2.1   Loading of Drugs

The entrapment efficiency of drugs and the drug-loading capacity depend on various factors such as solubility of the drugs in the matrix polymer and the presence of dispersion agent. The aforementioned factors are proportional to the interactions of the polymer with the drugs, attached functional groups such as ester and carboxylic acids, matrix composition as well as the molecular weights of the drugs [5–7]. The commonly used polymer is polyethylene glycol, which imparts a minimum loading capacity to the drug [8]. Mostly, the drug capacity is reached at the isoelectric point of the drug, particles, and macromolecules that are encapsulated in the layer of nanoparticles

[9]. One of the efficient methods for loading small molecules is the ionic interaction between the matrix and the drug [10, 11].

2.2.2 Drug Release

The uniform distribution of the drug in the nanosphere, where it is by the mechanism of diffusion and matrix degradation that releases the drug. When the diffusion rate is greater than the matrix erosion, the mechanism followed is the process of diffusion. Rapid as well as the primary release is related to the drug absorbed and weakly bound drug [12]. It has been reported that the mode of incorporation is related to the drug release profile. If the release of the drug is sustained with a minimal small burst, then the drug is loaded through the method of incorporation [13]. When there is polymer coating, the drug release is mediated through diffusion from the membrane. Interaction of the drug with the auxiliary ingredients forms a water-soluble complex along with the slow release of the drug with no burst effect [14]. Auxiliary ingredients such as ethylene oxide propylene block the chitosan, which further decreases the drug interaction because of PEO–PPO electrostatic interaction leading to elevation in the release of the drug [15].

2.3 Nanoparticle Delivery System

Nanospheres can be defined as a uniformly dispersed matrix system, whereas nanocapsules can be defined as a vesicular system in which the cavity consisting of the drug is surrounded by a polymeric membrane. By definition, nanoparticles are colloidal solid particles that constitute macromolecular compound ranging from 10 nm to 1000 nm [16]. Particles greater than 200 nm cannot be pursued; therefore, these should not be introduced as subjects of nanomedicines because of the lager width of microcapillaries. The desired drug can be disabled into the matrix polymer, which leads to entrapping, absorption, and attachment onto a nano-carried matrix. The properties of nanocarriers depend largely on the methods of preparation and the type of nanoparticles (whether nanospheres or nanocapsules) for having the efficient delivery of the encapsulated drug to impart the desired pharmaceutical effect [17–19].

2.3.1 Properties of Nanoparticles

The following are the properties of nanoparticles:
- Increase in bioavailability
- Increase in the loading capacity of the drug
- Reduced toxicity to the liver
- Improved stability of the drug by acting against enzymatic degradation
- Site specific as well as drug-target specific
- Adjuration of monoclonal antibodies enhances the specificity of drug
- Improved solubility of drugs, especially the poorly water soluble [20, 21]

2.3.2 Disadvantages of Nanoparticles

The following are the disadvantages of nanoparticles:
1. Leads to the formation of aggregates
2. Target specificity is poor
3. Immunogenicity is high
4. Expensive
5. Relatively lower half-life
6. Presence of organic solvent causes toxicity [21]

2.4 Nanoparticles: Classification

2.4.1 Depending on the Nature of Nanoparticles

2.4.1.1 Organic nanoparticles

Organic nanoparticles are defined as nanoparticles that are biodegradable in nature, and they amount to least toxicity in humans. These include liposomes and micelles containing a hollow core. These are often referred to as nanocapsules imparting sensitivity to electromagnetic radiation such light and to heat [22]. The above characteristics make them an optimum choice for specific target of drugs into the body. Various evaluation parameters such as capacity of drug loading, system of delivery, and stability mark their related application in various field. The evaluation of the efficiency of drugs in terms of morphology such as surface, size, and composition differentiates them from normal particles. These organic particles have shown various applications in the field of biomedical science. The examples of organic nanoparticles include ferritin, dendrimers, liposomes, and micelles [23].

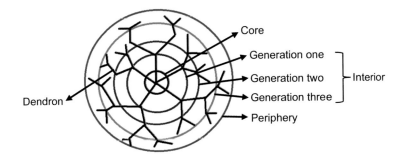

Figure 2.1   Dendrimers.

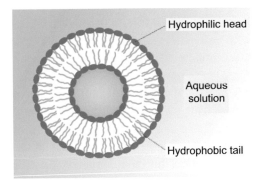

Figure 2.2   Liposomes [22].

2.4.1.2   Inorganic nanoparticles

Inorganic nanoparticles are defined as particles consisting of carbon. Inorganic nanoparticles can be further classified into metal and metal oxide nanoparticles [7].

Metal-based nanoparticles: Constructive as well as destructive methods are utilized for the synthesis of metal-based nanoparticles.

These methods can be employed for almost all types of nanoparticles [23]. Various metal-based nanoparticles include cobalt (Co), aluminum (Al), gold (au), silver (Ag), lead (Pb), cadmium (Cd), zinc (Zn), and iron (Fe). These inorganic nanoparticles include distinctive parameters and are of smaller size in the range of 10–100 nm. The various surface characteristics—volume ratio, density, shape and colour, crystalline and amorphous substances, and environmental parameters like air, heat, and humidity—also influence these nanoparticles [7].

Metal-oxide-based nanoparticles: These are modified metal nanoparticles that are usually altered to enhance the properties of metal nanoparticles. For instance, iron nanoparticles (Fe) generally get oxidized to iron oxide in the presence of oxygen, which leads to increase in reactivity compared to the former ones. Various metal-oxide-based nanoparticles are magnetite (Fe_3O_4), aluminum oxide (Al_2O_3), zinc oxide (ZnO), silicon dioxide (SiO_2), and titanium oxide (TiO_2) [7].

Carbon-based nanoparticles: Carbon-based nanoparticles consist of carbons; they can be further grouped into various types [7, 24].

- **Carbon nanotubes (CNTs):** CNTs consist of graphene nanofoil along with a carbon lattice in a honeycomb shape. These are wound into hollow cylinders, which lead to the formation of nanotubes. These nanotubes are of diameter less than 0.7 nm for single layer and about 100 nm for multilayer. Thus, the length varies from micrometers to millimeters [7].
- **Fullerenes:** These consist of C-60 molecules of carbon held by sp^2-hybridized carbon atoms and are spherical in shape. For a single layer of fullerenes, there exists about 28–1500 carbon atoms, while for multi-layered ones, they are 4–36 nm [7].
- **Graphene:** It is a carbon allotrope usually consisting of a hexagonal network in which the carbon atoms are arranged in a two-dimensional planar surface. Thickness varies up to 1 nm [7].
- **Carbon nanofibers:** They are similar to graphene nanofoils and CNTs, but they are wound into a conical or cup shape rather than cylindrical tubes [7].

- **Carbon black:** These are materials that are amorphous in nature. They are generally spherical, with diameter ranging from 20 to 79 nm. They can form aggregates because of the interaction between the particles, which are about 500 nm in size [7].

2.4.2 Depending on Physical and Chemical Basis

2.4.2.1 Carbon-based nanoparticles

The two main types of carbon-based nanoparticles are fullerenes and CNTs. Generally, fullerenes are nanomaterials made of allotropic forms of carbon. Strength testing, conductivity testing, affinity of electrons, and versatility are commercially interesting factors. These nanoparticles generally consist of pentagonal and hexagonal units of carbon where the carbon is under sp^2 hybridization.

Carbon nanotubes are defined as tubular and elongated structures having diameters of 1–2 nm [25, 26]. In addition, they are represented in the form of sheets, which can be single-walled nanotubes (SWNTs), double-walled nanotubes (DWNTs), and multi-walled nanotubes (MWNTs). Because of unique mechanical, chemical, and physical properties, these structures can be used in both nanocomposite and pristine forms for various applications such as filters [27], gas adsorbents [28], and as medium of support for various organic and inorganic catalysts [29].

2.4.2.2 Ceramic nanoparticles

Ceramic nanoparticles are defined as inorganic nonmetallic solids synthesized through the alternating process of heating and cooling. Ceramics can be produced in various forms such as polycrystalline, dense, amorphous, porous, and hollow [30]. Regarding utilization, they provide significant importance in dye-photodegradation, photolysis, imaging, and catalysis [31].

2.4.2.3 Semiconductor nanoparticles

Semiconductor nanoparticles have characteristics between metals and nonmetals. These have widened gaps and depict significant alteration along with bad tuning gap [35, 36]. Therefore, they are significant materials in the field of electronic devices and photo-optics [32, 33].

2.4.2.4 Polymeric nanoparticles

Polymeric nanoparticles are mainly organic nanoparticles of spherical or capsular shape [39]. These are majorly made of solid molecules on the surface. These nanocapsules contain a mass of solid encapsulated within a particle [40]. These have remarkable importance because of their functionalization [34, 35].

2.4.2.5 Lipid-based nanoparticles

Lipid-based nanoparticles consist of lipid moieties that are being used in various biomedical applications. The outermost part of the nanoparticles is mainly made of surfactants or emulsifiers [36]. Therefore, these lipid nanoparticles are utilized as remarkable carriers of drug and for targeted delivery of drugs along with their applications in cancer therapy [37–39].

2.4.2.6 Nanocarriers

Nanocarriers have optimized biological and physiochemical properties, which can be simply taken. These are molecules that have optimized biological and physiochemical properties, which are easily occupied by cells. These act as delivery tools for bioactive compounds. For instance, polymers, carbon materials, liposomes, magnetic nanoparticles, and silicon [40].

Liposomes: Liposomes are regarded as carriers of drugs. They are about 80–300 nm in size and are colloidal in nature. Liposomes are defined as spherical aggregates of steroids and phospholipids that spontaneously form and are dispersed in the aqueous medium through the method of sonication [41]. These aggregates generally lead to increase in drug solubility and improvement in physiochemical parameters such as pharmacological index of chemotherapeutic agents, which undergo metabolism. Liposomes also reduce adverse effects and lead to an increase in in vivo and in vitro invertible attachment of homing devices, which is useful in modified liposomal systems. Ligands, such as antibodies, are cleaved of in response to an environmental stimulus. In addition, gene-delivery carriers can be obtained from cationic liposomes [42].

2.5   Synthesis of Nanoparticles

Nanoparticles are synthesized by different methods classified into top-down or bottom-up methods [11]. The synthesis can be represented by the flow chart in Fig. 2.3.

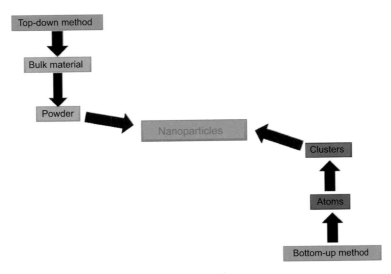

Figure 2.3   Process for synthesis.

2.5.1   Top-Down Method

The top-down synthesis is a destructive method, which helps in reducing the bulk material into nanoscale particles. Various methods included in the top-down synthesis are as follows [7]:

Nanolithography: It is defined as the fabricating study of the structures of nanometric scale, which ranges from 1 to 100 nm. Numerous nanolithography processes include electron beam processes, nanoimprint, optical processes, multiphoton processes, scanning probe lithography, and many more [43]. The major characteristic of the above method is to print the correct shape and structure on a surface that is selectively light sensitive, excluding the product portion as required for shape and structure design. The main benefit of the above process is the production of single nanoparticles from the group with desired size and shape. While disadvantages include necessity of equipment and high cost [44].

Sputtering: It is the process of deposition of nanoparticles on the surface through the ejection of particles upon collision with ions [45]. Generally, it is a process of thin-layer deposition of nanoparticles through the process of annealing. The various factors that affect the process are temperature, type of substrate, and thickness of layer, which affect the size and shape of the nanoparticles [46].

Mechanical milling: It is one of the widely used methods in the top-down synthesis of nanoparticles. This method is utilized for the post-annealing and milling processes during the synthesis in an inert atmosphere [47]. Factors that influence include shape and size of the particles. Fracture leads to decrease in particle size, while its increase causes cold welding.

Laser ablation: It is one of the common methods for the production of nanoparticles from solvents. The irradiation of a metal submerged in a liquid solution by a laser beam condenses a plasma plume, producing nanoparticles. It is regarded as one of the most reliable methods for providing a solution to all the conventional chemical reduction processes. It leads to the stable synthesis of nanoparticles especially in organic solvents as well as water, which does not require any agent for stability. Hence, it is referred to as a "green process" [7].

Thermal decomposition: It is referred to as an endothermic chemical decomposition method that produces heat, which breaks the chemical bonds within compounds [48]. The temperature at which the chemical begins to decompose is referred to as the decomposition temperature. The nanoparticles are produced through the decomposition of a metal at a temperature for the production of secondary products [7].

2.5.2   Bottom-Up Method

Unlike the top-down method, it is a constructible method, which helps in building the material from atom to the formation of clusters for the formation of nanoparticles. Various methods are as follows [7]:

Spinning: In this method, nanoparticles are produced by spinning carried out through an SDR. It consists of a revolving disk, which includes a reactor maintained at a temperature. The chamber is filled with nitrogen, which leads to the removal of oxygen and avoids chemical reactions [49]. The disk spins at different rpm, when the liquid material is pumped in. The spinning thus causes the atom to fuse collectively, which is then precipitated and dried [50]. The evaluation parameters include disk surface, feed location, rotation speed, and flow of liquid [7].

Sol–Gel–Sol: The colloidal mixture of solids is suspended into the liquid phase, which is then submerged into the solvent. It is one of the most preferred methods because of simplicity, and a high number of nanoparticles can be yielded by this method. It is defined as the wet chemical process, which contains a chemical mixture that acts as a precursor of the combined system of particles. The precursors that are generally used include metal oxides and chlorides [51]. These precursors then undergo dispersion method through shaking and then sonication method containing a solid and liquid phase. A phase separation technique is required for the recovery of nanoparticles, such as sedimentation, filtration, and centrifugation [52].

a. **Pyrolysis:** It is one of the most commonly used procedures for the large-scale production of nanoparticles. This involves the burning of precursor with the help of flame. The precursor can be either liquid or vapor, which is fed into the furnace at a very high pressure through a small hole [20]. The nanoparticles are then recovered through the combustion process or by product gases. Furthermore, the laser and plasma furnaces ensure successful evaporation [53]. The major advantage of this method is its effectiveness and cost friendliness as well as continuity with the production of high yield [7].

b. **Chemical vapor deposition:** This method aids in the deposition of a thin film of gases on the surface of reactants. This deposition is carried in a closed chamber through the combination of gases [15]. This reaction leads to the production of a thin film on the surface of the substrate, which undergoes a recovery process and utilized. One of the influencing factors is the substrate temperature. The advantages include its uniformity, purity, and absence of highly toxic byproducts [7].

c. **Biosynthesis:** It is an eco-friendly approach for the production of biodegradable nanoparticles free from harmful substances [54]. This process uses bacteria, fungi, and plant extracts along with the utilization of precursors for the production of nanoparticles. It provides effective biomedical applications [55].

Table 2.1 Categories of nanoparticles synthesized from various methods [7]

Classification	Nanoparticles	Method
Bottom up	• Carbon based • Metal oxide based	Sol–gel
	• Based on organic polymers	Spinning
	• Carbon based • Metal oxide based	Chemical vapors deposition
	• Carbon based • Metal oxide based	Pyrolysis
	• Based on organic polymer • Metal based	Biosynthesis
Top down	• Based on metal oxide • Polymer based	Mechanical milling
	• Metal based	Nanolithography
	• Carbon based • Metal oxide based	Laser ablation
	• Metal based	Sputtering
	• Carbon based • Metal oxide based	Thermal decomposition

2.6 Properties of Nanoparticles

The properties of nanoparticles are normally categorized into physical and chemical properties. The properties are given in Table 2.2.

Table 2.2 Chemical and physical properties of nanoparticles

Types	Nanoparticle	Properties	Reference
Carbon-based nanoparticles	Fullerenes	Transmission of light, more safe, and inert, superconductor	[30]
	Graphene	Electric conductor, high strength	[31]
	Carbon nanotubes	Tensile strength, thermal conductor, electrical conductor	[32]
	Carbon nanofibers	Mechanical properties, thermal stability, conductor of electricity	[33]
	Carbon black	Electricity conductor, greater surface area, resistant to ultraviolet degradation	[34]
Metal-based nanoparticles	Aluminum	More sensitivity for moisture, sunlight; highly reactive; greater surface area	[35]
	Iron	Sensitivity for air; instable and reactivity	[36]
	Silver	Ability of scattering the light, disinfectant, stable	[26]
	Gold	Reactive; visible light interaction	[37]
	Cobalt	Microwave absorption, toxic and instability	[38]
	Cadmium	Insolubility, semiconductor	[39]
	Lead	Highly stable but toxic, reactive	[40]
	Copper	Thermal conductivity, ductile, electrical conductivity	[41]
	Zinc	Antibacterial property, UV filtration, antifungal	[42]

(Continued)

Table 2.2 (*Continued*)

Types	Nanoparticle	Properties	Reference
	Titanium oxide	Inhibition of bacterial growth, greater surface area	[43]
	Iron oxide	Possess instability and reactivity	[44]
	Magnetite	Magnetic, highly reactive	[45]
Metal oxide–based nanoparticles	Silicon dioxide	Ability for functionalization of molecules, stability	[46]
	Zinc oxide	Antifungal and bacterial property, anti-corrosive	[47]
	Cerium oxide	Lower reduction potential, antioxidant properties	[48]
	Aluminum oxide	Greater reactivity, more sensitivity to heat, moisture, and light	[49]

2.6.1 Chemical Properties

The chemical properties of nanoparticles include reactivity properties, for instance target and stability factors. Sensitivity to atmosphere, moisture, light, and heat determines the applications. Antifungal, antibacterial, and toxicity properties are effective for their utilization in biomedical applications. Properties such as oxidation, reduction, corrosivity, and flammability determine the characteristics of nanoparticles [55].

2.6.2 Physical Characteristics

The physical characteristics of nanoparticles include color, absorption, light penetration, and UV absorption. Reflection properties and penetration of light influence the properties of nanoparticles. In addition, mechanical properties include flexibility, tensile strength, elasticity and ductility. While other properties include hydrophobicity, suspension, and diffusion characteristics.

Electrical and magnetic properties such as resistivity, conductivity, and semi-conductivity are aspects that affect the properties of nanoparticles [7].

2.7 Characterization

The application and potential of nanoparticles are determined through the unique characteristics they possess. Table 2.3 and Fig. 2.4 demonstrate the characterization of nanoparticles based on different techniques of measurement.

1. **Size:** The basis of nanoparticle measurement and characterization is particle. It decides the category in which the size range of nanoparticle falls, i.e., micro or nano. Electron microscopy is used to measure particle distribution and size, whereas transmission electron microscopy (TEM) and scanning electron microscopy (SEM) measure clusters and particles. The bulk samples are measured using laser diffraction in solid phases [48]. The liquid phase particle measurement is done using centrifugation techniques and photon-related spectroscopy. For gaseous phase particles, scanning mobility particle sizer (SMPS) is used as this technique provides quick and accurate measures in contrast to other methods.

2. **Surface area:** It is an important factor as the surface-area-to-volume ratio can play a crucial role in determining the properties and performance of nanoparticles. The Brunauer–Emmett–Teller (BET) analysis is used for measurement in which the analysis of surface area can be done by a simple titration (for particles in the liquid state). Since it is a process that requires intense labor, nuclear magnetic resonance (NMR) spectroscopy is employed. For the measurement of the surface area of particles in the gaseous phase, differential mobility analyzer (DMA) and SMPS are used.

3. **Composition:** The efficacy and performance of nanoparticles are the functions of their elemental and chemical purity. The higher the impurity (i.e., undesired or secondary elements), the lower the performance and the efficacy of a nanoparticle. This may also cause secondary or undesired reactions.

Table 2.3 Characterization methods of nanomaterials in different phases (solid, liquid, and gaseous) [11]

Characterization	Solid	Liquid	Gas
Size	Electron optical microscope and laser diffraction for bulk samples	Photon relationship spectroscopy and centrifugation	Scanning mobility, particle sizer, and optical particle counter
Surface area	Brunauer–Emmett–Teller isotherm	Normal titration and nuclear magnetic resonance spectroscopy experiments	Simple titration and nuclear magnetic resonance spectroscopy
Composition	X-ray photoelectron spectroscopy and chemical digestion followed by wet chemical analysis for bulk samples	Chemical digestion for mass spectrometry, atomic emission spectroscopy, and ion chromatography	Particles are collected for analysis by spectrometric or wet chemical techniques
Surface morphology	Image analysis of electron micrographs	Deposition on the surface for electron microscopy	Capture particles electrostatically or by filtration for imaging using electron microscopy
Surface charge	Zeta potential	Zeta potential	Differential mobility analyzer
Crystallography	Powder's X-ray or neutron diffraction	—	—
Concentration	—	—	Condensation particle counter

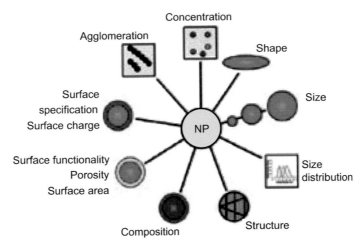

Figure 2.4   Characterization of nanoparticles [50].

X-ray photoelectron spectroscopy (XPS) is used to determine the composition of nanoparticles [7]. Other techniques such as ion chromatography, mass spectroscopy, and atomic emission spectroscopy determine the particle composition by chemically digesting the particle and then performing these wet chemical analysis techniques. For the particles present in the gaseous state, collection is done using electrostatic and spectrometric methods, filtration or techniques using wet chemicals [7].

4. **Morphology of surface:** Nanoparticles have different surface structures, and their shapes play an important role in the evaluation of their properties. Some of the nanoparticles have cylindrical, spherical, flat, tubular, conical, and irregular shapes with amorphous surface. The surface morphology is generally determined by TEM and SEM [54]. Nanoparticles in the liquid mixture phase are placed on a surface and analyzed, whereas nanoparticles in the gaseous state are captured electrostatically or by filtration for imaging using electron microscopy.

5. **Surface charge:** It demonstrates the interaction of nanoparticles with their target. Usually surface charge and the stability of a dispersion in a solution are determined using

a Zeta potentiometer [51]. For particles in the gaseous phase, a DMA is used [7].

6. **Crystallography:** It is the study of crystal structure and arrangement of atoms and molecules. It is done using neutron or electron diffraction or powder X-ray methods [55].

7. **Concentration:** For gaseous phase nanoparticles, concentration determines the volume of gas or air required for the method. The amount of nanoparticles per unit volume of gas or air, their distribution and size determine their performance and efficacy. These measurements are done preferably by using a condensation particle counter (CPC) [7].

2.8 Nanoparticles and Their Technological Enhancements

2.8.1 Biomedical and Nanomedical Applications

In the healthcare system, the application of nanotechnology and its benefits are known as nanomedicine. All around the world, these are being used to improve the health of patients suffering from different diseases such as chronic pain, menopause symptoms, fungal infections, kidney diseases, asthma, multiple sclerosis, hyperlipidemia, and cancers such as ovarian and breast cancers.

The current nanomedicines being used have an advantage of better drug delivery in comparison to the normal medical approaches, e.g., increasing the water solubility of poorly soluble drugs and also increasing the drug-retention time in the body. There are also off-target effects of drugs, for instance drugs that target cancer also affect other healthy cells in different organs. These difficulties can be overcome by the use of nanomedicines. They prolong the drug-retention time in the body and enhance the organ targeting [56].

2.8.2 Smart Drug-Delivery Technology

Since the past few years, there has been unprecedented growth in biomedical nanotechnology from the conventional drug-delivery system to smart drug-delivery systems with stimuli-responsive characteristics. Well-defined nanoplatforms can enhance the drug-targeting efficiency as well as the mean retention time and can decrease side effects, which are essential for improving patient passivity. In the academic field, various smart drug-delivery systems have been used for various interesting systems, such as stimuli-responsive liposomes, metal oxides polymeric nanoparticles, and exosomes [57].

2.8.3 Nanopharmaceuticals

Nanopharmaceuticals have the ability to identify diseases at much prior stages and then provide diagnostic applications for building up conventional measures using nanoparticles. In the pharmaceutical industry, a long-pending dispute is the difficulty of delivering the appropriate dose of a particle-active agent to specific disease site. Nanopharmaceuticals have an enormous potential in addressing this failure of traditional therapeutics, which offer site-specific targeting of active agents and ultimately reduce the toxic system side effects. Therefore, pharmaceutical companies are applying nanotechnology to enhance or supplement drug target discovery and drug delivery [58].

2.8.4 Cancer Treatments

About 15 million cases of cancer have been reported by the WHO. The spread of cancer to vital organs is estimated to be about 90%, and this process is referred to as metastasis [59]. The nanoparticles coated with polymeric micelles help in delivering the drugs to tumors. These coated iron oxide nanoparticles help in breaking the bacteria, which leads to the effective treatment of bacterial infections. The stimulation of immune response is mediated by the change in the surface of protein nanoparticles. The target treatment involves chemotherapy delivering the tumor-killing agent Tumor Necrosis Factor alpha (TNF-α) for the treatment of cancer. TNF binds to gold nanoparticles in conjunction with PEG (polyethylene glycol) thiol, thus hiding the TNF-bearing nanoparticles from the immune system.

The nanoparticles help in circulation through the bloodstream without being attacked. In addition, another technique for the destruction of cancer is mediated by heat. These nanoparticles are called Auro Shells, which absorb infrared light from the laser, thus turning light into heat. The company that developed this technique is Nanospectra Biosciences [60]. In addition, "nano trains" have been developed through connections between DNA strands, which are effective in delivering the chemotherapeutic drugs targeting cancer cells. For the enhancement of the drug-bearing nanoparticles to enter the tumor, a photosensitizing agent is utilized. Initially, accumulation in the tumor occurs, which is then illuminated by IR light. This agent then causes the blood vessels in the tumor to become porous, hence increasing the drug-loading capacity of nanoparticles, which bear its entry into the tumor. Recently, nanosponges have been developed that absorb toxins and help in their removal from the bloodstream. They are basically polymer nanoparticles coated with RBC membrane. This membrane then allows the passage of the nanosponges from the bloodstream and attracts the toxins to it.

2.8.5   Application in Diagnostic Technique

For the detection of cancer cells, researchers at the Worcester Polytechnic Institute have attached antibodies to carbon nanotubes in chips. This method can be utilized in a simple laboratory, which can help in early detection of cancer. These carbon nanotubes can be embedded into a gel with the aid of a sensor, which can be injected under the skin for monitoring the level of nitric oxide [61].

2.8.6   Sensors with the Aid of Nanotechnology

Sensors can help in the detection of a very small amount of chemical vapors. Certain detecting elements such as palladium, zinc oxide, and carbon nanotubes are used. After absorption, they lead to changes in electrical characteristics such as capacitance and resistance. Thereby, the detection of chemical vapors is allowed to change the electrical properties due to the smaller size of nanowires, tubes, and particles [62].

2.8.7 Nanotechnology in Food Science

Clay nanocomposites provide a barrier to gases such as carbon dioxide and are used in packaging films and light weight bottles. With the aid of silver nanoparticles, storage bins have been produced. Health risks can be minimized by the utilization of silver nanoparticles, which kill bacteria present in bins. In order to produce cost-effective sensors on surfaces such as plastic films for packaging food, a method that includes scattering carbon nanotubes on plastic exteriors has been demonstrated by researchers; the sensors can identify spoiled food. Silicate nanoparticles are utilized for providing a barrier to gases such as oxygen and moisture. In addition, they lead to minimal chances of spoilage of food. Zinc oxide incorporated into plastic packaging blocks UV rays and helps prevent bacterial infection. In addition, it provides stability and strength, thus providing bacterial protection when incorporated in plastic packages [63].

2.8.8 Nanoparticles in Ophthalmic Delivery

Nanoparticles help in overcoming the disadvantages of poorly soluble drugs in the lacrimal secretion of eye. They offer prolonged residence time to the nanosuspension in the sac, playing a vital role in the treatment of ocular diseases, thus maintaining the tonicity of eyes. The stability of cloricromens in the eye formulations also helps in enhancing bioavailability at the ocular lever through the utilization of solvent evaporation technique [64].

2.8.9 Nanotechnology in Heart Disease

The utilization of nanoparticles helps in the management of cardiovascular diseases. Nanoparticles help in

- Treatment of defective valves of heart
- Detection and treatment of arterial plaque
- Understanding of the functioning of heart tissue at the sub-cellular level

The delivery of cardiac stem cells helps in delivering nanoparticles to the injured tissue. To increase the amount, they are combined with nanovesicles, which attract them to the injured tissue. Researchers have used gold nanoparticles with a protein (collagen). Wrong levels of collagen in the heart valves affect their functioning. An increased amount of collagen causes stiffening of the valves, while a lesser amount helps in making the valves floppy. Gold nanoparticles when combined with collagen helps in changing the mechanical properties of valves, which repairs defective heart valves [65].

2.9   Conclusion

There have been rapid advancements in the field of nanotechnology, which has led to the enhancement of its performance and efficiency. It has led to a cleaner environment along with fresh air, better renewable resources, and water for the future development. In addition, investments have increased in the field of nanotechnology, especially in the field of research and development in various organizations and industries. Rapid research procedures have been carried out for the implementation of the technology. A wide range of applications have been utilized for increasing the efficiency and performance of objects along with cost reduction. Hence, the field of nanotechnology has the prospects of being a better and eco-friendly technology in the future.

References

1. Batista, Carlos, A. and Larson, Ronald, G. (2015). Nonadditivity of nanoparticle interactions. *Science,* **350**(6257) pp. 1242477.

2. The Royal Society and The Royal Academy of Engineering. (2004). Nanoscience and nanotechnologies: Opportunities and uncertainties. Available at https://royalsociety.org/~/media/Royal_Society_Content/policy/publications/2004/9693.pdf, Accessed May 3, 2018.

3. Giron, F., Pastó, A., Tasciotti, E., and Abraham, B. P. (2019). Nanotechnology in the treatment of inflammatory bowel disease. *Inflamm. Bowel Dis.,* **25**(12), pp. 1871–1880.

4. Khan, I., Saeed, K., and Khan, I. (2019). Nanoparticles: Properties, applications and toxicities. *Arab J. Chem.,* **12**, pp 908–931.

5. Yahya, I., Atif, R., Ahmed, L., Eldeen, T. S., Omara, A., and Eltayeb, M. (2019). Utilization of solid lipid nanoparticles loaded anticancer

agents as drug delivery systems for controlled release. *Int. J. Eng. Appl. Sci. Technol.*, **3**(12), pp. 7–16.

6. Tiwari, D. K., Behari, J., and Sen, P. (2008). Application of nanoparticles in wastewater treatment. *Nanobiotechnol. Bioformulat.*, **3**(3), pp. 417–433

7. Ealias, A. M. and Saravanakumar, M. P. (2017). A review on the classification, characterisation, synthesis of nanoparticles and their application. *IOP Conf. Ser. Mater. Sci. Eng.*, **263**, pp. 032019.

8. Nisini, R., Poerio, N., Mariotti, S., De Santis, F., and Fraziano, M. (2018). The multirole of liposomes in therapy and prevention of infectious diseases. *Front. Immunol.*, **9**, pp. 155.

9. Salavati-Niasari, M., Davar, F., and Mir, N. (2008). Synthesis and characterization of metallic copper nanoparticles via thermal decomposition. *Polyhedron*, **27**(17), pp. 3514–3518.

10. Tai, C. Y., Tai, C. T., Chang, M. H., and Liu, H. S. (2007). Synthesis of magnesium hydroxide and oxide nanoparticles using a spinning disk reactor. *Ind. Eng. Chem. Res.*, **46**(17), pp. 5536–5541.

11. Bhaviripudi, S., Mile, E., Steiner, S. A., Zare, A. T., Dresselhaus, M. S., Belcher, A. M., and Kong, J. (2007). CVD synthesis of single-walled carbon nanotubes from gold nanoparticle catalysts. *J. Am. Chem. Soc.*, **129**(6), pp. 1516–1517.

12. Ramesh, S. (2013). Sol–gel synthesis and characterization of nanoparticles. *J. Nanosci.*, 2013. DOI: 10.1155/2013/929321.

13. Mann, S., Burkett, S. L., Davis, S. A., Fowler, C. E., Mendelson, N. H., Sims, S. D., et al. (1997). Sol–gel synthesis of organized matter. *Chem. Mater.*, **9**(11), pp. 2300–2310.

14. Mohammadi, S., Harvey, A., and Boodhoo, K. V. (2014). Synthesis of TiO_2 nanoparticles in a spinning disc reactor. *Chem. Eng. J.*, **258**, pp. 171–184.

15. Adachi, M., Tsukui, S., and Okuyama, K. (2003). Nanoparticle synthesis by ionizing source gas in chemical vapor deposition. *Jap. J. Appl. Phys.*, **42**(1A), L77.

16. Kammler, H. K., Mädler, L., and Pratsinis, S. E. (2001). Flame synthesis of nanoparticles. *Chem. Eng. Technol. Ind. Chem. Plant Equip. Process Eng. Biotechnol.*, **24**(6), pp. 583–596.

17. Singh, R. and Lillard Jr, J. W. (2009). Nanoparticle-based targeted drug delivery. *Exp. Mol. Pathol.*, **86**(3), pp. 215–223.

18. Barratt, G. M. (2000). Therapeutic applications of colloidal drug carriers. *Pharmaceut. Sci. Tech. Today*, **3**, pp. 163–171.

19. Couvreur, P., Dubernet, C., and Puisieux, F. (1995). Controlled drug delivery with nanoparticles: Current possibilities and future trends. *Eur. J. Pharm. Biopharm.*, **41**(1), pp. 2–13.

20. Pitt, G. G., Gratzl, M. M., Kimmel, G. L., Surles, J., and Sohindler, A. (1981). Aliphatic polyesters II. The degradation of poly (DL-lactide), poly (ε-caprolactone), and their copolymers in vivo. *Biomaterials*, **2**(4), pp. 215–220.

21. Marsalek, R. (2014). Particle size and zeta potential of ZnO. *APCBEE Procedia*, **9**, pp. 13–17.

22. Amendola, V. and Meneghetti, M. (2009). Laser ablation synthesis in solution and size manipulation of noble metal nanoparticles. *Phys. Chem. Chem. Phys.*, **11**(20), pp. 3805–3821.

23. Shah, P. and Gavrin, A. (2006). Synthesis of nanoparticles using high-pressure sputtering for magnetic domain imaging. *J. Magn. Magn. Mater.*, **301**(1), pp. 118–123.

24. Lugscheider, E., Bärwulf, S., Barimani, C., Riester, M., and Hilgers, H. (1998). Magnetron-sputtered hard material coatings on thermoplastic polymers for clean room applications. *Surf. Coat. Tech.*, **108**, pp. 398–402.

25. Tenne, R. (2002). Fullerene-like materials and nanotubes from inorganic compounds with a layered (2-D) structure. *Colloid Surf. A Physicochem. Eng. Aspects*, **208**(1–3), pp. 83–92.

26. Huang, X., Boey, F., and Zhang, H. A. (2010). A brief review on graphene-nanoparticle composites. *Cosmos*, **6**(02), pp. 159–166.

27. De Volder, M. F., Tawfick, S. H., Baughman, R. H., and Hart, A. J. (2013). Carbon nanotubes: Present and future commercial applications. *Science*, **339**(6119), pp. 535–539.

28. Awasthi, K., Kumar, R., Raghubanshi, H., Awasthi, S., Pandey, R., Singh, D., et al. (2011). Synthesis of nano-carbon (nanotubes, nanofibres, graphene) materials. *Bull. Mater. Sci.*, **34**(4), pp. 607.

29. Fawole, O. G., Cai, X. M., and MacKenzie, A. R. (2016). Gas flaring and resultant air pollution: A review focusing on black carbon. *Environ. Pollut.*, **216**, pp. 182–197.

30. Geetha, P., Latha, M. S., Pillai, S. S., Deepa, B., Kumar, K. S., and Koshy, M. (2016). Green synthesis and characterization of alginate nanoparticles and its role as a biosorbent for Cr (VI) ions. *J. Mol. Struct.*, **1105**, pp. 54–60.

31. Harshiny, M., Iswarya, C. N., and Matheswaran, M. (2015). Biogenic synthesis of iron nanoparticles using *Amaranthus dubius* leaf extract as a reducing agent. *Powder Tech.*, **286**, pp. 744–749.

32. Syed, B., Prasad, N. M., and Satish, S. (2016). Endogenic mediated synthesis of gold nanoparticles bearing bactericidal activity. *J. Microsc. Ultrastruct.*, **4**(3), pp. 162–166.

33. Bau, V. M., Bo, X., and Guo, L. (2017). Nitrogen-doped cobalt nanoparticles/nitrogen-doped plate-like ordered mesoporous carbons composites as noble-metal free electrocatalysts for oxygen reduction reaction. *J. Energy Chem.*, **26**(1), pp. 63–71.

34. Osuntokun, J. and Ajibade, P. A. (2016). Morphology and thermal studies of zinc sulfide and cadmium sulfide nanoparticles in polyvinyl alcohol matrix. *Physica B Condensed Matter*, **496**, pp. 106–112.

35. Tyszczuk-Rotko, K., Sadok, I., and Barczak, M. (2016). Thiol-functionalized polysiloxanes modified by lead nanoparticles: Synthesis, characterization and application for determination of trace concentrations of mercury (II). *Micropor. Mesopor. Mater.*, **230**, pp. 109–117.

36. Ryu, C. H., Joo, S. J., and Kim, H. S. (2016). Two-step flash light sintering of copper nanoparticle ink to remove substrate warping. *Appl. Surf. Sci.*, **384**, pp. 182–191.

37. Bogutska, K. I., Sklyarov, Y. P., and Prylutskyy, Y. I. (2013). Zinc and zinc nanoparticles: Biological role and application in biomedicine. *Ukrainica Bioorganica Acta*, **1**, pp. 9–16.

38. Laad, M. and Jatti, V. K. S. (2018). Titanium oxide nanoparticles as additives in engine oil. *J. King Saud Univ. Eng. Sci.*, **30**(2), pp. 116–122.

39. Ruales-Lonfat, C., Barona, J. F., Sienkiewicz, A., Bensimon, M., Vélez-Colmenares, J., Benítez, N., et al. (2015). Iron oxides semiconductors are efficients for solar water disinfection: A comparison with photo-Fenton processes at neutral pH. *Appl. Catal. B Environ.*, **166**, pp. 497–508.

40. Carlos, L., Einschlag, F. G., González, M. C., and Mártire, D. O. (2013). Applications of magnetite nanoparticles for heavy metal removal from wastewater. In *Waste Water: Treatment Technologies and Recent Analytical Developments*, pp. 63–77. IntechOpen.

41. Kaynar, Ü. H., Şabikoğlu, I., Kaynar, S. Ç., and Eral, M. (2016). Modeling of thorium (IV) ions adsorption onto a novel adsorbent material silicon dioxide nano-balls using response surface methodology. *Appl. Rad. Isotopes*, **115**, pp. 280–288.

42. Bajpai, S. K., Jadaun, M., and Tiwari, S. (2016). Synthesis, characterization and antimicrobial applications of zinc oxide nanoparticles loaded gum acacia/poly (SA) hydrogels. *Carbohydr. Polym.*, **153**, pp. 60–65.

43. Kim, S. J. and Chung, B. H. (2016). Antioxidant activity of levan coated cerium oxide nanoparticles. *Carbohydr. Polym.*, **150**, pp. 400–407.

44. Munuswamy, D. B., Madhavan, V. R., and Mohan, M. (2015). Synthesis and surface area determination of alumina nanoparticles by chemical combustion method. *Int. J. Chem. Tech. Res.*, **8**, pp. 413–419.

45. http://50.87.149.212/sites/default/files/images/knowledge-base/tox/NanoparticleCharacteristics_0.jpg

46. D'Amato, R., Falconieri, M., Gagliardi, S., Popovici, E., Serra, E., Terranova, G., et al. (2013). Synthesis of ceramic nanoparticles by laser pyrolysis: From research to applications. *J. Anal. Appl. Pyrolysis*, **104**, pp. 461–469.

47. Kuppusamy, P., Yusoff, M. M., Maniam, G. P., and Govindan, N. (2016). Biosynthesis of metallic nanoparticles using plant derivatives and their new avenues in pharmacological applications: An updated report. *Saudi Pharm. J.*, **24**(4), pp. 473–484.

48. Hasan, S. (2015). A review on nanoparticles: Their synthesis and types. *Res. J. Recent Sci.*, **4**, pp. 9–11.

49. Yadav, T. P., Yadav, R. M., and Singh, D. P. (2012). Mechanical milling: A top down approach for the synthesis of nanomaterials and nanocomposites. *Nanosci. Nanotechnol.*, **2**(3), pp. 22–48.

50. Pimpin, A. and Srituravanich, W. (2012). Review on micro- and nanolithography techniques and their applications. *Eng. J.*, **16**(1), pp. 37–56.

51. Asuri, P., Karajanagi, S. S., Dordick, J. S., and Kane, R. S. (2006). Directed assembly of carbon nanotubes at liquid–liquid interfaces: Nanoscale conveyors for interfacial biocatalysis. *J. Am. Chem. Soc.*, **128**(4), pp. 1046–1047.

52. Narayanan, R., Zhu, G., and Wang, P. (2007). Stabilization of interface-binding chloroperoxidase for interfacial biotransformation. *J. Biotechnol.*, **128**(1), pp. 86–92.

53. Qhobosheane, M., Zhang, P., and Tan, W. (2004). Assembly of silica nanoparticles for two-dimensional nanomaterials. *J. Nanosci. Nanotechnol.*, **4**(6), pp. 635–640.

54. Shipway, A. N., Katz, E., and Willner, I. (2000). Nanoparticle arrays on surfaces for electronic, optical, and sensor applications. *ChemPhysChem*, **1**(1), pp. 8–52.

55. Greulich, C., Diendorf, J., Simon, T., Eggeler, G., Epple, M., and Köller, M. (2011). Uptake and intracellular distribution of silver nanoparticles in human mesenchymal stem cells. *Acta Biomaterialia*, **7**(1), pp. 347–354.

56. Hanley, C., Thurber, A., Hanna, C., Punnoose, A., Zhang, J., and Wingett, D. G. (2009). The influences of cell type and ZnO nanoparticle size on immune cell cytotoxicity and cytokine induction. *Nanoscale Res. Lett.*, **4**(12), pp. 1409.

57. Faunce, T. A. and Vines, T. (2009). Assessing the safety and cost-effectiveness of early nanodrugs. *J. Law Med.*, **16**, pp. 822–845.

58. Benson, H. A., Sarveiya, V., Risk, S., and Roberts, M. S. (2005). Influence of anatomical site and topical formulation on skin penetration of sunscreens. *Ther. Clin. Risk Manag.*, **1**(3), pp. 209.

59. Gadad, A. P., Kumar, S. V., Dandagi, P. M., Bolmol, U. B., and Pallavi, N. P. (2014). Nanoparticles and their therapeutic applications in pharmacy. *Int. J. Pharm. Sci. Nanotechnol.*, **7**(3), pp. 2515–2516.

60. Joshi, A., Singh, N., and Verma, G. (2016). Preparation and applications of self-assembled natural and synthetic nanostructures. In *Fabrication and Self-Assembly of Nanobiomaterials*, pp. 29–55. Elsevier, Inc.

61. Belloni, J., Mostafavi, M., Remita, H., Marignier, J. L., and Delcourt, M. O. (1998). Radiation-induced synthesis of mono-and multi-metallic clusters and nanocolloids. *New J. Chem.*, **22**(11), pp. 1239–1255.

62. Iravani, S. (2011). Green synthesis of metal nanoparticles using plants. *Green Chem.*, **13**(10), pp. 2638–2650.

63. Bello, S. A., Agunsoye, J. O., and Hassan, S. B. (2015). Synthesis of coconut shell nanoparticles via a top down approach: Assessment of milling duration on the particle sizes and morphologies of coconut shell nanoparticles. *Mater. Lett.*, **159**, pp. 514–519.

64. Khan, I., Ali, S., Mansha, M., and Qurashi, A. (2017). Sonochemical assisted hydrothermal synthesis of pseudo-flower shaped bismuth vanadate ($BiVO_4$) and their solar-driven water splitting application. *Ultrasonics Sonochem.*, **36**, pp. 386–392.

65. Khan, I., Yamani, Z. H., and Qurashi, A. (2017). Sonochemical-driven ultrafast facile synthesis of SnO_2 nanoparticles: Growth mechanism structural electrical and hydrogen gas sensing properties. *Ultrasonics Sonochem.*, **34**, pp. 484–490.

Multiple Choice Questions

1. Particles less than 100 nm are referred to as
   - (a) Liposomes
   - (b) Nanoparticles
   - (c) Microparticles
   - (d) Emulsions

2. The following are not examples of metal-based nanoparticles:
   - (a) Silver
   - (b) Cadmium
   - (c) Titanium oxide
   - (d) Gold

3. Carbon nanoparticles that consist of c-60 molecules of sp^2 hybridization are called
   - (a) Fullerenes
   - (b) Carbon nanofiber
   - (c) Carbon black
   - (d) Graphene

4. The spherical aggregates of steroids and phospholipids that spontaneously form and are dispersed in the aqueous medium are called
   - (a) Nanocarriers
   - (b) Liposomes
   - (c) Polymeric nanoparticles
   - (d) Ceramics nanoparticles

5. Following are not examples of the top-down method of synthesis:
   - (a) Sputtering
   - (b) Laser ablation
   - (c) Thermal decomposition
   - (d) Spinning

6. Which method is utilized for the determination of surface charge in gaseous phase?
   - (a) Condensation particle counter
   - (b) Differential mobility analyzer
   - (c) Zeta potentiometer
   - (d) None of the above

7. Bottom-up method is referred to as
   - (a) Constructive method
   - (b) Destructive method
   - (c) Both
   - (d) None of the above

8. The tubular and elongated structures of diameter 1–2 nm are called
   - (a) Semiconductor nanoparticles
   - (b) Ceramics nanoparticles
   - (c) Carbon nanotubes
   - (d) Carbon nanofiber

9. Which of the following is not an advantage of nanoparticles?
   - (a) High immunogenicity
   - (b) Greater bioavailability
   - (c) Increased drug loading
   - (d) None

10. The loading capacity of a drug is directly proportional to the interactions of the polymer with the drug.
    - (a) True
    - (b) False
    - (c) Not given
    - (d) None

Answer Key

1. (b)   2. (c)   3. (a)   4. (b)   5. (d)   6. (c)   7. (a)
8. (c)   9. (a)   10. (a)

Short Answer Questions

1. Explain in brief the delivery system of nanoparticles along with their properties.
2. Classify nanoparticles based on their chemical nature.
3. Discuss carbon-based nanoparticles.
4. Describe in brief various applications of nanoparticles.
5. Enlist the top-down methods of synthesis.

Long Answer Questions

1. Describe in detail the various method of synthesis of nanoparticles.
2. What are nanoparticles and classify them in detail?
3. Write a note on the various characterization parameters of nanoparticles.
4. Discuss nanoparticles and their associated technological advancements.

Chapter 3

Nanographenes for Renewable Energy

Parth Malik,[a] Showkat Hassan Mir,[b] Rachna Gupta,[c] and Tapan K. Mukherjee[d]

[a]*School of Chemical Sciences, Central University of Gujarat, Gandhinagar, India*
[b]*Department of Chemistry, IIT Kanpur, Kanpur, UP, India*
[c]*Department of Biotechnology, Visva-Bharati, Santiniketan (Bolpur), West Bengal, India*
[d]*Department of Internal Medicine, University of Utah, Salt Lake City, Utah, USA*
tapan400@gmail.com

The past few years have witnessed significant advancements in the development of revolutionary products having manifold improved mechanical strength, electrical conductivity, transparency besides shape- and size-driven modulations. Among many such materials, nanoscale graphene flakes (also known as nanographene, which is graphene with limited lateral dimensions) have emerged as a promising agent owing to remarkable manipulation of the graphene physics and chemistry. The marketable extent of nanographene is presently on the right track to challenge most advanced conventional materials (such as carbon fibers and carbon

Nanotechnology: Principles and Applications
Edited by Rakesh K. Sindhu, Mansi Chitkara, and Inderjeet Singh Sandhu
Copyright © 2021 Jenny Stanford Publishing Pte. Ltd.
ISBN 978-981-4877-43-5 (Hardcover), 978-1-003-12026-1 (eBook)
www.jennystanford.com

nanotubes, CNTs) for future applications within the domain of flexible electronics, composites, and energy storage. The most distinctive structural attributes of these materials comprise their long- and short-range orders in self-assembled configurations apart from multiple functional groups at the edge of basil planes. With the versatility, flexibility, and dynamic self-adjusting ability of graphene, the formation of its nanoscale architectures further advances these potentials to a great extent. A high surface area (SA) of these nanoscale materials confers them remarkable cyclic durability and considerable charge storage capacity capable of being used as lithium-ion battery (LIB) anodes. Pertaining to their energy storage abilities, it is worth noting that these materials are semiconductors having tunable electronic structures expressible within the organic solvents. Significant improved functional aspects of these nanostructures are due to their elongated structure having numerous oxygen-containing functional groups and optimal interlayer spacing (between two consecutive layers, allowing for Li adsorption and diffusion). Inclusion of these entities has significantly improved the efficiency of energy conversion, charge storage, and photo-absorption. With these incentives, this chapter focuses on the possible improvement in renewable energy utilization through nanographene inclusion or involvement.

3.1   Introduction

Since the inceptive finding of graphene by Geim and Novoselov in 2004, many researchers have made provisions and optimum fits to accommodate graphene and its derivatives within their application-oriented research, from the basic to applied domains [1]. Several remarkable physical and chemical properties of graphene consolidate a significant platform toward realistic quantum research, substantiated from its versatile surface functionalities [2, 3]. Natively composed of two-dimensional single-layered carbon atoms, graphene serves as a building block for extending the dimensions of constitutional derivatives, such as fullerenes (0-dimensional), nanotubes (one-dimensional), and graphite (three-dimensional) [4, 5]. Perhaps it is only the two-dimensional structure that confers an electronic behavior with linear dispersion closer to Fermi energy echelon. Moreover, it acts as the most suitable candidate for electronic applications, owing to its large charge carrier mobility $(2 \times 10^5 \text{ cm}^2/\text{V·s})$ [6]. The much-envisaged application prospects of graphene arise from its inherent stable nature, versatility of being functionalized in manifold ways, and the engineered surface features, encompassed via high aspect ratios and absorptive capabilities [1, 7–10]. Besides these, it also possesses strong intrinsic breaking strength (42 N/m), high optical transmittance (97.7%), and uncharacteristically remarkable thermal conduction [11–14]. Such characteristics have been the basis of tremendous advancements in making stimulus-sensitive electronic gadgets, drug-delivery biological and electromechanical sensing, energy storage, and others [15–21].

Bestowed with an infinite network (monolayer benzene rings) and missing band gap, graphene finds immense suitability in charge storage, electronic modulation, and efficient energy conversion [22]. The planar structure and hexagonal honeycomb-like pattern are the prime factors impairing its steadfast usage, often culminating in aggregation and poor aqueous solubility. Thereby, to overcome these hindrances, the properties of graphene are tuned via size and shape control, bringing together numerous defects and edges in order to reduce the infinite network organization [23–25]. Such modifications are the fundamental genesis of nano-sized graphene, i.e., nanographene (NG) containing nonzero band gap with confined size and edge state that makes them superior to graphene. This NG size ranges from 1 to 100 nm, where infinite π-electron system is disordered and exhibits properties intermediate of bulk graphene and large polycyclic aromatic hydrocarbon (PAH) molecules [26]. Apart from this, improvement in synthesis approaches for controllable preparation of NG having defined size, shape, and edge state in high quality and quantities is one of the major challenges in the latest science and technology uptake themes [27, 28]. The bottom-up methodology has been studied with greater interest toward size-controllable NG generation, wherein chemical vapor deposition (CVD) is used for synthesis via concurrent participation of organic precursors and appropriate doping agents [29, 30]. Contrary to this, the top-down methods (thermal processing, arc discharge, and plasma treatment) lack a precise control of growing size extents and wide-ranged geometry. Several studies have reported the NG preparation through varied shape, size, constituents, and edges. For instance, Halleux et al. reported planar NG C_{42} significantly for the first time using hexa-peri-hexabenzocoronene processing (Fig. 3.1) [31]. Mullen et al. successfully prepared the series of planar NG containing 48, 72, 96, 132, 150, and 222 carbons using the Diels–Alder reaction followed by cyclo-trimerization. The process used oligophenylene precursor aligned in a planar configuration using oxidative cyclodehydrogenation to the hexagonally positioned PAH molecules [32]. The resultant PAH molecules emerged as the smallest NG with sp^2 carbon framework of 1 nm. Another study by Itami et al. reported NG preparation and subsequent $C_{80}H_{30}$ formation through nonhexagonal ring introduction within an NG subunit, suggesting significant breakthrough in graphene chemistry. Such methodology enabled NG comprising approximately 80 C, joined in a 26-ring framework having 30 H positioned over rims [33]. Similarly, Suzuki et al. used the Scholl reaction to develop two aromatic saddles, i.e., tetrabenzo-8-circulene (TB8C) and its octamethyl derivative (OM-TB8C). This mechanism for TB8C synthesis optimized the research methodology for preparing other curved NG and graphene molecules [34].

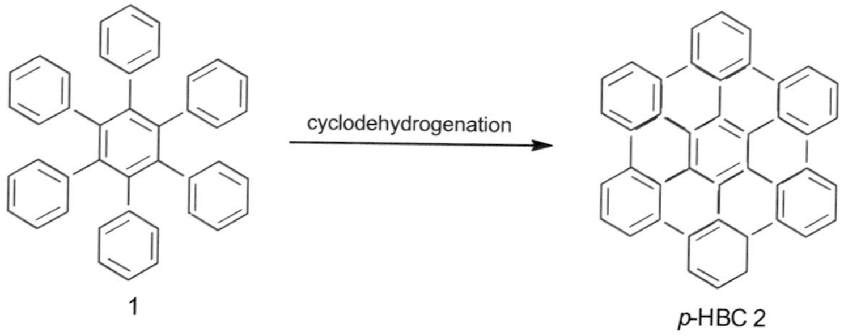

Figure 3.1 Synthesis of NG C_{42} using the hexa-peri-hexabenzocoronene process.

Furthermore, recently a synthesis approach has been developed toward modifying the NG structure and size, advancing its potential toward nanoelectronics, optoelectronics, and spintronics development [35–37]. With these advancements, the present-day research is focused on the NG formation over the insulated substrate in order to simplify the transfer process for graphene device fabrication. In one such study, Ismach et al. utilized Cu as the sacrificial layer for graphene development on quartz via CVD, while Rummeli et al. used catalyst-free CVD at low temperature for controllable graphene growth on MgO substrate as high temperature is required for Cu residues removal [38, 39]. Similarly, Lee et al. used glass as substrate for 15 nm NG scalable growth via temperature-sensitive annealing, through introducing an Ni layer [40].

Recently, among the rigorously studied carbon materials, CNTs, activated carbons, nanohorns, and nanographenes have attracted significant interest for energy storage applications, manifested by high pore volume, SA, low density, and tunable pore morphology [41]. The H_2 molecules are capable of being adsorbed on the surface of carbon, facilitating a further increment in the presence of metal nanoparticles (NPs) such as Pt or Ni [42, 43]. For example, Li et al. described the octagonal vacancies at the center of TB8C and OM-TB8C, having the capability of metal binding and further enhanced H_2 storage. Their outcomes exhibited TB8C as superior carbon material than pristine graphene. Furthermore, fabrication of NG with amino groups enabled enhanced adsorption for metal atoms because of the increment in the central electron density. Therefore, the covalent binding of metals to NG considerably improved the native physical and chemical properties, exhibiting significant energy storage potential. The H atoms mainly adsorbed to the edge sites (rather than inner ones) and the curved surface than flat one. Thus, such H_2 adsorption indicates its efficient modulation through the NG structure, so its size and structure should be controlled to enhance energy storage.

3.2   Major Aspects of Graphene Chemistry

Due to the manifold unique physical properties, graphene has rapidly emerged as a superior material in diversified domains, lending manifold fascinations to be pursued within the cross-linked domains of physics, chemistry, and material sciences. At present, graphene is the thinnest nanomaterial known on earth with remarkable electrical, thermal, mechanical, optical, and electronic attributes. A high specific SA, chemical inertness, optical transmittance, restoring ability, porous texture, biological compatibility, and tunable band gap are some eye-catching features of graphene materials chemistry

[44, 45]. Some of the exciting graphene properties include half-integer room-temperature quantum Hall effect, long-range ballistic transport, charge carriers prevalence as massless relative quasiparticles with quantum confinement (QC) enabled finite band gap, and Coulomb blockade effect. A characteristic aspect of graphene is the linear dependence of its electronic energy on the wave vector junctions in the Brillouin zone. Massless relativistic particles, therefore, can be predicted for their behavior using Dirac equation rather than more conventional Schrodinger equation. Physically, graphene is characterized as a flat single graphite sheet, having two-dimensional structure with a honeycomb crystal plane packed C atom monolayer [46]. Constitutional hexagonal lattice has 0.14 nm as characteristic C–C bond length, imparting the flexibility of being a fundamental building block to make several other graphene-based nanomaterials (NMs) (Fig. 3.2). For instance, wrapping up of graphene forms zero-dimensional fullerenes and rolling can provide one-dimensional nanotubes, while stacking can provide three-dimensional graphite. This is the reason why graphene is frequently addressed as the mother of all graphitic carbon-based NMs. Energy storage applications of graphene exhibit its most significant advances through the devices equipped for hydrogen storage and supercapacitor kind of functioning.

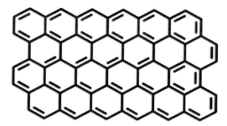

Figure 3.2 Schematic representation of *sp²*-hybridized pure carbon hexagonal lattice.

Another prominent domain evaluated with respect to graphene chemistry is its suitability as the novel substrate for Raman enhancement. Select graphene physical characteristics, making its feasible application toward the design of quick fast and robust electronic devices, such as field-effect transistor (FET), charge storage and energy sensing, are provided in Table 3.1.

Table 3.1   Some extraordinary physical properties of graphene

S. No.	Characteristic property	Typical value for graphene
1.	Young's modulus	11,000 GPa
2.	Fracture ability	125 GPa
3.	Thermal conductivity	5000 W/m·K
4.	Charge carrier motility	2×10^5 cm^2/V·s
5.	Specific SA	2630 m^2/g

It has been more than six decades since graphene or two-dimensional graphite came into academic domain. The very first line of understanding regarding graphene was its non-feasible free-state existence, owing to the existence of thermodynamically unstable curved structures such as soot, fullerenes, and nanotubes. It was perhaps more than seven decades back that Peierls and Landau defied a nonexistence of two-dimensional crystals, due to an inherent thermodynamic unstable nature [47, 48]. Therefore, it was gradually agreed upon that graphene exists as a functional constituent of larger three-dimensional structures, grown epitaxially on monocrystals surface with similar crystal lattices [49]. Without such a three-dimensional support, there was a consensus regarding nonexistence of two-dimensional material, until in 2004, when graphene was accidentally discovered using micromechanical cleavage [1]. This method utilizes a cellophane tape to remove graphene layers from a graphite flake, followed by pressing against a substrate. Subsequent removal of this tape provides a single sheet, with popular reference as being "Scotch tape" or "Peel-off" approach. Single-layer graphene consists of a continuous two-dimensional crystals with a high structural quality [1, 50]. So defining two-dimensional crystals presumes importance before discussing the energy-efficient applications of graphene. In most general words, a two-dimensional crystal refers to a single atomic plane, while more than 100 layers could comprise a thin film of three-dimensional crystal. The most robust distinction of two-dimensional and three-dimensional crystals can be made through analysis of electronic spectra, wherein graphene and its bilayers exhibit uncomplicated electronic attributes. These moieties are zero-band-gap semiconductors with one type of electron and hole. Exceeding these layers to three and more enhances the complex nature of electronic spectra, through an emergence of manifold charge carriers, arising from non-gapped valence and conduction bands. Perhaps, it was due to such electronic properties that the three-dimensional limit was considered apt for 10 or more layers. Most established research groups currently utilize the morphologies using micromechanical cleavage of bulk graphite (in view of missing quality grade graphite wafers).

3.3   Synthesis Strategy of Graphene for Advancing Renewable Energy Utilization

Several methods are practiced to prepare graphene sheets through chemical processing of graphite, such as mechanical cleavage, CNTs unzipping, chemical exfoliation, solvothermal reduction, epitaxial growth on SiC and metal surfaces, hydrocarbon CVD over metal surfaces, and reduction of graphene oxide (GO, produced from graphite oxide). The last among these methods is extensively used for the sake of usage compatibility of reducing agents; this method provides only chemically modified graphene. This method is easy to scale up for obtaining graphene on a large scale, wherein the concurrent approach is widely used to make chemically derived graphene. The initial step involves graphite oxidation to graphite oxide (famously termed Hummer's or modified Hummer's method), using vigorous acids and oxidizing agents [51, 52]. Retrieved graphite oxide is promptly exfoliated as individual GO sheets using aqueous ultrasonication. Inherently the oxidized form of graphene, GO possesses characteristic –OH and epoxy functionalities on the hexagonally arranged C atom network with –COOH groups flanking the edges (Fig. 3.3). The native molecule is highly hydrophilic and persists stably in the aqueous colloids, owing to the prevalence of manifold O comprising functional groups and electrostatic repulsions [53]. Inherently an electrical insulator, harvested GO can be chemically reduced to obtain electrically conducting graphene. Harvested GO readily exfoliates in the course of aqueous sonication.

Plenty of reducing moieties have been investigated to obtain graphene from GO, notable among which include hydrazine and its higher derivatives, hydroquinone, borohydrides of alkali metals, ethylene glycol, and several others. The scalability of this process has been significantly encouraging, with few-layer graphene synthesis being reported using mono, di-tri, and polysaccharides as reducing agents. These agents are easily available and, therefore, enable a greener prospect to provide GO. Apart from these entities, some other notable methods to prepare GO from graphene are microwave treatment, laser exposure, sonochemical and hydrothermal approaches [54–56].

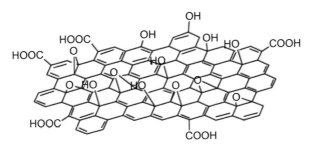

Figure 3.3   Schematic representation of graphene oxide (GO).

3.4   Potential Areas where Graphene Can Maximize Renewable Energy Output

The emergence of NMs has catalyzed the feasibility of manifold novel technologies useful for the society. Presently, we are at the initial stage of the development of a whole new regime of two-dimensional materials. Among the several two-dimensional NMs investigated for their better structural and energy-driven applications, some entities include graphene [1, 44], hexagonal boron nitride (BN) [57], transition metal dichalcogenides [58–61], MXenes [62, 63], bismuth telluride (Bi_2Te_3), bismuth selenide (Bi_2Se_3), silicones [64], phosphorene [65, 66], antimonene [67], and graphyne [68, 69]. Graphene is perhaps the most researched one out of these, due to its unique properties. It is being investigated with considerable interest for its manifold unconventional attributes such as carrier mobility, mechanical strength, thermal conductivity (1500 and 5000 W/m·K), and SA (2630 m^2/g), till date unnoticed in any other two-dimensional material [70–73]. These uncommon and splendid material properties of graphene owe their existence to a constitutional strong bonding between the hexagonal geometry of carbons [73] (Fig. 3.4).

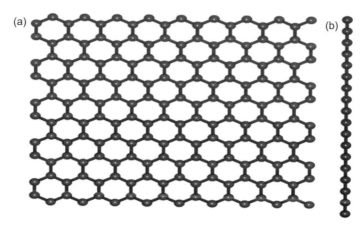

Figure 3.4 Graphene morphologies as viewed from top (a) and side (b). The figure is created using VESTA tool [75].

Several known graphene forms are graphene flakes (GFs), nanoribbons, and large-area sheets (henceforth referred to as sheets) [74–79]. These chemically alike but physically distinct forms differ in lateral dimensions. For instance, nanoflakes have restricted lateral dimensions (from several nano- to micrometers), while nanoribbons have at least one dimension, one order larger than the other. Similarly, sheets possess extended (macroscopic) dimensions [80]. Owing to this remarkable structural flexibility, these different graphene forms are well-equipped to meet specific requirements, such as graphene sheets higher suitability in making wafer-scale thin films (foundations for sensors, transparent conductive electrodes, and displays) [81]. Contrary to this, reduced manufacturing cost of GFs easily challenges several conventional materials (in particular being CNTs and carbon fibers for flexible electronics, energy storage, composites). GFs are also investigated for their use in conductive ink, with wide-ranged terminologies, including graphene nano-/micro-sheets, platelets, powder, and quantum dots [82, 83]. Herein we summarize the recent research findings pertaining to energy transformation attributes of graphene and graphene-based materials, such in Li$^+$ batteries (LIBs), renewable energy, and in photovoltaics (solar cells).

3.4.1   Batteries

Increasing dependence on fossil fuels for energy has elevated the risks of environmental degradation, greenhouse phenomenon, and global warming [84]. Furthermore, a rapidly growing energy demand across the globe has mandated a shift toward renewable and sustainable energy resources. As a consequence, several research groups are intensively working to optimize the energy yield of potential alternative energy sources [85]. Recent focus has pointed out the use of non-fossil fuels instead of low emission carbon fuels, forming the basis of efficient energy conversion in several energy storage mechanisms [86–91]. Due to the extensive use of batteries in portable electronic devices, they have garnered tremendous research interests [85]. Similarly, LIB is the most popular rechargeable battery, due to its suitability in portable electronics, such as mobile phones, calculators, tablets, electric vehicles, and several others [92–94].

To improve LIB performance and life cycle, the electrodes that are essential components of the battery assembly are being investigated for better performance. In this regard, carbon-based materials are swiftly emerging as amicable replacements [95, 96]. Graphite is an extensively utilized anodic agent in LIBs owing to compatible electrochemical and physical characteristics [97, 98]. However, a low theoretical capacity ($\sim$372 mAh/g) and rate ability limit the performance of LIBs [98–100]. Interestingly, the theoretical studies have revealed that lowering anode material dimensions, using NMs, can improve the existing energy storage extents, forming basis for investigating suitability of low-dimensional carbon materials as anode material in LIBs [98, 101]. In the recent years, the graphene and graphene-based LIBs have become promising technology because of presumably high energy density, wide range of working potentials, improved recharge ability, reduced spontaneous discharging, and light weight [102–104].

For adequate power generation (over the coming generations), it is necessary to upgrade the existing working mechanism of LIBs with reference to performance and reliability [85]. An LIB is an advanced technology working on the basis of Li^+ transport through electrolyte during the charging/discharging cycle. During discharging, Li-atoms in the anode get ionized and subsequently migrate to cathode through the electrolyte (Fig. 3.5) [50]. Although

LIBs comprise important device for present-day rechargeable and portable electronics, they suffer from certain lacunae, such as limited life cycle, usage and handling safety, and high expenditure. For overcoming such limitations, several C materials are being studied as possible LIB component replacements [105, 106]. Among these alternatives, graphene, GO, and the functionalization-driven compounds attract significant attention, in view of higher conductivity, larger SA (>2000 m^2/g), and extraordinary charge transport ability. Henceforth, the advances in graphene-based LIB research are discussed with an emphasis on major modifications and improvements [85].

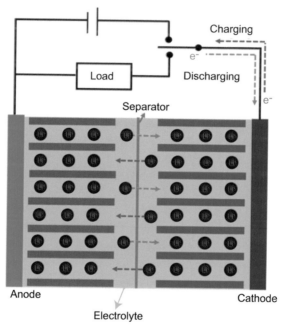

Figure 3.5 Typical representation of an Li$^+$ battery [105].

Because of the high conductive nature of carbon-based materials, they are considered a promising electrode material in LIBs. For efficient Li$^+$ movement between electrodes as well as exchange of electrons, a good conductive supporting network of C atoms is needed. In this direction, graphene by virtue of its superior electrical properties and large Li$^+$ diffusivity considerably enhances the rate capability, making it potential for LIBs [85]. Furthermore, a long-range π-conjugation in graphene confers it a large ionic mobility, making it suitable for ballistic electron transport [85, 107]. Besides being used as a cathode, graphene capably serves as a conducting agent, as metal oxide constituted anode. Theoretically, it has been shown to have ~764 mAh/g capacity, nearly twice than the conventional graphite electrodes [54]. Moreover, the functionalized graphene, for instance graphene oxide (GO), comprises –OH, –COOH, and epoxy groups on the surface, altering the concentration of which can easily convert GO to a tunable electronic material. The large active sites in GO make it feasible for electronic modulation of sensitive matter. Studies have shown better electrochemical performance of LIBs through metal-oxide-modified graphene/GO-based materials [108, 109].

The electrochemical performance of LIBs made using porous NFs as anode materials was investigated by Zho et al. [110]. It was observed that electrodes made from porous graphene exhibited large discharge capacity (830.4 mAh/g at 0.01 A/g current density). Interestingly, despite increasing current density to 20 A/g, the LIB displayed 211.6 mAh/g discharge capacity. Better cycling stability (nearly 100% retention ability) was noticed for graphene electrode, subsequent to 10^4 cycles. Such splendid electrochemical characteristics could be attributed to lose porous NG morphology, facilitating easier Li^+ diffusion and electron mobility.

Another area of research in this dimension involves improvement in capacity of graphene as LIB component, through integration with other carbon-based materials. Yoo et al. reported improved nanographene-specific capacity on CNTs and C_{60} inclusion from 540 to 730 mAh/g and 784 mAh/g, respectively [111]. The enhancement in specific capacity originated from the significant d-spacing enhancement, attributable to C_{60} and CNT, π-electron network. The scientists concluded that inclusion of graphene and other carbon nanostructures created additional sites for Li^+ accommodation, with an increment in d-spacing of graphene layers, enabling improved specific capacity. The performance of LIB can also be improved by introducing defects in graphene nanosheets, based on the hypothesis that the defects thereof create additional sites for Li, thus enhancing the charge capacity. To study the effects of a specific method on corresponding graphene's electronic properties, Pan et al. synthesized different grades of reduced graphene oxide (rGO) (such as hydrazine reaction, pyrolytic de-oxidation, and electron bean irradiation) [112]. It was noted that the electron-beam-driven rGO possessed the highest charge and discharge capacities of 1054 mAh/g and 2042 mAh/g, respectively. Apart from this, it was also noted that sheets of pyrolytic rGO displayed 82% retention capacity (after 15 cycles) than the 745 for electron-beam-irradiated GO. Henceforth, the researchers correlated the intensity ratio (I_D/I_G) of Raman D and G bands and observed a reversible capacity, viz-à-viz effect of defects on electrochemical performance. Further, with the help of Raman spectroscopy, high-resolution transmission electron microscopy (HRTEM), and electron paramagnetic resonance (EPR), it was concluded that electron beam rGO possesses maximum defects (comprising disordered carbon, topological and edge distortions, etc.). Though the interfacial defects between electrolyte and nanosheets were mitigated toward reversible Li$^+$ storage, the edge and internal defect contributed more rigorously toward reversible responses. Moreover, the decrease in cycling capacity may also be related to structural vulnerability, conferred by the constituted defects.

In addition, heteroatom doping in graphene can also be used to improve the LIB performance. Doping can play a critical role to enhance Li$^+$ storage capacity, via permitting an intercalation [85]. Jiang et al. synthesized the N-doped graphene using graphite oxide, GO and rGO, and melamine [113]. They reported that rGO has the highest specific capacity (~1250.8 mAh/g) among the three doped electrodes. A native high capacity was attributed to large SA (687.7 m^2/g), with higher 30-cycle capacity for N-doped rGO. Not only heteroatom doping, graphene micropores also contributed to a better working performance. N-doped holey graphene hollow microspheres (NHGHSs) and NGHSs (nitrogen-doped graphene hollow microspheres) were synthesized by Jiang et al. [60]. It was observed that NHGHSs expressed a large specific SA (948 m^2/g), larger than extensive graphene materials. The NHGHSs expressed a reversible charge storage capacity (≈1563 mAh/g) at 1/2 C low rate, which decreased to ≈254 mAh/g at 20 C. The electrodes also displayed a large energy density (637.4 Wh/kg) with 182.4 W/kg power density. Besides a low electrolyte resistance allowed easier electron transfer and Li^+ diffusion. The improved electrochemical functioning of the doped entities was attributed to its large conductivity and the extent of N contribution, resulting in enhanced adsorption energy as well as lesser Li^+ penetration energy barrier. Furthermore, Reddy et al. investigated the N-doped graphene (produced on a copper collector via CVD) capacitive properties [114, 115]. By creating larger surface defects, the scientists observed that N-doped graphene's reversible discharge capacity nearly doubles in comparison to pure graphene. It was inferred that the presence of pyridinic N-atoms and topological defects was the main reason for the improvement. In addition, direct contact between doped electrode and current collector also decreased the electronic resistance [85].

Besides doping with heteroatoms, structural modifications of graphene present another viable mechanism to improve the life cycle of LIBs. For example, Liu et al. [116] reported the superior performance of LIBs fabricated from N-doped monolithic films of graphene (having porous carbon framework) as electrode material. Though with such films, LIBs showed low reversible capacity (493 mAh/g), a high cycling stability (98% after 5×10^3 cycles) was noticed. Notably, a higher rate capability was accomplished with quasi-graphene films, compared to commercial graphite. The LIBs exhibited high energy density (72.5 Wh/kg) and a 10.7 W/g power density. Such an electrochemical performance could be attributed to the characteristic structural attributes expressing in a synergistic manner, arising from monolithic structure as well as N-doping.

Co-doping graphene is another approach to improve LIB performance, wherein N and S inclusion revealed (1016 mAh/g) initial reversible capacity at 100 mA/g current density, via producing greater active sites as well as S and N synergistic effects [117]. Importantly, the co-doped graphene showed better charge storage activity than pure graphene through lowering the O accommodating residual functional groups [85]. Another study by Liu et al. illustrated that N and Cl co-doping enabled high specific capacity (1.2 Ah/g) at 100 mA/g current density and 1.01 Ah/g rate capability [118]. Interestingly, high retention capacity of 95% after 1800 cycles was exhibited by the electrode. In this case, the p-type dopant (low dopant level 2.8%) enhanced the graphene conductivity and carrier mobility, while N-type doping increases electron mobility and number of active sites.

Over the last several years, several investigators used graphene and its derivatives for making LIB cathode. Graphene and $LI_3V_2(PO_4)/LiMn_2O_4/LiFePO_4$-modified GO were investigated as negative electrode agents in LIBs, revealing high rate ability and cycling stability [119–122] N-doped graphene, due to its enhanced Li^+-absorbing capability (conferred by pyrrolic- and pyrimidinic-N), functioned as suitable cathode material. As mentioned earlier, N-doping improves conductivity and boosts graphene wettability by electrolyte [123]. Table 3.2 summarizes some specific materials used in making LIB anode.

Table 3.2 Electrochemical performance of graphene and its derivatives in LIBs [85]

	Specific material	Specific capacity (mAh/g)	Capacity retention
Anode	Functionalized graphene paper	450 @ 300 mA/g	—
	N-doped graphene sheet	1250.8 @ 100 mA/g	87.9% per 30 cycles
	N,S co-doped graphene	1690 @100 mA/g	92% per 200 cycles
	B-doped graphene	1016 @ 100 mA/g	77.5% per 50 cycles
	GO paper	702 @ 50 mA/g	73.11% per 55 cycles
	N-doped holey graphene	1563 @ 0.5 C	104% per 10^5 cycles
	rGO	545 @ 0.372 mA/g	98.5% per 50 cycles
	Graphene/Co_3O_4	1287.7 @ 200 mA/g	85.5% per 100 cycles
	Graphene/CuO	583.5 @ 67 mA/g	75.5% per 50 cycles

	Specific material	Specific capacity (mAh/g)	Capacity retention
Cathode	Graphene	195 @ 50 mA/g	79.5% per 300 cycles
	Graphene/LiFePO$_4$	172 @ 0.06 C	97.5% per 50 cycles
	N-doped graphene	330 @ 50 mA/g	61.2% per 200 cycles
	GO/LiMn$_2$O$_4$	149 @ 0.05 C	87% per 100 cycles
	N-doped graphene/ LiFePO$_4$	163.1 @ 0.1 C	98.2% per 150 cycles

3.4.2 Renewable Fuels

Today energy crisis is an alarming issue; the world is facing, in lieu of ever-increasing energy demand, depletion of conventional energy resources [124]. Electricity, which is the most commonly used energy form, is chiefly generated via combustion of fossil fuels such as coal, petroleum, and natural gas. Such utilization of precious fossil fuels frequently results in the emission of considerable poisonous gases, polluting waterbodies and ecosystems, besides emitting greenhouse gases such as CO_2.

Λ more critical factor regarding the usage of fossil fuels is their non-conventional nature and their faster consumption compared to their restoring pace. Therefore, production and storage of clean and renewable energy sources are major challenges humans are facing at present [124, 125]. Recently, immense efforts have been made for using graphene in energy-related devices for an improved performance, stability, and cost effectiveness [126–130]. Considering the rapidly growing attention in this acutely emerging global problem, we discuss here some of the latest advances in graphene utilization for fuel cells.

Fuel cells (FCs) are one of the most well-organized and eco-friendly energy transformation devices. The typical working principle of an FC involves converting a fuel's chemical potential into electrical energy using chemical reactions mediated along electrode–electrolyte interface [130]. Major reactions, therein, involve typical fuel oxidation (on anode) and concurrent oxygen reduction (on cathode) [131]. Several modules of FCs are known, such as PEMFC, DMFC, and DFAFC, depending on the electrolyte used, with each configuration having its own merits and demerits. Typical representation of PEMFC is depicted in Fig. 3.6, having hydrogen split into its constituents at the anode. The H⁺ diffuses through the cell toward cathode, subsequent of which, electrons flow from anode to complete the circuit. The positively and negatively charged species combine at the cathode to give off water as by-product [131]. For an efficient chemical potential to electrical energy transformation, there is an indispensable requirement of a catalyst to initiate the reaction. Therefore, the FC performance implicitly depends on the catalyst efficacy [124]. Among the several available metal catalysts, Pt is most widely used for FCs. However, high price, below par utilization of Pt catalyst loading per unit area, and self-poisoning of Pt due to CO adsorption on surface restrict its practical applications [131]. Also, using Pt in DMFCs as a cathodic catalyst, the anode to cathode methanol crossover decreases the oxygen reduction reaction (ORR) performance because of mixing potentials formed due to simultaneous oxygen reduction and methanol oxidation at the cathode. According to a report of the US Department of Energy (DOE), 27–43% of the entire PEMFC stack cost is attributable to Pt catalysts [132].

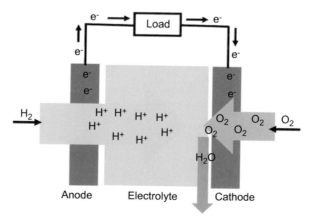

Figure 3.6 An example of PEMFC [131].

To avoid these limitations, different types of carbon supports such as activated carbon, graphite nanofibers, and graphene nanosheets have been investigated as catalyst [124, 133, 134]. The features needed for a catalyst support are (i) high SA for catalyst dispersion, (ii) chemical stability, and (iii) high electrical conductivity [135, 136].

Table 3.3 The electrochemical SA (ECSA), power/current density, and graphene-based material as electrode in FCs [124]

Type of cell	Type of electrocatalyst	Electrochemical SA (ECSA) m^2/g	Maximum current/ power density
DMFC	Pt/G	44.6	199.6 mA/mgPt
DMFC	Pt/C	30.1	101.2 mA/mgPt
	Pt/rGO	—	4.2 mA/cm^2
	Pt/NG	2.5	135 mA/mg
	PtPd/G	49.8	394 mA/mgPt
PEMFC	Pt/G	65	790 mW·cm^2
PEMFC	PdPt/G	49.8	394 mA/mgPt
PEMFC	Pt$_3$Cr/G	55	985 mW·cm^2
	PtCo/G	57	875 mW·cm^2
	Pt/NG	—	390 mW·cm^2
DFAFC	Pt/G	—	91 mW·cm^2

To enhance the Pt catalyst efficiency and activity, Pt NPs supported on graphene nanosheets have been examined to possess greater electrochemical SA, ORR activity, and greater stability compared to commercial Pt catalyst (from the E-TEK division, BASF Fuel Cell). Better performance of ORR is because of the inherent small size and lower aggregation of Pt NPs (put out of action on graphene sheet) [137]. Table 3.3 summarizes some catalyst support applications of graphene in fuel cell devices. Studies have shown that methanol and ethanol oxidations are catalyzed more efficiently by Pt/graphene as compared to commercially available Pt/C and Pt/MWCNTs [138, 139]. In one such attempt, Xin et al. synthesized methanol reduction catalyst by depositing Pt NPs on graphene sheet (Pt/G) through NaBH$_4$-mediated GO and H$_2$PtCl$_6$ suspension, reduction [140]. The researchers observed better stability for Pt/G catalyst, which also exhibited greater catalyzing ability compared to Pt/C catalyst. This was attributed to (i) additional exposure of Pt active sites due to graphene sheet rolling, (ii) increase in interaction between Pt and graphene nanosheet, (iii) partial decomposition of surface functional groups that resulted in less defects on graphene, and (iv) uniform distribution (morphology) of Pt nanoparticles on graphene surface [140, 141]. In another elegant effort, Lee et al. observed PtRu/G catalyst to exhibit improved action and superior forbearance to CO poisoning in methanol oxidation than PtRu/MWCNT catalyst [142]. The better activity of PtRu/G catalyst was ascribed to large graphene sheet specific area, promising finer NP dispersion. Several other bimetallic catalysts are PtNi, PtAu, PdPt, etc., in combination with graphene, wherein enhanced catalytic activity, improved stability, and reduced Pt usage have been observed [143–146]. The results indicated that electrocatalyst dispersion on graphene surface is an effective strategy to obtain higher electrocatalytic performance in FCs. In one significant attempt, Jafri et al. constructed PEMFCs with high power densities (4400 W/cm^2 and 3900 W/cm^2) using graphene and nitrogen-doped graphene (NG) nanoplatelets, acting as sustaining moiety for Pt NPs [147]. As apparent, the NG nanoplatelets showed better performance due to pyrrolic N defects, serving as anchoring sites for carbon–catalyst binding, as well as increased electrical conductivity. To counter CO deactivation, Dong et al. [148] developed graphene-supported Pt-uthenium NPs, bettering the electrocatalytic activity alongside lesser overpotential and better reversibility than graphite-supported alternatives. Besides its use as catalyst support for fuel cells, graphene and related materials have also drawn immense significance as metal-free catalyst [124]. For instance, Gong et al. observed mediation of ORR with higher catalytic activity for vertically aligned nitrogen-doped carbon nanotubes (VA-NCNTs), resulting in lower overpotential, less crossover sensitivity, and significantly long-lasting operational stability, compared to commercially available Pt/C catalyst. The improved activity of VA-NCNTs as ORR catalyst encouraged the investigation of NG as catalyst material [149]. Quantum mechanical calculations revealed that in VA-NCNTs, the C atoms in the N vicinity have higher positive charge capacity to equilibrate the N (strong) electron affinity. Additionally, charge delocalization induced by N can change the O$_2$ chemisorption mode and electrochemical potential, which in turn weakens the

O–O bond and hence facilitates the ORR [124, 149]. It was inferred that heteroatom (nitrogen) doping to CNTs efficiently generates metal-free active sites for electrochemical ORR. Discovering this phenomenon in VA-NCNTs was important in the sense that the same principle may be applied to other carbon NMs [122]. Indeed, it was revealed that nitrogen-doped graphene sheets generated in the presence of ammonia by the CVD technique show better ORR performance [124]. Also, nitrogen-doped graphene films produced by other methods such as edge functionalization, thermal treatment with ammonia, solvothermal treatment with tetrachloromethane and lithium nitride, and nitrogen plasma treatment of graphene also exhibit good electrocatalytic activity for ORR [150–153].

The NG is regarded as a compatible metal NP support. It strengthens metal–graphene binding and increases the number of chemically active sites for catalytic reactions, besides increasing graphene's electrical conductivity [154]. Importantly, the N-doping increases the number of active sites, useful for a homogeneous dispersion of metal NPs. These properties of NG are already illustrated in hybrid catalysts, such as Pt/NG [155] and Co_3O_4/NG for ORR [156]. The hybrid catalyst Co_3O_4/NG was found to exhibit high ORR activity due to NG. The response was observed as equivalent to Pt with higher alkaline stability. Further, graphene and its derivatives (not having N as doping agent) also exhibit enhanced ORR as demonstrated by P-doped graphite layers, B-doped CNTs (BCNTs), and S-doped graphene [157–159]. The intramolecular charge transfer imparted by heteroatom doping has been studied to impart oxygen-reducing electrocatalytic activity to graphene. For instance, it was reported that when functionalized with poly(diallyldimethylammonium chloride) (PDDA), graphene exhibits [160] superior ORR activity, due to PDDA and graphene charge transfer. This intramolecular charge transfer imparts a simple and cost-effective methodology for metal-free ORR catalyst development, depending on the C materials, such as graphene. Continued research in this growing field may provide a thriving energy conversion technology.

3.4.3 Photovoltaic Cells

To address the energy crisis, solar energy presents a neat and clean approach, having the special quality of being quickly regenerated or replaced [161]. From the past several decades, the emergence of new technologies and devices to produce, store, and utilize solar energy has enabled manifold novel mechanisms to produce green energy. Solar cells, indeed, provide a competent method to convert sunlight into energy using photoelectric effect [162]. The first-generation solar cells are produced from ultrapure Si [161]. Unfortunately, due to the requirement of high energy for ultrapure Si production, the electricity generated from these cells is much more expensive than that from fossil fuels [163]. Unlike Si-based solar cells, dye-sensitized solar cells (DSSCs) and perovskite solar cells (PSCs) are cost-effective and environment-friendly photovoltaic devices because of simple fabrication process and use of inexpensive material [164, 165].

Due to high optical transparency, carrier motility, and significant mechanical strength, graphene-based materials are widely studied for photovoltaic (PV) attributes [70, 71, 161]. GO and rGO are practically the most investigated graphene derivatives. Despite such extraordinary properties of graphene, its application in PV devices remains hindered demanding optimization of its optical response. For instance, no photoluminescence (PL) is observed for graphene sheets, rendering it infeasible to applied conditions. Contrary to this, zero-dimensional graphene quantum dots are fluorescent, well suited for QC and edge effects (energy traps), which impart them photoluminescence attributes. Scientific research on graphene-based provisions has witnessed an outstanding growth, owing to which, it is important to appreciate the significance of graphene and its derivatives (thereof) in PV devices as one of the major technological developments [162, 166]. Therefore, we describe the advances of graphene or graphene derivatives in PV devices (DSSCs and PSCs), mostly studied at present.

3.4.3.1 Graphene-based materials for DSSCs

Introduced in 1991, DSSCs soon gathered enough robustness that conferred them potential replacements of Si-based solar cells [167, 168]. DSSCs, also called Grätzel cells in honor of their inventor, comprise a photo electrode, a counter electrode, photosensitizer, and electrolyte (Fig. 3.7) [161]. DSSCs are fabricated at a sufficiently low price, exhibiting comparatively better efficiency. DSSCs also work better during darker conditions (at dawn and dusk) and in cloudy weather (unlike solar cells). This capability of DSSCs to utilize diffuse light makes them better choice for indoor applications such as windows and sunroofs [169, 170].

A typical DSSC comprises a transparent anode made of indium tin oxide (ITO) or fluorine-doped indium tin oxide (FTO) with a thin coating of TiO_2 NPs, which confers porous texture along with a large SA. This plate (which acts as an electrode) is immersed in a photosensitive dye solution, known as molecular sensitizer. The other plate (counter electrode) consists of Pt with a thin iodide electrolyte layer spread over. The two plates are annealed together in a bid to avoid any electrolyte leakage [171]. Unlike conventional solar cells whose working principle is based on the p-n junction, the principle of DSSCs is based on photogeneration of electrons from dye, similar to photosynthesis [162]. The light-enabled excitation of dye sensitizer results in electron transfer to TiO_2 NPs, restoring the used dye's ground state. This dye configuration is reached through electron transfer from electrolyte, which is further regenerated by the reduction of electrolyte oxidant at the counter electrode [172].

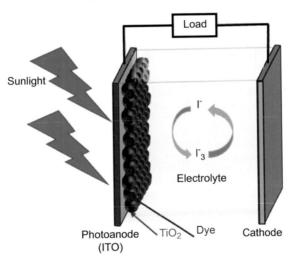

Figure 3.7 Schematic representation of DSSC [131].

The high conductivity and catalytic nature of graphene make it highly suitable for electrolyte redox reaction at the counter electrode of DSSCs [162]. However, pristine graphene suffers from the lack of active sites for electrocatalysis, necessitating the use of functionalized graphene such as GO, rGO, and three-dimensional graphene scaffold, in the counter electrode [173, 174]. The oxygen atoms serve as active sites as in catalytic reactions. With the use of rGO, the power conversion efficiency (PCE) of DSSC attains a 6.81% value with 1.2 $\Omega \cdot cm^2$ R_{ct} [175]. It was found that the high conductivity and low R_{ct} value enhance the electrolyte reduction at GO/iodide interface and decrease charge recombination, resulting in an increment of open circuit voltage (V_{oc}) and PCE. Three-dimensional graphene has also been used as counter electrode in DSSCs and has shown reasonable PCE (7.63%) contrary to 8.485 for Pt, due to larger conductivity, SA, and better electrolyte–electrode interactivities [176–178]. By hybridizing and doping with polymers, metals, and metal oxides, graphene delivers better performance as an electrode in DSSCs. GO decorated with Pt nanoparticles fared better compared to sputter Pt as counter electrode. DSSCs based on Pt-graphene exhibited better PCE (8.56%) than that of a GO-only device (4.48%) with a small R_{ct} value of 0.61 $\Omega \cdot cm^2$ [179]. The PCE of DSSCs with TiO_2/graphene photoanode achieved a 6.97% value through considerable increase in current density (45%), better than a TiO_2-only device (5.01%). The role of graphene in TiO_2 scaffold is like a two-dimensional bridge, which enables a superior optical scattering besides restricting the carrier recombination, on account of better charge transport [180, 181]. One study also utilized a composite of three-dimensional hybridized graphene (CNT-TiO_2) as photoanode, wherein a PCE of 6.11% was noticed, depicting a 31% enhancement compared to only a TiO_2-controlled device [182]. The photoanode of a DSSC is made using dye-sensitized TiO_2 coating over transparent conducting glass or plastic substrate [161]. To reduce the charge recombination, it is essential to enhance the dye loading, via amplifying the electrolyte–dye interface, through enhanced semiconductor layer conductivity.

Large SA-driven interacting properties of graphene impart it the potential of being used as semiconductor layer electron promoter (as transparent electrodes and contributory in sensitizer dye) [16–19, 161, 183]. ITO is widely used as transparent electrode although it is unstable at high temperature and is also brittle. Because of these problems, graphene has become a promising alternative material, which is not only cheap but also possesses better properties than ITO as photoelectrode material [13]. The first transparent and conductive electrode for DSSCs based on graphene was fabricated by Wang et al. [184]. This electrode comprised graphene films made through graphite oxide exfoliation and thermal oxidation of output platelets. Using graphene film as anode in DSSC with 1.8 kΩ/sq R_s and 72% transmittance, 0.26% efficiency was achieved. This low efficiency was attributed to the low quality of graphene film [184]. An improvement in this direction was proposed by Hang et al., wherein graphene films were prepared on SiO_2 substrate by employing ambient pressure in the CVD process [185]. Using such films, a power conversion efficiency of 4.25% was achieved, comparable with the use of FTO as counter electrode [185].

Graphene is also included within the semiconductor, enhancing the charge collection and improving the working efficacy of DSSCs [186–188]. Moreover, by forming a graphene–TiO_2 composite porous network, the scattering of light at the photoanode can be increased. One study in this regard reported a 39% higher efficacy than with P25 TiO_2 (titania photocatalyst intentionally made for high activity in photocatalytic reactions, with a typical efficacy of 4–7%) [188]. In another study, Kim et al. investigated the UV-assisted photocatalytic reduction of GO/TiO_2 NPs, wherein the prepared graphene/TiO_2 composite functioned as the interfacial layer within the nanocrystalline material and FTO gap [189]. Due to an inherently low graphene/TiO_2 roughness, it yielded superior adhesion compared to normal TiO_2 and FTO interfacial layer. The solar cells so constructed showed 5.26% PCE, larger than the one obtained without the blocking layer (4.89%). A similar methodology was used by Chen et al. wherein coated graphene on FTO was used to stop charge recombination between FTO and TiO_2, marginally improving the 5.8% working efficacy to 8.13% [186].

Graphene also finds utility as a current collector in DSSCs [190, 191]. In one such attempt, Yang et al. incorporated two-dimensional graphene into TiO_2 nanostructures, resulting in increased light collection and restricted charge recombination [190]. Similarly, Bavir and Fattah [191] used graphene and TiO_2 composite as photoelectrode to substitute TiO_2–ZnO composite. DSSCs made with graphene–TiO_2 photoelectrode exhibited higher efficiency (9.3%) than that of TiO_2–ZnO (6.5%). Table 3.4 provides the major attributes of DSSCs, photovoltaic aspects, made with graphene incorporated into the semiconducting layers.

Table 3.4 Performance of DSSCs with graphene/graphene-based material as photoanode [161, 191, 192]

Graphene type	J_{sc} (mA/cm^2)	V_{oc} (V)	FF	Efficiency (%)
rGO	1.01	0.70	0.36	0.26
Graphene	NA	NA	NA	2.0
Underlayer with rGO	23.3	0.73	0.55	9.2
Scaffold layer with T-CRGO	8.4	0.75	0.68	4.3
Graphene on Al_2O_3	10.2	0.78	0.68	5.4
CTAB functionalized graphene	12.8	0.82	0.62	6.5
Scaffold layer with T-CRGO	16.3	0.69	0.62	7.0
	22.47	0.744	0.71	11.2

3.5 Application of Graphene Materials for Perovskite Solar Cells

Perovskite solar cells (PSCs) have also generated tremendous research interest owing to their comparable PCE with the traditional solar cells, such as Si, GaAs, and CdTe [194]. Besides these, other notable attributes of PSCs include their flexibility, light weight, low production cost, and tuneability with flexible plastic substrates for large-area generation. The improvement in the PCE of PSCs has risen rapidly from 3.8% (in 2009) to 22.1% (in 2016) (Fig. 3.8) [194, 195]. It is significant to note here that 16.0 ± 0.4% efficiency is achieved for a small-module PSC with a 16.29 cm^2 aperture [196]. By contrast, the PCE of the first-generation solar cells based on silicon wafers lies between 15% and 20% with a comparatively higher aperture of 1.6 m^2 [197]. This indicates PSC efficacy for high energy conversion through non-C energy generation.

So far, electrode modification, metal-doped cathode buffer layer, surface-engineered perovskite layer, or introducing graphene-based materials within device layers have been studied to monitor the functioning of PSCs [198–204]. In particular, introducing graphene in PSCs has garnered significant interest due to the promising device designs, low production cost, and high chemical stability [194, 205, 206]. Interestingly, greater than 18% efficiency of PSCs (the highest till date) was achieved using graphene-based materials [153]. This clearly shows the potential of graphene for PSCs development.

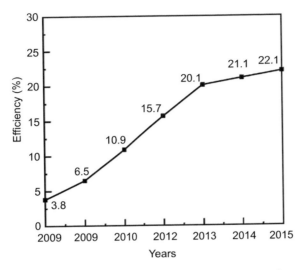

Figure 3.8 Variance in PSC efficiencies from 2009 to 2016 [193].

Over the last few years, graphene has been successfully used as a conductive electrode in PSC applications [154–158]. For the first time, Yan et al. reported graphene-based PSCs produced using CVD lamination [156]. However, graphene films produced through this method showed 1050 ± 150 Ω/sq single-layer sheet resistance. Due to this, a ZonylFS300 (a fluorosurfactant) doped 20 nm layer of poly-(3,4-ethylenedioxythiophene): poly(styrenesulfonate) (PEDOT:PSS) together with D-sorbitol was spin coated on the surface of graphene [194] to (i) lower the resistance of graphene films, (ii) function as lamination adhesion layer, and (iii) increase hole doping to graphene electrode. After optimizing the processing conditions, the efficiency of the device with double-layer graphene films reached 12.37%, relatively high when compared to semitransparent free PSCs [207–211]. Inception of the superior performance and large short-circuit current density (J_{sc}) was attributed to a lesser sheet resistance and higher graphene electrode conductivity, caused via PEDOT:PSS spin coating. The transmittance of graphene films in the visible region (>90%) also enhanced the device performance [209]. In a significant attempt, Choi et al. studied the efficacy of single-layer coated glass substrate (as transparent anode). It was observed that accommodating a 2 nm (molybdenum trioxide) hole transporting material (HTM) on graphene electrode (using interfacial engineering), a significantly high PCE (17.1%) was achieved for graphene-based PSC. A considerable high device performance was due to MoO_3 inclusion as HTM that provided hydrophobicity conferring adequate energy-induced inter-alignment of MoO_3/graphene electrode and PEDOT:PSS [207]. In yet another attempt, Choi et al. used a similar structure on an elastic polyethylene naphthalate (PEN) substrate and studied the operational stability against repeating bending [211, 212]. The device with graphene-coated electrode displayed 16.8% PCE (strain-free conditions), remaining within approximately 90% of the initial extent post-device bending over 10^3 times at 6 mm, 4 mm, and 2 mm R values. This was sheerly due to the graphene-based electrode that no injury was noticed in graphene or perovskite thin films in bending configurations. On the contrary, ITO-based flexible PSCs displayed rapid PCE decrement, after being bent higher than 250 times at $R = 4$ mm. This happened due to cracks that occurred on brittle ITO surface, subsequently being diffused into perovskite thin film [213–216].

Thus, graphene is a highly preferred material as conductive electrode and carrier transport in PSCs. So far, CVD is the most widely used approach to make graphene as it generates large-area graphene films. Besides, it also enables facile transfer of graphene on the target substrate, either via roll-to-roll or spin-coating mechanisms, for making conductive electrode. It is also suggested that when graphene is used on top of perovskite/HTM, the stability of PSCs can be further improved.

3.6   Future Directions

The turning point of graphene chemistry is characterized by its native ease of functionalization where nearly every intended application could be accomplished through the fitting of structure–activity relationship (SAR). Charge carrier and transport attributes are its fundamental characteristics implicated in the better functioning of electronic devices. Such abilities are the outcomes of robust graphene surface functionalities and its manifold derivatives (designed through covalent and ionic interactions), which confer improved chemical interactions and involvements toward industrial potential. Energy-efficient applications of graphene are attributed to its robust functionalization, doping, photosensitivity, catalytic and supramolecular chemistries. Except the last, all these attributes are derived out of the electronic activities of graphene and its native configuration of carbon atoms, whereas the supramolceular aspects are the outcomes of the molecular behavior of graphene derivatives. Nevertheless, it has not been long since the chemistry of graphene received the right scientific impetus, with many chemical reactions still not being able to be controlled precisely, owing to an incomplete understanding of the underlying mechanisms. For instance, defect-instigated chemical reactions happen either on basal planes or along the graphene sheet edges. Thereby, it is a challenge to optimize a methodology that controls the reactions within specified domains of a graphene sheet.

A daunting task in conceptualizing the electronic impacts of graphene is the modulation of its band gap, which has emerged as a significant challenge. Though several methods are proposed for controlling the graphene band gap, yet the accuracy of their experimental implementation needs an urgent revamp. The band gap of graphene can be tuned through varying its electronic structure via constitutional rearrangement of its sp^2-conjugated C network. The chemical approaches such as functionalization, doping, and photochemical reactions are not much successful toward controlling its compositions and the structural morphologies of sheet-like assembly at the atomic level. Plentiful improvements in the existing attributes are possible through improving the experimental conduct of the CVD process, wherein better controls over functional constitution, defects, number of layers, and sheet sizes could be exercised. Graphene sheets possess the unique ability of being modifiable into fibers, membranes, and three-dimensional frameworks. Optimization of performance could be accomplished through controlling the size, shape, and functional groups of chemically modified graphene, which could enable a better functional understanding of controlling self-assembly driven molecular activities.

3.7 Conclusion

We have presented an outlook on graphene structural chemistry and constitutional aspects responsible for its application in improving the energy outputs of conventional devices. The robust functionalization attributes of graphene and its derivatives thereof have enriched the realm of multiple applications. The flexibility and surface versatility have rightly emerged as bestowed artefacts through improved catalysis; energy conversion and storage have been made possible in reasonably shorter durations. Terminal and surface chemistries are the dual assets of multifunctional usability modes of graphene, presenting unique mechanisms to engineer the surface hydrophobicity into distributed philic–phobic configurations. The attributes such as deposition and adsorption of metals at the designated sites and in the needful extents, optimized through computational mechanisms, have emerged as vigorous performance-modulating prospects. In a nutshell, understanding the fundamentals of graphene chemistry remains a prerequisite for making and configuring graphene materials, vis-à-vis practical feasibilities toward improved energy provision.

References

1. Novoselov, K. S., Geim, A. K., Morozov, S. V., Jiang, D., Zhang, Y., Dubonos, S. V., Grigorieva, I. V., and Firsov, A. A. (2004). Electric field effect in atomically thin carbon films, *Science*, **306**, 666–669.

2. Novoselov, K. S., Morozov, S. V., Mohinddin, T. M. G., Ponomarenko, L. A., Elias, D. C., Yang, R., Barbolina, I. I., Blake, P., Booth, T. J., Jiang, D., Giesbers, J., Hill, E. W., and Geim, A. K. (2007). Electronic properties of graphene, *Phys. Status Solidi B*, **244**, 4106–4111.

3. Katsnelson M. I., Novoselov, K. S., and Geim, A. K. (2006). Chiral tunnelling and the Klein paradox in graphene, *Nat. Phys.*, **2**, 620–625.

4. Kroto, H. W., Heath, J. R., O'Brien, S. C., Curl, R. F., and Smalley, R. E. (1985). C60: Buckminsterfullerene, *Nature*, **318**, 162–163.

5. Iijima, S. (1991). Helical microtubules of graphitic carbon, *Nature*, **354**, 56–58.

6. Du, X., Skachko, I., Barker, A., and Andrei, E. Y. (2008). Approaching ballistic transport in suspended graphene, *Nat. Nanotechnol.*, **3**, 491–495.

7. Jiang, Z., Zhang, Y., Stormer, H. L., and Kim, P. (2007). Quantum Hall states near the charge-neutral Dirac point in graphene, *Phys. Rev. Lett.*, **99**, 106802.

8. Zhang, Y. B., Tan, Y. W., Stormer, H. L., and Kim, P. (2005). Experimental observation of the quantum Hall effect and Berry's phase in graphene, *Nature*, **438**, 201–204.

9. Novoselov, K. S., Jiang, Z., Zhang, Y., Morozov, S. V., Stormer, H. L., Zeitler, U., Maan, J. C., Boebinger, G. S., Kim, P., and Geim, A. K. (2007). Room temperature Quantum Hall effect in graphene, *Science*, **315**, 1379.

10. Novoselov, K. S., Geim, A. K., Morozov, S. V., Jiang, D., Katsnelson, M. I., Grigorieva, I. V., Dubonos, S. V., and Firsov, A. A. (2005). Two-dimensional gas of massless Dirac fermions in graphene, *Nature*, **438**, 197–200.

11. Lee, C., Wei, X., Kysar, J. W., and Hone, J. (2008). Measurement of the elastic properties and intrinsic strength of monolayer graphene. *Science*, **321**, 385–388.

12. Balandin, A. A., Ghosh, S., Bao, W., Calizo, I., Teweldebrhan, D., Miao, F., and Lau, C. N. (2008). Superior thermal conductivity of single-layer graphene, *Nano Lett.*, **8**, 902–907.

13. Bonaccorso, F., Sun, Z., Hasan, T., and Ferrari, A. C. (2010). Graphene photonics and optoelectronics, *Nat. Photon.*, **4**, 611–622.

14. Meng, J., Shi, D., and Zhang, G. (2014). A review of nanographene: Growth and applications, *Mod. Phys. Lett. B*, **28**, 1430009.

15. Westervelt, R. M. (2008). Graphene nanoelectronics, *Science*, **320**, 324–325.

16. Dua, V., Surwade, S. P., Ammu, S., Agnihotra, S. R., Jain, S., Roberts, K. E., Park, S., Ruoff, R. S., and Manohar, S. K. (2010). Graphene sensors, *Angew. Chem. Int. Ed.*, **49**, 2154–2157.

17. Dreyer, D. R. and Bielawski, C. W. (2011). Carbocatalysis: Heterogeneous carbons finding utility in synthetic chemistry, *Chem. Sci.*, **2**, 1233–1240.

18. Schedin, F., Geim, A. K., Morozov, S. V., Hill, E. W., Blake, P., Katsnelson, M. I., and Novoselov, K. S. (2007). Detection of individual gas molecules adsorbed on graphene, *Nat. Mater.*, **6**, 652–655.

19. Scheuermann, G. M., Rumi, L., Steurer, P., Bannwarthand, W., and Mulhaupt, R. (2009). Palladium nanoparticles on graphite oxide and its functionalized graphene derivatives as highly active catalysts for the Suzuki–Miyaura coupling reaction, *J. Am. Chem. Soc.*, **131**, 8262–8270.

20. Sun, Y., Wu, Q., and Shi, G. (2011). Graphene based new energy materials, *Energ. Environ. Sci.*, **4**, 1113–1132.

21. Yang, W. R., Ratinac, K. R., Ringer, S. P., Thordarson, P., Gooding, J. J., and Braet, F. (2010). Carbon nanomaterials in biosensors: Should you use nanotubes or graphene? *Angew. Chem. Int. Ed.*, **49**, 2114–2138.

22. Schwierz, F. (2010). Graphene transistors, *Nat. Nanotechnol.*, **5**, 487–496.

23. Bacon, M., Bradley, S. J., and Nann, T. (2014). Graphene quantum dots, *Part. Part. Syst. Charact.*, **31**, 415–428.

24. Kim, S., Hwang, S. W., Kim, M. K., Shin, D. Y., Shin, D. H., Kim, C. O., Yang, S. B., Park, J. H., Hwang, E., Choi, S. H., Ko, G., Sim, S., Sone, C., Choi, H. J., Bae, S., and Hong, B. H. (2012). Anomalous behaviors of visible luminescence from graphene quantum dots: Interplay between size and shape, *ACS Nano*, **6**, 8203–8208.

25. Banhart, F., Kotakoski, J., and Krasheninnikov, A. V. (2011). Structural defects in graphene, *ACS Nano*, **5**, 26–41.

26. Fujii, S. and Enoki, T. (2013). Nanographene and graphene edges: Electronic structure and nanofabrication, *Acc. Chem. Res.*, **46**, 2202–2210.

27. Novoselov, K. S., Falko, V. I., Colombo, L., Gellert, P. R., Schwab, M. G., and Kim, K. (2012). A roadmap for graphene, *Nature*, **490**, 192–200.

28. Allen, M. J., Tung, V. C., and Kaner, R. B. (2010). Honeycomb carbon: A review of graphene, *Chem. Rev.*, **110**, 132–145.

29. Englert, J. M., Hirsch, A., Feng, X. L., and llen, K. M. (2011). Chemical methods for the generation of graphenes and graphene nanoribbons, *Angew. Chem. Int. Ed.*, **50**, 17–24.

30. Cui, H., Zhou, Z., and Jia, D. (2017). Heteroatom-doped graphene as electrocatalysts for air cathodes, *Mater. Horiz.*, **4**, 7–19.

31. Halleux, A., Martin, R. H., and King, G. S. D. (1958). Syntheses in the series of highly condensed aromatic polycyclic derivatives. The hexabenzo-1.12; 2.3; 4.5; 6.7; 8.9; 10.11-coronene, tetrabenzo-4.5; 6.7; 11.12; 13,14-peropyrene and tetrabenzo- 1,2; 3.4; 8.9; 10,11-bisanthène, *Helv. Chim. Acta.*, **41**, 1177.

32. Simpson, D. C., Brand, J. D., Berresheim, A. J., Przybilla, L., Rader, H. J., and Müllen, K. (2002). Synthesis of a giant 222 carbon graphite sheet, *Chem. Eur. J.*, **8**, 1424–1429.

33. Kawasumi, K., Zhang, Q., Segawa, Y., Scott, L. T., and Itami, K. (2013). A grossly warped nanographene and the consequences of multiple odd-membered-ring defects, *Nat. Chem.*, **5**, 739–744.

34. Sakamoto, Y. and Suzuki, T. (2013). Tetrabenzo[8]circulene: Aromatic saddles from negatively curved graphene, *J. Am. Chem. Soc.*, **135**, 14074–14077.

35. Ball, M., Zhong, Y., Wu, Y., Schenck, C., Ng, F., Steigerwald, M., Xiao, S., and Nuckolls, C. (2015). Contorted polycyclic aromatics, *Acc. Chem. Res.*, **48**, 267–276.

36. Sun, Z., Zeng, Z., and Wu, J. (2013). Benzenoid polycyclic hydrocarbons with an open-shell biradical ground state, *Chemistry: An Asian J.*, **8**, 2894–2904.

37. Morita, Y., Suzuki, S., Sato, K., and Takui, T. (2011). Synthetic organic spin chemistry for structurally well-defined open-shell graphene fragments, *Nat. Chem.*, **3**, 197–204.

38. Kim, H., Song, I., Park, C., Son, M., Hong, M., Kim, Y., Kim, J. S., Shim, H. J., Baik, J., and Cho, H. C. (2013). Copper-vapor-assisted chemical vapor deposition for high-quality and metal-free single-layer graphene on amorphous SiO_2 substrate, *ACS Nano*, **7**, 6575–6582.

39. Rummeli, M. H., Bachmatiuk, A., Scott, A., Börrnert, F., Warner, J. H., Hoffman, V., Lin, J. H., Cuniberti, G., and Büchner, B. (2010). Direct low-temperature nanographene CVD synthesis over a dielectric insulator, *ACS Nano*, **4**, 4206–4210.

40. Lee, C. M. and Choi, J. (2011). Direct growth of nanographene on glass and post deposition size control, *Appl. Phys. Lett.*, **98**, 183106.

41. Stein, A., Wang, Z., and Fierke, M. A. (2009). Functionalization of porous carbon materials with designed pore architecture, *Adv. Mater.*, **21**, 265–293.

42. Li, Y. W. and Yang, R. T. (2007). Hydrogen storage on platinum nanoparticles doped on superactivated carbon, *J. Phys. Chem. C*, **111**, 11086–11094.

43. Jeon, N. J., Noh, J. H., Yang, W. S., Kim, Y. C., Ryu, S., Seo, J., and Seok, S. I. (2015). Compositional engineering of perovskite materials for high-performance solar cells, *Nature*, **517**, 476–480.

44. Geim, A. K. and Novoselov, K. S. (2007). The rise of graphene, *Nat. Mater.*, **6**, 183–191.

45. Rao, C. N. R., Sood, A. K., Subrahmanyam, K. S., and Govindaraj, A. (2009). Graphene: The new two-dimensional nanomaterial, *Angewandte Chemie*, **48**, 7752–7777.

46. Geim, A. K. (2009). Graphene: Status and prospects, *Science*, **324**, 1530–1534.

47. Meyer, J. C., Geim, A. K., Katsnelson, M. I., Novoselov, K. S., Booth, T. J., and Roth, S. (2007). The structure of suspended graphene sheets, *Nature*, **446**, 60–63.

48. Landau, L. D. and Lifshitz, E. M. (1980). *Statistical Physics*, Part I, Sections 137 and 138 (Pergamon, Oxford).

49. Evans, J. W., Thiel, P. A., and Bartelt, M. C. (2006). Morphological evolution during epitaxial thin film growth: Formation of 2D islands and 3D mounds, *Surf. Sci. Rep.*, **61**, 1–28.

50. Novoselov, K. S., Jiang, D., Schedin, F., Booth, T. J., Khotkevich, V. V., Morozov, S. V., and Geim, A. K. (2005). Two-dimensional atomic crystals, *Proc. Natl. Acad. Sci. U. S. A.*, **102**, 10451–10453.

51. Cao, X., Yin, Z., and Zhang, H. (2014). Three-dimensional graphene materials: Preparation, structures and application in supercapacitors. *Energ. Environ. Sci.*, **7**, 1850–1865.

52. Kovtyukhova, N. I. (1999). Layer by layer assembly of ultrathin composite films from micron-sized graphite oxide sheets and polycations, *Chem. Mater.*, **11**, 771–778.

53. Li, D., Müller, M. B., Gilje, S., Kaner, R. B., and Wallace, G. G. (2008). Processable aqueous dispersions of graphene nanosheets, *Nat. Nanotechnol.*, **3**, 101–105.

54. Chen, W., Yan, L., and Bangal, P. R. (2010). Preparation of graphene by the rapid and mild thermal reduction of graphite oxide, *J. Phys. Chem. Lett.*, **1**, 2633–2636.

55. Vinodgopal, K., Neppolian, B., Lightcap, I. V., Grieser, F., Ashokkumar, M., and Kamat, P. V. (2010). Sonolytic design of graphene-Au nanocomposites, simultaneous and sequential reduction of graphene oxide and Au (III), *J. Phys. Chem. Lett.*, **1**, 1987–1993.

56. Zhou, Y., Bao, Q., Tang, L. A. I., and Loh, K. P. (2009). Hydrothermal dehydration for the green reduction of exfoliated graphene oxide to graphene and demonstration of tunable optical limiting properties, *Chem. Mater.*, **21**, 2950–2956.

57. Pacile, D., Meyer, J. C., Girit, C. O., and Zettl, A. (2008). The two-dimensional phase of boron nitride: Few-atomic-layer sheets and suspended membranes, *Appl. Phys. Lett.*, **92**, 1–4.

58. Zhang, J., Wang, Q., Wang, L. H., Li, X. A., and Huang, W. (2015). Layer-controllable WS2-reduced graphene oxide hybrid nanosheets with high electrocatalytic activity for hydrogen evolution, *Nanoscale*, **7**, 10391–10397.

59. Johari, P. and Shenoy, V. B. (2011). Tunable dielectric properties of transition metal dichalcogenides, *ACS Nano,* **5**, 5903–5908.

60. Tsai, C., Chan, K., Nørskov, J. K., and Abild-Pedersen, F. (2015). Rational design of MoS$_2$ catalysts: Tuning the structure and activity via transition metal doping, *Catal. Sci. Technol.,* **5**, 246–253.

61. Wang, Q. H., Kalantar-Zadeh, K., Kis, A., Coleman, J. N., and Strano, M. S. (2012). Electronics and optoelectronics of two-dimensional transition metal dichalcogenides, *Nat. Nanotechnol.,* **7**, 699–712.

62. Naguib, M., Halim, J., Lu, J., Cook, K. M., Hultman, L., Gogotsi, Y., and Barsoum, M. W. (2013). New two-dimensional niobium and vanadium carbides as promising materials for Li-ion batteries, *J. Am. Chem. Soc.,* **135**, 15966–15969.

63. Chaudhari, N. K., Jin, H., Kim, B., San Baek, D., Joo, S. H., and Lee, K. (2017). MXene: An emerging two-dimensional material for future energy conversion and storage applications, *J. Mater. Chem. A,* **5**, 24564–24579.

64. Vogt, P., Capiod, P., Berthe, M., Resta, A., De Padova, P., Bruhn, T., Le Lay, G., and Grandidier, B. (2014). Synthesis and electrical conductivity of multilayer silicene, *Appl. Phys. Lett.,* **104**, 021602.

65. Mir, S. H. (2019). Exploring the electronic, charge transport and lattice dynamic properties of two-dimensional phosphorene, *Physica B: Condensed Matter,* **572**, 88–93.

66. Liu, H., Neal, A. T., Zhu, Z., Luo, Z., Xu, X., Tománek, D., and Ye, P. D. (2014). Phosphorene: An unexplored 2D semiconductor with a high hole mobility, *ACS Nano,* **8**, 4033–4041.

67. Mir, S. H., Yadav, V. K., and Singh, J. K. (2020). Unraveling the stacking effect and stability in nanocrystalline antimony through DFT, *J. Phys. Chem. Solids,* **136**, 109156.

68. Peng, Q., Dearden, A. K., Crean, J., Han, L., Liu, S., Wen, X., and De, S. (2014). New materials graphyne, graphdiyne, graphone, and graphane: Review of properties, synthesis, and application in nanotechnology, *Nanotechnol. Sci. Appl.,* **7**, 1.

69. Arguilla, M. Q., Jiang, S., Chitara, B., and Goldberger, J. E. (2014). Synthesis and stability of two-dimensional Ge/Sn graphane alloys, *Chem. Mater.,* **26**, 6941–6946.

70. Bolotin, K. I., Sikes, K. J., Jiang, Z., Klima, M., Fudenberg, G., Hone, J., Kim, P., and Stormer, H. L. (2008). Ultrahigh electron mobility in suspended graphene, *Solid State Commun.,* **146**, 351–355.

71. Lee, J. U., Yoon, D., and Cheong, H. (2012). Estimation of Young's modulus of graphene by Raman spectroscopy, *Nano Lett.*, **12**, 4444–4448.

72. Bosak, A., Krisch, M., Mohr, M., Maultzsch, J., and Thomsen, C. (2007). Elasticity of single-crystalline graphite: Inelastic X-ray scattering study, *Phys. Rev. B*, **75**, 153408.

73. Das, S., Pandey, D., Thomas, J., and Roy, T. (2019). The role of graphene and other 2D materials in solar photovoltaics, *Adv. Mater.*, **31**, 1802722.

74. Chae, S., Bratescu, M. A., and Saito, N. (2017). Synthesis of few-layer graphene by peeling graphite flakes via electron exchange in solution plasma, *J. Phys. Chem. C*, **121**, 23793–23802.

75. https://jp-minerals.org/vesta/en/download.html

76. Ghaemi, F., Abdullah, L. C., Rahman, N. M. A. N. A., Najmuddin, S. U. F. S., Abdi, M. M., and Ariffin, H. (2017). Synthesis and comparative study of thermal, electrochemical, and cytotoxicity properties of graphene flake and sheet, *Res. Chem. Intermed.*, **43**, 4981–4991.

77. An, H., Lee, W. G., and Jung, J. (2012). Synthesis of graphene ribbons using selective chemical vapor deposition, *Curr. Appl. Phys.*, **12**(4), 1113–1117.

78. Liu, F., Shen, X., Wu, Y., Bai, L., Zhao, H., and Ba, X. (2016). Synthesis of ladder-type graphene ribbon oligomers from pyrene units, *Tetrahedr. Lett.*, **57**, 4157–4161.

79. Bhaviripudi, S., Jia, X., Dresselhaus, M. S., and Kong, J. (2010). Role of kinetic factors in chemical vapor deposition synthesis of uniform large area graphene using copper catalyst, *Nano Lett.*, **10**, 4128–4133.

80. Yan, Z., Lin, J., Peng, Z., Sun, Z., Zhu, Y., Li, L., Xiang, C., Samuel, E. L., Kittrell, C., and Tour, J. M. (2012). Toward the synthesis of wafer-scale single-crystal graphene on copper foils, *ACS Nano*, **6**, 9110–9117.

81. Bianco, A., Cheng, H. M., Enoki, T., Gogotsi, Y., Hurt, R. H., Koratkar, N., Kyotani, T., Monthioux, M., Park, C. R., Tascon, J. M., and Zhang, J. (2013). All in the graphene family: A recommended nomenclature for two-dimensional carbon materials, *Carbon*, **65**, 1–6.

82. Li, X., Zhu, Y., Cai, W., Borysiak, M., Han, B., Chen, D., Piner, R. D., Colombo, L., and Ruoff, R. S. (2009). Transfer of large-area graphene films for high-performance transparent conductive electrodes, *Nano Lett.*, **9**, 4359–4363.

83. Yang, W. and Wang, C. (2016). Graphene and the related conductive inks for flexible electronics, *J. Mater. Chem. C*, **4**, 7193–7207.

84. Kairi, M. I., Dayou, S., Kairi, N. I., Bakar, S. A., Vigolo, B., and Mohamed, A. R. (2018). Toward high production of graphene flakes: A review on recent developments in their synthesis methods and scalability, *J. Mater. Chem. A*, **6**, 15010–15026.

85. Kang, M. G., Kim, M. S., Kim, J., and Guo, L. J. (2008). Organic solar cells using nanoimprinted transparent metal electrodes, *Adv. Mater.*, **20**, 4408–4413.

86. Kumar, R., Sahoo, S., Joanni, E., Singh, R. K., Tan, W. K., Kar, K. K., and Matsuda, A. (2019). Recent progress in the synthesis of graphene and derived materials for next generation electrodes of high-performance lithium ion batteries, *Prog. Energy Combust. Sci.*, **75**, 100786.

87. Zhang, Q., Uchaker, E., Candelaria, S. L., and Cao, G. (2013). Nanomaterials for energy conversion and storage, *Chem. Soc. Rev.*, **42**(7), 3127–3171.

88. Linares, N., Silvestre-Albero, A. M., Serrano, E., Silvestre-Albero, J., and García-Martínez, J. (2014). Mesoporous materials for clean energy technologies, *Chem. Soc. Rev.*, **43**, 7681–717.

89. Chen, K., Wang, Q., Niu, Z., and Chen, J. (2018). Graphene-based materials for flexible energy storage devices, *J. Energy Chem.*, **27**, 12–24.

90. Zhou, M., Xu, Y., and Lei, Y. (2018). Heterogeneous nanostructure array for electrochemical energy conversion and storage, *Nano Today*, **20**, 33–57.

91. Cui, K. and Maruyama, S. (2019). Multifunctional graphene and carbon nanotube films for planar heterojunction solar cells, *Prog. Energy Combust. Sci.*, **70**, 1–21.

92. Wang, S., Wu, Z. S., Zheng, S., Zhou, F., Sun, C., Cheng, H. M., and Bao, X. (2017). Scalable fabrication of photochemically reduced graphene-based monolithic micro-supercapacitors with superior energy and power densities, *ACS Nano*, **11**, 4283–4291.

93. Gaines, L. (2018). Lithium-ion battery recycling processes: Research towards a sustainable course, *Sustain. Mater. Technol.*, **17**, 00068.

94. Zubi, G., Dufo-López, R., Carvalho, M., and Pasaoglu, G. (2018). The lithium-ion battery: State of the art and future perspectives, *Renew. Sustain. Energy Rev.*, **89**, 292–308.

95. Barai, A., Uddin, K., Dubarry, M., Somerville, L., McGordon, A., Jennings, P., and Bloom, I. (2019). A comparison of methodologies for the non-invasive characterisation of commercial Li-ion cells, *Prog. Energy Combust. Sci.*, **72**, 1–31.

96. Kim, S. H., Lee, D. H., Park, C., and Kim, D. W. (2018). Nanocrystalline silicon embedded in an alloy matrix as an anode material for high energy density lithium-ion batteries, *J. Power Sources*, **395**, 328–335.

97. Kwasi-Effah, C. C. and Rabczuk, T. (2018). Dimensional analysis and modelling of energy density of lithium-ion battery, *J. Energy Storage*, **18**, 308–315.

98. Kim, J. H., Jung, M. J., Kim, M. J., and Lee, Y. S. (2018). Electrochemical performances of lithium and sodium ion batteries based on carbon materials, *J. Industr. Eng. Chem.*, **61**, 368–380.

99. Mao, C., An, S. J., Meyer III, H. M., Li, J., Wood, M., Ruther, R. E., and Wood III, D. L. (2018). Balancing formation time and electrochemical performance of high energy lithium-ion batteries, *J. Power Sources*, **402**, 107–115.

100. Sahoo, M. and Ramaprabhu, S. (2018). One-pot environment-friendly synthesis of boron doped graphene-SnO_2 for anodic performance in Li ion battery, *Carbon*, **127**, 627–635.

101. Xiao, Z., Ning, G., Ma, X., Li, W., and Xu, C. (2018). MnO-encapsulated graphene cubes derived from homogeneous $MnCO_3$-C cubes as high-performance anode material for Li ion batteries, *Carbon*, **139**, 750–758.

102. Fang, S., Shen, L., and Zhang, X. (2017). Application of carbon nanotubes in lithium-ion batteries. In: Peng, H., Li, Q., and Chen, T., Eds., *Industrial Applications of Carbon Nanotubes*. Boston: Elsevier, 251–276.

103. Hassoun, J., Panero, S., Reale, P., and Scrosati, B. (2009). A new, safe, high-rate and high-energy polymer lithium-ion battery, *Adv. Mater.*, **21**, 4807–4810.

104. Kim, T., Wentao, S., Dae-Yong, S., Luis, K. O., and Yabing, Q. (2019). Lithium-ion batteries: Outlook on present, future, and hybridized technologies, *J. Mater. Chem. A*, **7**, 2942–2964.

105. Zang, X., Wang, T., Han, Z., Li, L., and Wu, X. (2019). Recent advances of 2D nanomaterials in the electrode materials of lithium-ion batteries, *Nano: Brief Reports Rev.*, **14**, 1930001.

106. Thackeray, M. M., Wolverton, C., and Isaacs, E. D. (2012). Electrical energy storage for transportation-approaching the limits of, and going beyond, lithium-ion batteries, *Energ. Environ. Sci.*, **5**, 7854–7863.

107. Deng, D. (2015). Li-ion batteries: Basics, progress, and challenges, *Energy Sci. Eng.*, **3**, 385–418.

108. Tse, W. K., Hwang, E. H., and Das Sarma, S. (2008). Ballistic hot electron transport in graphene, *Appl. Phys. Lett.*, **93**, 023128.

109. Wang, G., Wang, B., Wang, X., Park, J., Dou, S., Ahn, H., and Kim, K. (2009). Sn/graphene nanocomposite with 3D architecture for enhanced reversible lithium storage in lithium ion batteries, *J. Mater. Chem.*, **19**, 8378–8384.

110. Raccichini, R., Varzi, A., Passerini, S., and Scrosati, B. (2014). The role of graphene for electrochemical energy storage, *Nat. Mater.*, **14**, 271.

111. Zhao, D., Wang, L., Yu, P., Zhao, L., Tian, C., Zhou, W., Zhang, L., and Fu, H. (2015). From graphite to porous graphene-like nanosheets for high rate lithium-ion batteries, *Nano Res.*, **8**, 2998–3010.

112. Yoo, E., Kim, J., Hosono, E., Zhou, H. S., Kudo, T., and Honma, I. (2008). Large reversible Li storage of graphene nanosheet families for use in rechargeable lithium ion batteries, *Nano Lett.*, **8**, 2277–2282.

113. Pan, D., Wang, S., Zhao, B., Wu, M., Zhang, H., Wang, Y., and Jiao, Z. (2009). Li storage properties of disordered graphene nanosheets, *Chem. Mater.*, **21**, 3136–3142.

114. Jiang, M. H., Cai, D., and Tan, N. (2017). Nitrogen-doped graphene sheets prepared from different graphene-based precursors as high capacity anode materials for lithium-ion batteries, *Int. J. Electrochem. Sci.*, **12**, 7154–7165.

115. Jiang, Z. J. and Jiang, Z. (2014). Fabrication of nitrogen-doped holey graphene hollow microspheres and their use as an active electrode material for lithium ion batteries, *ACS Appl. Mater. Interfaces*, **6**, 19082–19091.

116. Reddy, A. L. M., Srivastava, A., Gowda, S. R., Gullapalli, H., Dubey, M., and Ajayan, P. M. (2010). Synthesis of nitrogen-doped graphene films for lithium battery application, *ACS Nano*, **4**, 6337–6342.

117. Liu, X., Chao, D., Li, Y., Hao, J., Liu, X., Zhao, J., Lin, J., Fan, H. J., and Shen, Z. X. (2015). A low-cost and one-step synthesis of N-doped monolithic quasi-graphene films with porous carbon frameworks for Li-ion batteries, *Nano Energy*, **17**, 43–51.

118. Cai, D., Wang, C., Shi, C., and Tan, N. (2018). Facile synthesis of N and S co-doped graphene sheets as anode materials for high-performance lithium-ion batteries, *J. Alloys Compd.*, **731**, 235–242.

119. Liu, H., Tang, Y., Zhao, W., Ding, W., Xu, J., Liang, C., Zhang, Z., Lin, T., and Huang, F. (2018). Facile synthesis of nitrogen and halogen dual-doped porous graphene as an advanced performance anode for lithium-ion batteries, *Adv. Mater. Interfaces*, **5**, 1701261.

120. Wang, L., Wang, H., Liu, Z., Xiao, C., Dong, S., Han, P., Zhang, Z., Zhang, X., Bi, C., and Cui, G. (2010). A facile method of preparing mixed

conducting LiFePO$_4$/graphene composites for lithium-ion batteries, *Solid State Ionics*, **181**, 1685–1689.

121. Zhou, X., Wang, F., Zhu, Y., and Liu, Z. (2011). Graphene modified LiFePO$_4$ cathode materials for high power lithium ion batteries, *J. Mater. Chem.*, **21**, 3353–3358.

122. Ding, Y., Jiang, Y., Xu, F., Yin, J., Ren, H., Zhuo, Q., Long, Z., and Zhang, P. (2010). Preparation of nano-structured LiFePO$_4$/graphene composites by co-precipitation method, *Electrochem. Commun.*, **12**, 10–13.

123. Liu, H., Gao, P., Fang, J., and Yang, G. (2011). Li$_3$V$_2$(PO$_4$)$_3$/graphene nanocomposites as cathode material for lithium ion batteries, *Chem. Commun.*, **47**, 9110–9112.

124. Xiong, D., Li, X., Bai, Z., Shan, H., Fan, L., Wu, C., Li, D., and Lu, S. (2017). Superior cathode performance of nitrogen-doped graphene frameworks for lithium ion batteries, *ACS Appl. Mater. Interfaces*, **9**, 10643–10651.

125. Sahoo, N. G., Pan, Y., Li, L., and Chan, S. H. (2012). Graphene-based materials for energy conversion, *Adv. Mater.*, **24**, 4203–4210.

126. Xie, G., Zhang, K., Guo, B., Liu, Q., Fang, L., and Gong, J. R. (2013). Graphene-based materials for hydrogen generation from light-driven water splitting, *Adv. Mater.*, **25**, 3820–3839.

127. Wang, X., Zhi, L., and Müllen, K. (2008). Transparent, conductive graphene electrodes for dye-sensitized solar cells, *Nano Lett.*, **8**, 323–327.

128. Qu, L., Liu, Y., Baek, J. B., and Dai, L. (2010). Nitrogen-doped graphene as efficient metal-free electrocatalyst for oxygen reduction in fuel cells, *ACS Nano*, **4**, 1321–1326.

129. Yoo, E., Kim, J., Hosono, E., Zhou, H. S., Kudo, T., and Honma, I. (2008). Large reversible Li storage of graphene nanosheet families for use in rechargeable lithium ion batteries, *Nano Lett.*, **8**, 2277–2282.

130. Murugan, A. V., Muraliganth, T., and Manthiram, A. (2009). Rapid, facile microwave-solvothermal synthesis of graphene nanosheets and their polyaniline nanocomposites for energy storage, *Chem. Mater.*, **21**, 5004–5006.

131. Liu, J., Xue, Y., Zhang, M., and Dai, L. (2012). Graphene-based materials for energy applications, *MRS Bull.*, **37**, 1265–1272.

132. Liu, M., Zhang, R., and Chen, W. (2014). Graphene-supported nanoelectrocatalysts for fuel cells: Synthesis, properties, and applications, *Chem. Rev.*, **114**, 5117–5160.

133. James, B. D., Kalinoski, J. A., and Baum, K. N. (2009). Mass production cost estimation for direct H2 PEM fuel cell systems for automotive applications: 2008 Update, *Contract No. GS-10F-0099J*.

134. Bang, J. H., Han, K., Skrabalak, S. E., Kim, H., and Suslick, K. S. (2007). Porous carbon supports prepared by ultrasonic spray pyrolysis for direct methanol fuel cell electrodes, *J. Phys. Chem. C*, **111**, 10959–10964.

135. Wang, C., Waje, M., Wang, X., Tang, J. M., Haddon, R. C., and Yan, Y. (2004). Proton exchange membrane fuel cells with carbon nanotube based electrodes, *Nano Lett.*, **4**, 345–348.

136. Chen, Z., Higgins, D., Yu, A., Zhang, L., and Zhang, J. (2011). A review on non-precious metal electrocatalysts for PEM fuel cells, *Energ. Environ. Sci.*, **4**, 3167–3192.

137. Shao, Y., Yin, G., and Gao, Y. (2007). Understanding and approaches for the durability issues of Pt-based catalysts for PEM fuel cell, *J. Power Sources*, **171**, 558–566.

138. Kou, R., Shao, Y., Wang, D., Engelhard, M. H., Kwak, J. H., Wang, J., Viswanathan, V. V., Wang, C., Lin, Y., Wang, Y., and Aksay, I. A. (2009). Enhanced activity and stability of Pt catalysts on functionalized graphene sheets for electrocatalytic oxygen reduction, *Electrochem. Commun.*, **11**, 954–957.

139. Liu, S., Wang, J., Zeng, J., Ou, J., Li, Z., Liu, X., and Yang, S. (2010). Green electrochemical synthesis of Pt/graphene sheet nanocomposite film and its electrocatalytic property, *J. Power Sources*, **195**, 4628–4633.

140. Bong, S., Kim, Y. R., Kim, I., Woo, S., Uhm, S., Lee, J., and Kim, H. (2010). Graphene supported electrocatalysts for methanol oxidation, *Electrochem. Commun.*, **12**, 129–131.

141. Xin, Y., Liu, J. G., Zhou, Y., Liu, W., Gao, J., Xie, Y., Yin, Y., and Zou, Z. (2011). Preparation and characterization of Pt supported on graphene with enhanced electrocatalytic activity in fuel cell, *J. Power Sources*, **196**, 1012–1018.

142. Li, Y., Tang, L., and Li, J. (2009). Preparation and electrochemical performance for methanol oxidation of Pt/graphene nanocomposites, *Electrochem. Commun.*, **11**, 846–849.

143. Lee, S. H., Kakati, N., Jee, S. H., Maiti, J., and Yoon, Y. S. (2011). Hydrothermal synthesis of PtRu nanoparticles supported on graphene sheets for methanol oxidation in direct methanol fuel cell, *Mater. Lett.*, **65**, 3281–3284.

144. Rao, C. V., Reddy, A. L. M., Ishikawa, Y., and Ajayan, P. M. (2011). Synthesis and electrocatalytic oxygen reduction activity of graphene-supported Pt_3Co and Pt_3Cr alloy nanoparticles, *Carbon*, **49**, 931–936.

145. Zhang, K., Yue, Q., Chen, G., Zhai, Y., Wang, L., Wang, H., Zhao, J., Liu, J., Jia, J., and Li, H. (2011). Effects of acid treatment of Pt-Ni alloy nanoparticles@ graphene on the kinetics of the oxygen reduction reaction in acidic and alkaline solutions, *J. Phys. Chem. C*, **115**(2), 379–389.

146. Zhang, H., Xu, X., Gu, P., Li, C., Wu, P., and Cai, C. (2011). Microwave-assisted synthesis of graphene-supported Pd_1Pt_3 nanostructures and their electrocatalytic activity for methanol oxidation, *Electrochimica Acta*, **56**, 7064–7070.

147. Venkateswara Rao, C., Cabrera, C. R., and Ishikawa, Y. (2011). Graphene-supported Pt-Au alloy nanoparticles: A highly efficient anode for direct formic acid fuel cells, *J. Phys. Chem. C*, **115**, 21963–21970.

148. Jafri, R. I., Rajalakshmi, N., and Ramaprabhu, S. (2010). Nitrogen doped graphene nanoplatelets as catalyst support for oxygen reduction reaction in proton exchange membrane fuel cell, *J. Mater. Chem.*, **20**, 7114–7117.

149. Dong, L., Gari, R. R. S., Li, Z., Craig, M. M., and Hou, S. (2010). Graphene-supported platinum and platinum–ruthenium nanoparticles with high electrocatalytic activity for methanol and ethanol oxidation, *Carbon*, **48**, 781–787.

150. Gong, K., Du, F., Xia, Z., Durstock, M., and Dai, L. (2009). Nitrogen-doped carbon nanotube arrays with high electrocatalytic activity for oxygen reduction, *Science*, **323**, 760–764.

151. Jeon, I. Y., Yu, D., Bae, S. Y., Choi, H. J., Chang, D. W., Dai, L., and Baek, J. B. (2011). Formation of large-area nitrogen-doped graphene film prepared from simple solution casting of edge-selectively functionalized graphite and its electrocatalytic activity, *Chem. Mater.*, **23**, 3987–3992.

152. Geng, D., Chen, Y., Chen, Y., Li, Y., Li, R., Sun, X., Ye, S., and Knights, S. (2011). High oxygen-reduction activity and durability of nitrogen-doped graphene, *Energ. Environ. Sci.*, **4**, 760–764.

153. Deng, D., Pan, X., Yu, L., Cui, Y., Jiang, Y., Qi, J., Li, W. X., Fu, Q., Ma, X., Xue, Q., and Sun, G. (2011). Toward N-doped graphene via solvothermal synthesis, *Chem. Mater.*, **23**, 1188–1193.

154. Shao, Y., Wang, J., Wu, H., Liu, J., Aksay, I. A., and Lin, Y. (2010). Graphene based electrochemical sensors and biosensors: A review,

Electroanalysis: Int. J. Devoted Fundamental Practical Aspects Electroanalysis, **22**, 1027–1036.

155. Zhang, L. S., Liang, X. Q., Song, W. G., and Wu, Z. Y. (2010). Identification of the nitrogen species on N-doped graphene layers and Pt/NG composite catalyst for direct methanol fuel cell, *Phys. Chem. Chem. Phys.*, **12**, 12055–12059.

156. Jafri, R. I., Rajalakshmi, N., and Ramaprabhu, S. (2010). Nitrogen doped graphene nanoplatelets as catalyst support for oxygen reduction reaction in proton exchange membrane fuel cell, *J. Mater. Chem.*, **20**, 7114–7117.

157. Liang, Y., Li, Y., Wang, H., Zhou, J., Wang, J., Regier, T., and Dai, H. (2011). Co_3O_4 nanocrystals on graphene as a synergistic catalyst for oxygen reduction reaction, *Nat. Mater.*, **10**, 780–786.

158. Liu, Z. W., Peng, F., Wang, H. J., Yu, H., Zheng, W. X., and Yang, J. (2011). Phosphorus-doped graphite layers with high electrocatalytic activity for the O_2 reduction in an alkaline medium, *Angewandte Chemie Int. Ed.*, **50**, 3257–3261.

159. Yang, L., Jiang, S., Zhao, Y., Zhu, L., Chen, S., Wang, X., Wu, Q., Ma, J., Ma, Y., and Hu, Z. (2011). Boron-doped carbon nanotubes as metal-free electrocatalysts for the oxygen reduction reaction, *Angewandte Chemie Int. Ed.*, **50**, 7132–7135.

160. Yang, Z., Yao, Z., Li, G., Fang, G., Nie, H., Liu, Z., Zhou, X., Chen, X. A., and Huang, S. (2012). Sulfur-doped graphene as an efficient metal-free cathode catalyst for oxygen reduction, *ACS Nano*, **6**, 205–211.

161. Wang, S., Yu, D., Dai, L., Chang, D. W., and Baek, J. B. (2011). Polyelectrolyte-functionalized graphene as metal-free electrocatalysts for oxygen reduction, *ACS Nano*, **5**, 6202–6209.

162. Patil, K., Rashidi, S., Wang, H., and Wei, W. (2019). Recent progress of graphene-based photoelectrode materials for dye-sensitized solar cells, *Int. J. Photoenergy*, 2019. doi.org/10.1155/2019/1812879.

163. Mahmoudi, T., Wang, Y., and Hahn, Y. B. (2018). Graphene and its derivatives for solar cells application, *Nano Energy*, **47**, 51–65.

164. Wei, W., Wang, H., and Hu, Y. H. (2014). A review on PEDOT-based counter electrodes for dye-sensitized solar cells, *Int. J. Energy Res.*, **38**, 1099–1111.

165. Grätzel, M. (2003). Dye-sensitized solar cells, *J. Photochem. Photobiol. C: Photochem. Rev.*, **4**, 145–153.

166. Wei, W., Wang, H., and Hu, Y. H. (2013). Unusual particle-size-induced promoter-to-poison transition of ZrN in counter electrodes for dye-sensitized solar cells, *J. Mater. Chem. A*, **1**, 14350–14357.

167. Teymourinia, H., Salavati-Niasari, M., Amiri, O., and Farangi, M. (2018). Facile synthesis of graphene quantum dots from corn powder and their application as down conversion effect in quantum dot-dye-sensitized solar cell, *J. Mol. Liq.*, **251**, 267–272.

168. Gong, J., Sumathy, K., Qiao, Q., and Zhou, Z. (2017). Review on dye-sensitized solar cells (DSSCs): Advanced techniques and research trends, *Renew. Sust. Energ. Rev.*, **68**, 234–246.

169. Saadi, S. and Nazari, B. (2019). Recent developments and applications of nanocomposites in solar cells: A review, *J. Comp. Compd.*, **1**, 48–58.

170. Wang, D. H., Kim, J. K., Seo, J. H., Park, I., Hong, B. H., Park, J. H., and Heeger, A. J. (2013). Transferable graphene oxide by stamping nanotechnology: Electron-transport layer for efficient bulk-heterojunction solar cells. *Angewandte Chemie Int. Ed.*, **52**, 2874–2880.

171. O'regan, B. and Grätzel, M. (1991). A low-cost, high-efficiency solar cell based on dye-sensitized colloidal TiO$_2$ films, *Nature*, **353**, 737–740.

172. Wu, J., Lan, Z., Lin, J., Huang, M., Huang, Y., Fan, L., and Luo, G. (2015). Electrolytes in dye-sensitized solar cells, *Chem. Rev.*, **115**, 2136–2173.

173. Roy-Mayhew, J. D. and Aksay, I. A. (2014). Graphene materials and their use in dye-sensitized solar cells, *Chem. Rev.*, **114**, 6323–6348.

174. Trancik, J. E., Barton, S. C., and Hone, J. (2008). Transparent and catalytic carbon nanotube films, *Nano Lett.*, **8**, 982–987.

175. Jang, S. Y., Kim, Y. G., Kim, D. Y., Kim, H. G., and Jo, S. M. (2012). Electrodynamically sprayed thin films of aqueous dispersible graphene nanosheets: Highly efficient cathodes for dye-sensitized solar cells, *ACS Appl. Mater. Interfaces*, **4**, 3500–3507.

176. Ju, M. J., Jeon, I. Y., Kim, J. C., Lim, K., Choi, H. J., Jung, S. M., Choi, I. T., Eom, Y. K., Kwon, Y. J., Ko, J., and Lee, J. J. (2014). Graphene nanoplatelets doped with N at its edges as metal-free cathodes for organic dye-sensitized solar cells, *Adv. Mater.*, **26**, 3055–3062.

177. Nechiyil, D., Vinayan, B. P., and Ramaprabhu, S. (2017). Tri-iodide reduction activity of ultra-small size PtFe nanoparticles supported nitrogen-doped graphene as counter electrode for dye-sensitized solar cell, *J. Colloid Interface Sci.*, **488**, 309–316.

178. Yu, K., Wen, Z., Pu, H., Lu, G., Bo, Z., Kim, H., Qian, Y., Andrew, E., Mao, S., and Chen, J. (2013). Hierarchical vertically oriented graphene as a

catalytic counter electrode in dye-sensitized solar cells, *J. Mater. Chem. A*, **1**, 188–193.

179. Song, M., Ameen, S., Akhtar, M. S., Seo, H. K., and Shin, H. S. (2013). HFCVD grown graphene like carbon–nickel nanocomposite thin film as effective counter electrode for dye sensitized solar cells, *Mater. Res. Bull.*, **48**, 4538–4543.

180. Dao, V. D., Hoa, N. T. Q., Larina, L. L., Lee, J. K., and Choi, H. S. (2013). Graphene-platinum nanohybrid as a robust and low-cost counter electrode for dye-sensitized solar cells, *Nanoscale*, **5**, 12237–12244.

181. Zhao, J., Wu, J., Zheng, M., Huo, J., and Tu, Y. (2015). Improving the photovoltaic performance of dye-sensitized solar cell by graphene/titania photoanode, *Electrochimica Acta*, **156**, 261–266.

182. Yang, N., Zhai, J., Wang, D., Chen, Y., and Jiang, L. (2010). Two-dimensional graphene bridges enhanced photoinduced charge transport in dye-sensitized solar cells, *ACS Nano*, **4**, 887–894.

183. Yen, M. Y., Hsiao, M. C., Liao, S. H., Liu, P. I., Tsai, H. M., Ma, C. C. M., Pu, N. W., and Ger, M. D. (2011). Preparation of graphene/multi-walled carbon nanotube hybrid and its use as photoanodes of dye-sensitized solar cells, *Carbon*, **49**, 3597–3606.

184. Cao, X., Yin, Z., and Zhang, H. (2014). Three-dimensional graphene materials: Preparation, structures and application in supercapacitors, *Energ. Environ. Sci.*, **7**, 1850–1865.

185. Wang, X., Zhi, L., and Müllen, K. (2008). Transparent, conductive graphene electrodes for dye-sensitized solar cells, *Nano Lett.*, **8**, 323–327.

186. Bi, H., Sun, S., Huang, F., Xie, X., and Jiang, M. (2012). Direct growth of few-layer graphene films on SiO_2 substrates and their photovoltaic applications, *J. Mater. Chem.*, **22**, 411–416.

187. Chen, T., Hu, W., Song, J., Guai, G. H., and Li, C. M. (2012). Interface functionalization of photoelectrodes with graphene for high performance dye-sensitized solar cells, *Adv. Funct. Mater.*, **22**, 5245–5250.

188. Wang, H., Leonard, S. L., and Hu, Y. H. (2012). Promoting effect of graphene on dye-sensitized solar cells, *Ind. Eng. Chem. Res.*, **51**, 10613–10620.

189. Tang, Y. B., Lee, C. S., Xu, J., Liu, Z. T., Chen, Z. H., He, Z., Cao, Y. L., Yuan, G., Song, H., Chen, L., and Luo, L. (2010). Incorporation of graphenes in nanostructured TiO_2 films via molecular grafting for dye-sensitized solar cell application, *ACS Nano*, **4**, 3482–3488.

190. Kim, S. R., Parvez, M. K., and Chhowalla, M. (2009). UV-reduction of graphene oxide and its application as an interfacial layer to reduce the back-transport reactions in dye-sensitized solar cells, *Chem. Phys. Lett.*, **483**, 124–127.

191. Yang, W., Xu, X., Gao, Y., Li, Z., Li, C., Wang, W., Chen, Y., Ning, G., Zhang, L., Yang, F., and Chen, S. (2016). High-surface-area nanomesh graphene with enriched edge sites as efficient metal-free cathodes for dye-sensitized solar cells, *Nanoscale*, **8**, 13059–13066.

192. Bavir, M. and Fattah, A. (2016). An investigation and simulation of the graphene performance in dye-sensitized solar cell, *Opt. Quant. Electron.*, **48**, 559.

193. Rosli, N. N., Ibrahim, M. A., Ludin, N. A., Teridi, M. A. M., and Sopian, K. (2019). A review of graphene based transparent conducting films for use in solar photovoltaic applications, *Renew. Sust. Energ. Rev.*, **99**, 83–99.

194. Low, F. W. and Lai, C. W. (2018). Recent developments of graphene-TiO_2 composite nanomaterials as efficient photoelectrodes in dye-sensitized solar cells: A review, *Renew. Sust. Energ. Rev.*, **82**, 103–125.

195. Lim, E. L., Yap, C. C., Jumali, M. H. H., Teridi, M. A. M., and Teh, C. H. (2018). A mini review: Can graphene be a novel material for perovskite solar cell applications, *Nano Micro Lett.*, **10**, 27.

196. Correa-Baena, J. P., Abate, A., Saliba, M., Tress, W., Jacobsson, T. J., Grätzel, M., and Hagfeldt, A. (2017). The rapid evolution of highly efficient perovskite solar cells, *Energ. Environ. Sci.*, **10**, 710–727.

197. Green, M. A., Hishikawa, Y., Warta, W., Dunlop, E. D., Levi, D. H., Hohl-Ebinger, J., and Ho-Baillie, A. W. (2017). Solar cell efficiency tables, *Prog. Photovolt. Res. Appl.*, **25**, 668–676.

198. Albrecht, S. and Rech, B. (2017). Perovskite solar cells: On top of commercial photovoltaics, *Nat. Energy*, **2**, 1–2.

199. Cheng, N., Liu, P., Qi, F., Xiao, Y., Yu, W., Yu, Z., Liu, W., Guo, S. S., and Zhao, X. Z. (2016). Multi-walled carbon nanotubes act as charge transport channel to boost the efficiency of hole transport material free perovskite solar cells, *J. Power Sources*, **332**, 24–29.

200. Zhang, C., Luo, Y., Chen, X., Chen, Y., Sun, Z., and Huang, S. (2016). Effective improvement of the photovoltaic performance of carbon-based perovskite solar cells by additional solvents, *Nano Micro Lett.*, **8**, 347–357.

201. Giordano, F., Abate, A., Baena, J. P. C., Saliba, M., Matsui, T., Im, S. H., Zakeeruddin, S. M., Nazeeruddin, M. K., Hagfeldt, A., and Graetzel,

M. (2016). Enhanced electronic properties in mesoporous TiO_2 via lithium doping for high-efficiency perovskite solar cells, *Nat. Commun.*, **7**, 10379.

202. Zhao, X., Shen, H., Zhang, Y., Li, X., Zhao, X., Tai, M., Li, J., Li, J., Li, X., and Lin, H. (2016). Aluminum-doped zinc oxide as highly stable electron collection layer for perovskite solar cells, *ACS Appl. Mater. Interfaces*, **8**, 7826–7833.

203. Li, X., Dar, M. I., Yi, C., Luo, J., Tschumi, M., Zakeeruddin, S. M., Nazeeruddin, M. K., Han, H., and Grätzel, M. (2015). Improved performance and stability of perovskite solar cells by crystal crosslinking with alkylphosphonic acid ω-ammonium chlorides, *Nat. Chem.*, **7**, 703.

204. Liu, C., Ding, W., Zhou, X., Gao, J., Cheng, C., Zhao, X., and Xu, B. (2017). Efficient and stable perovskite solar cells prepared in ambient air based on surface-modified perovskite layer, *J. Phys. Chem. C*, **121**, 6546–6553.

205. Zhang, Y., Wang, J., Xu, J., Chen, W., Zhu, D., Zheng, W., and Bao, X. (2016). Efficient inverted planar formamidinium lead iodide perovskite solar cells via a post improved perovskite layer, *RSC Adv.*, **6**, 79952–79957.

206. Batmunkh, M., Shearer, C. J., Biggs, M. J., and Shapter, J. G. (2015). Nanocarbons for mesoscopic perovskite solar cells, *J. Mater. Chem. A*, **3**, 9020–9031.

207. Agresti, A., Pescetelli, S., Taheri, B., Del Rio Castillo, A. E., Cinà, L., Bonaccorso, F., and Di Carlo, A. (2016). Graphene–perovskite solar cells exceed 18% efficiency: A stability study, *Chem. Sus. Chem.*, **9**, 2609–2619.

208. Sung, H., Ahn, N., Jang, M. S., Lee, J. K., Yoon, H., Park, N. G., and Choi, M. (2016). Transparent conductive oxide-free graphene-based perovskite solar cells with over 17% efficiency, *Adv. Energy Mater.*, **6**, 1501873.

209. Heo, J. H., Shin, D. H., Kim, S., Jang, M. H., Lee, M. H., Seo, S. W., Choi, S. H., and Im, S. H. (2017). Highly efficient $CH_3NH_3PbI_3$ perovskite solar cells prepared by $AuCl_3$-doped graphene transparent conducting electrodes, *Chem. Eng. J.*, **323**, 153–159.

210. You, P., Liu, Z., Tai, Q., Liu, S., and Yan, F. (2015). Efficient semitransparent perovskite solar cells with graphene electrodes, *Adv. Mater.*, **27**, 3632–3638.

211. Liu, Z., You, P., Xie, C., Tang, G., and Yan, F. (2016). Ultrathin and flexible perovskite solar cells with graphene transparent electrodes, *Nano Energy*, **28**, 151–157.

212. Yoon, J., Sung, H., Lee, G., Cho, W., Ahn, N., Jung, H. S., and Choi, M. (2017). Superflexible, high-efficiency perovskite solar cells utilizing graphene electrodes: Towards future foldable power sources, *Energ. Environ. Sci.*, **10**, 337–345.

213. Guo, F., Azimi, H., Hou, Y., Przybilla, T., Hu, M., Bronnbauer, C., Langner, S., Spiecker, E., Forberich, K., and Brabec, C. J. (2015). High-performance semitransparent perovskite solar cells with solution-processed silver nanowires as top electrodes, *Nanoscale*, **7**, 1642–1649.

214. Li, Z., Kulkarni, S. A., Boix, P. P., Shi, E., Cao, A., Fu, K., Batabyal, S. K., Zhang, J., Xiong, Q., Wong, L. H., and Mathews, N. (2014). Laminated carbon nanotube networks for metal electrode-free efficient perovskite solar cells, *ACS Nano*, **8**, 6797–6804.

215. Kim, B. J., Kim, D. H., Lee, Y. Y., Shin, H. W., Han, G. S., Hong, J. S., Mahmood, K., Ahn, T. K., Joo, Y. C., Hong, K. S., and Park, N. G. (2015). Highly efficient and bending durable perovskite solar cells: Toward a wearable power source, *Energ. Environ. Sci.*, **8**, 916–921.

216. Heo, J. H., Lee, M. H., Han, H. J., Patil, B. R., Yu, J. S., and Im, S. H. (2016). Highly efficient low temperature solution processable planar type $CH_3NH_3PbI_3$ perovskite flexible solar cells, *J. Mater. Chem. A*, **4**, 1572–1578.

Multiple Choice Questions

1. Formation of wrinkles in free-standing graphene layers is due to
   - (a) Covalent interactions
   - (b) Ion-hydrophilic interactions
   - (c) π–π and hydrophobic interactions
   - (d) Van der Waals and London dispersive forces

2. The most effective mechanism to modify the structure and properties of graphene is
   - (a) Tuning the band gap
   - (b) Chemical functionalization
   - (c) Dispersion in amphipathic solvents
   - (d) Doping with compatible moieties

3. Most chemical reactions occur at the edges of graphene sheets due to
   - (a) Large surface area and robust edge functionalization
   - (b) Sheet-like morphology of native state

    (c) Giant π-conjugation, little structure of curvature, and missing dangling bonds

    (d) Both (b) and (c)

4. The mechanism being profusely used to develop new graphene-based catalysts is

    (a) Multiple strategies of functionalization

    (b) Chemisorption

    (c) Ability of dynamic self-adjustment

    (d) Physisorption

5. Hydrogenation of graphene changes its hybridization from

    (a) sp^2 to sp^3 (b) sp^3 to sp^3d

    (c) sp^2 to sp^3d^2 (d) None of these

6. The mass of carbon comprised in graphene oxide is nearly

    (a) 30%   (b) 25%

    (c) 65%   (d) 45%

7. How many chemical routes are known for the oxidation of graphene?

    (a) Two   (b) Four

    (c) Three   (d) Only one

8. Most widely used nanoscale form of graphene is

    (a) Graphite   (b) Carbon nanotubes

    (c) Fullerenes   (d) Both (b) and (c)

9. Despite expected flat topology, the reason for ripples on graphene surface is

    (a) Physical non-symmetry   (b) Contractile motions

    (c) Thermal fluctuations   (d) None of these

10. The correct order of the discovery years for graphene, fullerenes, and carbon nanotubes is

    (a) 2004, 1991, and 1985   (b) 2004, 1985, and 1991

    (c) 1991, 1985, and 2004   (d) None of these

11. The most common anode-grade material used in rechargeable lithium-ion batteries is

    (a) Graphene   (b) Graphite

    (c) Carbon nanotubes   (d) Both (b) and (c)

12. The differences of lithium to carbon binding stoichiometries for graphite and monolayer graphene are

(a) 1:6 and 1:3, respectively   (b) 1:3 and 1:6, respectively

(c) 1:4 and 2:5, respectively   (d) None of these

13. Reduced graphene oxide platelets exhibit 540 mAh/g specific capacity for anodes. Following the introduction of CNTs and C_{60} in these anodes, the specific capacity increased to

(a) 730 mAh/g and 784 mAh/g, respectively

(b) 784 mAh/g and 730 mAh/g, respectively

(c) Both increased to 784 mAh/g

(d) None of these

14. The better electrode performance of metal–graphene hybrids is due to

(a) Structural flexibility of reduced graphene oxide

(b) Superior conductivity of reduced graphene oxide

(c) Greater super-flexibility of graphene oxide

(d) Both (a) and (b)

15. Which nanocomposites upon being combined with graphene oxide derived graphene materials as fibers have shown significant improvements in elastic modulus, tensile strength, electrical conductivity, and thermal stability?

(a) Ceramic nanocomposites   (b) Metal nanocomposites

(c) Polymer nanocomposites   (d) Both (a) and (c)

16. The features conferring amphiphilic character to a graphene oxide sheet are

(a) Hydrophobic edges and hydrophilic basal plane

(b) Hydrophobic edges and hydrophobic basal plane

(c) Hydrophilic edges and hydrophobic basal plane

(d) Hydrogen bonding at edges and hydrophobic interactions at basal plane

17. The reason for a graphene layer being buckled is

(a) One-sided hydrogenation of graphene sheet

(b) Fully hydrogenated graphene (every C being covalently bonded to a H atom)

(c) Elongation of C–C bonds in hydrogenated graphene

(d) None of these

18. The doping regime specifically used to achieve superior graphene performance in catalysis and sensing is

(a) Nitrogen doping

(b) Boron-doped graphene

(c) Surface transfer doping

(d) Substitutional doping

19. The optical transmittance of graphene is

   (a) 96%

   (b) 97.7%

   (c) 96.7%

   (d) 100%

20. The electronic band structure of graphene near the Fermi level shows dispersion relation of which type?

   (a) Quadratic

   (b) Linear

   (c) A combination of quadratic and linear behavior

   (d) None of the above

21. In which category, chemical vapor deposition comes?

   (a) Bottom-up approach

   (b) Top-down approach

   (c) Thermal exfoliation

   (d) Chemical exfoliation

22. The electrons in graphene behave as

   (a) Massless Dirac fermions

   (b) Heavy mass Dirac fermions

   (c) Ordinary electrons

   (d) Bosons

23. The Peierls and Landau theory predicts the nonexistence of

   (a) Only zero-dimensional nanomaterials

   (b) Only one-dimensional nanomaterials

   (d) Only two-dimensional nanomaterials

   (d) All types of nanomaterials

24. What type of charged particles transport through electrolyte during charging/discharging of LIBs?

   (a) Electrons

   (b) Holes

   (c) Electrons during charging and holes during discharging

   (d) Li^+

25. Major redox reactions involving in a typical FC occur at

   (a) Oxidation at anode and reduction at cathode

   (b) Reduction at anode and oxidation at cathode

   (c) Oxidation and reduction reactions occur in electrolyte between the electrodes

   (d) Depends on how FCs are designed

26. DSSCs generate electrons from a dye similar to
    (a) Electrons in semiconductors
    (b) Electrons in p-n junctions
    (c) Photosynthesis
    (d) None of the above

27. In a DSSC, the dye is used to
    (a) Generate electrons photoelectrons
    (b) Capture electrons photoelectrons
    (c) Transport of electrons
    (d) Transport of holes

28. The highest power conversion efficiency of PSCs obtained is
    (a) 22.1%   (b) 21.1%
    (c) 50%   (d) 20.1%

Answer Key

1. (c)	2. (b)	3. (c)	4. (b)	5. (a)	6. (d)	7. (c)
8. (b)	9. (c)	10. (b)	11. (b)	12. (a)	13. (a)	14. (d)
15. (c)	16. (c)	17. (b)	18. (d)	19. (b)	20. (b)	21. (a)
22. (a)	23. (c)	24. (d)	25. (a)	26. (c)	27. (a)	28. (a)

Short Answer Questions

1. What are the different techniques used to improve the performance of graphene? Give examples.
2. What are the basic differences between LIBs and fuel cells?
3. What is the basic principle of DSSCs?
4. Describe the application of graphene in PSC.
5. What is the role of dye in a DSSC?
6. How graphene as electrode material can improve the working efficiency of lithium-ion batteries?
7. How preparation of graphene from the graphene oxide emerges as the most efficient method?
8. How graphene functionalization proves to be an asset in its industrial applications?

9. Compare the catalytic potential of graphene and carbon nanotubes with respect to organic reactions.

Long Answer Questions

1. Describe how graphene can help to improve the performance of LIBs?

2. What is the working principle of FCs? Explain how graphene is a promising cathodic material in FCs to replace Pt?

3. Explain how Si as doping agent can alter the electronic properties of graphene in varying extents.

4. Explain the significance of graphene characteristics toward the design of nanocomposites.

5. Comment on the potential usefulness of graphene in meeting energy crisis with respect to developing countries.

Chapter 4

Nanotechnology: Applications, Opportunities, and Constraints in Agriculture

Priya Chugh and Saleem Jahangir Dar

Department of Botany, Punjab Agricultural University, Ludhiana, India

priya-bot@pau.edu, saleemjahangir23@gmail.com

Agriculture is the backbone of most developing countries, with more than 60% of the population reliant on it for their livelihood. With the changing climatic conditions, global agricultural systems are facing numerous, unprecedented challenges. To address these challenges, significant technological advancements and innovations have been made in recent years in the field of agriculture for sustainable production and food security. Nanotechnology is the art and science of manipulating matter at the nanoscale (≤ 100 nm). The applications of nanotechnology in agriculture include nano-fertilizers to increase plant growth and yield, nano-pesticides for pest and disease management, and nano-sensors for monitoring soil quality and plant health. The integration of biology and nanotechnology has greatly increased the potential to sense and identify environmental

Nanotechnology: Principles and Applications

Edited by Rakesh K. Sindhu, Mansi Chitkara, and Inderjeet Singh Sandhu

Copyright © 2021 Jenny Stanford Publishing Pte. Ltd.

ISBN 978-981-4877-43-5 (Hardcover), 978-1-003-12026-1 (eBook)

www.jennystanford.com

conditions or impairments. This collective approach also enhances the efficiency and sustainability of agricultural practices by requiring less input and generating less waste as compared to conventional methods. The employment of engineered nanoparticles (ENPs), whether carbon or metal based, has wide application in developing smart field systems to monitor soil health, plant pathogen detection, genetic transformation, nanocoating of seeds to improve germination, nanoparticles-based gene carrier, crop's nutrient management, nano-herbicide for weed control, soil remediation, and recycling agricultural waste. Despite the potential benefits of nanotechnology in agriculture, their relevance has not reached up to the field conditions. There is still a huge untapped potential in nanoparticles, which need to be explored. Besides offering benefits, the elevating concerns about fate, transport, bioavailability, nanoparticle toxicity, and inappropriateness of regulatory framework limit the complete acceptance and inclination to adopt nanotechnologies in the agricultural sector and pose potential challenges associated with them.

4.1 Introduction

The Indian economy and population rely wholly on the agricultural sector. Due to the Green Revolution in the 1960s, India became self-sufficient in food production, but still food security remains a big concern and for that the government has invested a huge amount of money. Agricultural has been proven to be the prime field for maintaining sustainability. Hence, agriculture needs extensive research and efficient machinery to enhance productivity. In the recent decades, the agricultural sector has faced several challenges such as declining farm income, fading of natural resources, emergence of novel pests and diseases, and increased global temperatures through climate change. The population explosion has further aggravated pressure on this sector to meet the growing food demand. Due to these reasons, pivotal attention should be given on research, technology generation, and dissemination with human resources development. Toward this end, the conventional research methods need to be replaced by novel science and technology approaches, which must be time and cost-effective. To replenish the declining agricultural sector, modern technologies such as biotechnology and nanotechnology can become farmer friendly by enhancing and improving the quantity and quality of food. There is a proper belief that through these modern technological approaches, world food demands can be met with the mountain of economical, health, and environmental benefits.

4.2 Nanotechnology

Nanotechnology involves the manipulation of atoms and molecules to fabricate materials, devices, and systems. It is a branch of applied science and technology. Nanotechnology operates at the nanoscale (100 nm or less) by applying both physical and chemical methods for nanoparticle blend. Metal-originated nanoparticles such as gold and silver nanoparticles are popular in plant science for various applications, but their cost of synthesis is high, and hazardous chemicals are used (Fig. 4.1) [1]. A number of novel methods that are more "ecofriendly," such as plant extracts, are now used [2]. The traditional way of synthesis often relies on chemical reduction in a liquid phase

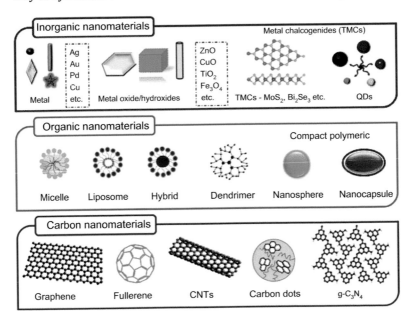

Figure 4.1 Types of nanomaterials (modified from Ref. [45]).

enabling great control over structure and yield. Reducing agents include sodium carbonate hydroxylamine, hydrogen peroxide, cellulose, citric acid, and sodium hydroxide. For stabilization, some agents are frequently added to promote diffusion and uniform distribution of particle size. Several physical techniques can also be employed for NM synthesis, such as chemical vapor deposition (CVD), laser ablation, supercritical fluids, sonochemical reduction, and gamma radiation.

4.3 Nano-farming: Novel Approach in Agriculture

Nano-farming is the term used for the invasion of nanotechnology in the field of agriculture. The role of nanotechnology in other fields provides sufficient shreds of evidence that it could be a novel approach in agriculture. The terminology "nanotechnology" was introduced by Norio Taniguchi in 1974, one of the professors at the Tokyo University of Science [3]. The agricultural sector is a very dynamic field and faces unprecedented challenges such as reduced yield, nutrient deficiency, and various environmental stresses. Precision agriculture means everything that makes agriculture farming highly accurate and controlled when it comes to the growing of crops and maintaining livestock. Particularly, it is related to site-specific crop management with an extensive collection of pre-production and post-production aspects of farming, varied from field to horticultural crops [4]. It involves integration of information technology, i.e., wireless networking (GPS) and use of sensor technology for controlled monitoring, assessing, and scheming farming practices. The advancement in genetically modified crops achieved based on nanomaterials-based target deliverance of CRISPR/Cas [(clustered regularly interspaced short palindromic repeats)/(CRISPR-associated protein)] mRNA and SgRNA is one of the best example [5].

The expansion of nano-sensors has a prevalent scenario for the scrutiny of environment-related stress and improving the defending functions of plants against diseases [6]. Hence, the persistent improvements in nanotechnology with special emphasis on the identification of problems and development methods, using collaborative approaches for agricultural growth, have great potentials to provide extensive benefits.

Figure 4.2 Application of nanotechnology (modified from Ref. [46]).

The key challenges of agriculture to be tackled by nanotechnology are as follows:

- Food security for the growing population
- Cultivable areas with low productivity
- Poor agricultural input efficiency
- Unsustainable farm execution
- Large barren and unproductive areas
- The decline of cultivable lands
- Wastage of agricultural produce
- Stumpy shelf life
- Impairment in processing and packaging (post-harvest losses)
- Diseases prone and vulnerable to climate change

4.4 Mechanism of Uptake and Diffusion of Nanoparticles in the Biological System

The uptake of nanoparticles in plants depends on several factors such as chemical composition, size, charge and surface adsorption, their intake routes, interactions with the ecosystem (soil texture, water availability, biota), cell structured limitation, and cell physiological mechanism. The first route for applying nanoparticles is either through base, i.e., roots, or through the upper portion, i.e., vegetative part of plants/leaves. The uptake of nanoparticles occurs passively through natural openings [7]. This is accomplished by nanoscale or microscale exclusion size, such as stigma, bark, stomata, and hydathodes texture over the shoot. The deep knowledge and understanding of nanoparticles–plant interaction as well as anatomical and physiological characteristics of plants need to be gathered. There are only certain openings through which nanoparticles get access, while the whole plant system is a reserve system with physical barrier. For example, shoot surface is generally covered by a cuticle made from biopolymers and waxes, which form a lipophilic barrier to plant's primary organs (Fig. 4.3) [8] and the only way is through natural openings. Trichomes, which are hairy structures, entrap nanoparticles on the surface of plants and thus enhance the steadiness time of applied chemicals/material on tissues, which can affect dynamics on the surface of the plant. The damages and wounds in both aerial and hypogeal parts form other viable routes for nanoparticle intake in plants [9]. After penetration, nanoparticles follow two routes: One is apoplastic where transportation takes place on the outer surface of the cell membrane through the cell walls and extracellular spaces. It promotes radial movement through which the nanoparticles move to the root's cylinder and then to vascular tissue and promoting upward movement [10]. The second route is symplastic, which includes the transportation of solutes and water between the cytoplasm of neighboring cells and the strainer plate pores. The barrier to apoplastic movement is the presence of casparian strip, which prevents the movement of material within the endodermis of roots [11]. To overcome this barrier, the symplastic pathway is preferred, but the only hurdle to this pathway is the presence of a rigid cell membrane that prevents this intracellular delivery of nanoparticles in plants [12]. There are certain nanoparticles with an average diameter between 3 nm and 50 nm and carbon nanotubes that can simply undergo the cell membrane in various plant species [13]. Endocytosis is the preferred method for cell internalization, while pore formation, membrane translocation, and carrier proteins are also utilized by cells [14, 15]

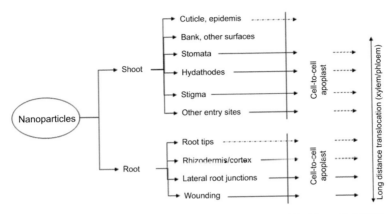

Figure 4.3 Routes for nanoparticles internalization (modified from Ref. [45]).

4.5 Nano-products in Precision Agriculture

4.5.1 Seed Germination

A seed is a basic unit of the fate of productivity. Traditionally, seed quality is gritty by their germination percentage. Despite recording higher germination percentage (80–90%) in the laboratory, seeds show lower germination rates in fields. Recent studies indicated the role of nanoparticles in enhancing seed germination and vitality. Seed germination and seedling vigor of old rice seeds enhanced by the use of silver nanoparticles as a nano-priming agent at 5 and 10 ppm concentration as compared to unprimed control and conventional hydropriming [16]. Another application of nano-priming was reported in the form of enhanced α-amylase activity leading to higher sugar content required for germination and upregulation of aquaporin (AQ) genes for more uptake of water during germination. Further studies revealed that the use of zinc oxide nanoparticle

(0.10 g/kg dose) in common beans enhanced shoot growth [17]. The use of zinc oxide nanoparticles at low concentration improved seed germination of mung bean (*Vigna radiata*) and chickpeas (*Cicer arietinum*) [18], and another report proved enhanced growth and photosynthesis by the use of manganese nanoparticles in mung bean [19]. The effect of nano and non–nano-TiO_2 was studied on the growth of naturally aged spinach seeds, and results revealed that nano-TiO_2-treated seeds produced 73% more dry weight, three times higher photosynthetic rate, and 45% increase in chlorophyll a formation as compared to the control over a germination period of 30 days [20]. The nano-TiO_2 enhances the seed stress resistance and promotes capsule penetration for the intake of oxygen and water for germination. Photo-sterilization and photo-generation of active oxygen such as superoxide and hydroxide anions could also be the reason for increased growth rate as they enhanced photosynthesis. Carbon nanotubes penetrate the seed coat and act as a gate to channelize the water from the substrate in the seeds. It serves as a new pore for water permeation as evident by enhanced germination in tomato seeds. This method served best in the rainfed ecosystem [21]. An increase was observed in yield component, iron content, and chlorophyll content in black-eyed peas due to foliar application of 500 mg/L iron nanoparticles as compared to the control, which was sprayed with regular iron salt [22].

4.5.2 Nano-fertilizers

Precise fertilizer management is an important prerequisite for sustainable agricultural development. Supplementation of essential nutrients at the appropriate time is foreseeable for the improvement in crop yield and productiveness. Intensive farming to fulfill the demands of the increasing population has decreased soil fertility and productivity of 40% world's agricultural land. The only traditional approach to improve soil fertility is the usage of a huge amount of fertilizers, resulting in adverse effect on soil's nutrient equilibrium [23]; the water environment was seriously affected due to leaching of toxic materials, which further contaminated drinking water [24]. Additionally, all the chemically composed conventional fertilizers persist in the soil or may enter other ecological compartments (i.e., may enter the food chain in different ecosystems), resulting in severe environmental pollution. So nano-fertilizer provides a platform to normalize the discharge of nutrients, depending on the requirement of the crop and also undertaking its efficient usage. Nano-formulation of nitrogenous fertilizers syncs with the release of nitrogen depending on their uptake demands as compared to conventional fertilizers, which degrade by 50–70% due to leaching or evaporation, hence reducing the efficiency and elevating the cost [25].

4.5.3 Shelf Life Extension for Fruits and Vegetables

The shelf life of fruits and vegetables is important from the commercial point of view. As per statistics, the total annual production of fruits and vegetables worldwide is about 3.7 billion tons [26], and they are the richest source of nutrients in the human diet [27]. Fruits and vegetables continue their biological process even after harvesting. Due to browning, off-flavor softening, and wilting, reduction in nutritional content occurs, which shortens their shelf life or makes them undesirable for consumption. The conventional methods are not up to the mark to inhibit this process. Therefore, nano-level detection technologies that enlighten the mechanism of change in the properties of food during storage and processing have gained much curiosity in the area of preservation of fruits and vegetables [28]. Considering the mechanical properties of nanoparticles, the shelf life of fruits and vegetables can be extended by using nanomaterials based on more stable packaging material. Nano-crystalline cellulose strengthens the hardness and elasticity of fruits and vegetables [29]. Another property of nanomaterials is photocatalysis, which enhances shelf life. Nano-TiO_2 is the most frequently used photocatalyst that delays the decay of fruits and vegetables by accelerated oxidative rotting of ethylene and other gases during storage [30]. The antibacterial properties of nanomaterials play a significant role in preventing bacterial infections. Antibacterial nanomaterials, by inhibiting the reproduction of microorganisms and killing individual microorganisms, can extend the shelf life of fruits and vegetables.

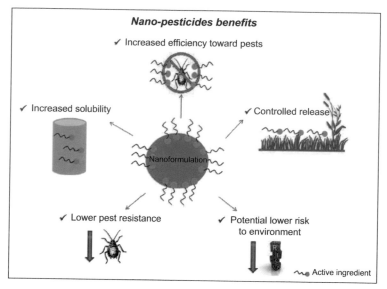

Figure 4.4 Effects of nano-pesticides. *Source*: Potential benefits of nano-pesticides (© Anita Jemec/University of Ljubljana).

4.5.4 Nano-pesticides

Pesticides consistently consist of a few substances, where the actual active ingredient (which is a drug) is blended with other compounds (auxiliary compounds). It allows easy absorption, equal distribution, and stability to the active ingredient after use. Nanomaterials are used in pesticides, but they themselves do not act as the active agent; rather, they work as the auxiliary compound. This is to stabilize the active agent and its prescribed release (e.g., as nanocapsule) (Table 4.1). Nano-copper and nano-silver are the best examples of nanomaterials as active agents. The insolubility of an active ingredient in water makes them difficult to use, which has been overcome by nanomaterials as the accompanying/auxiliary compound. In this way, the active constituent can be better protected against early degradation and also its release and distribution can be controlled (Fig. 4.4). This also reduces the amount of pesticides used and also improvisation in the uptake and effectiveness of the pest-control agent. Further, the integration of nanomaterials into active ingredients is less injurious to the environment and reduces the growth of resistance of pests to pesticides. The nanomaterial is used as a mode of transporter for the actual active constituent in emulsions, capsules, gels, fibers, and particles.

Table 4.1 Types of nano-pesticides with their active ingredient, benefits, and examples

Type	Relation to nano	Benefits	Example of active ingredient
Nanoemulsions	The active ingredient is located in nanoscale oil droplets that float in an aqueous solution.	Improved distribution and increased uptake by organism	Neem oil (parasitic worms and insects), Permethrin (ticks, mosquitoes)
Nanocapsule	The active ingredient is packed in nanoscale solution.	Increase uptake and thus increased effectiveness against pests	Lansiumamide B (antibacterial to treat tobacco bacterial wilt)
Nanogel	The active ingredient is distributed in a gel, which consists of nanoscale building blocks.	Easier handling, decreased evaporation of active ingredient	Copper (antifungal activity), essential oils (various pests)
Nanofibers	The active ingredient is incorporated in a nanoscaled fiber.	Improved distribution, slow release of active ingredient	Thiametoxam (insecticide)
Liposomes	Liposomes are in nano-sized particles.	Slow release of active ingredient	Etofenprox (insecticide), Pyrifluquinazon (insecticide)

4.5.5   Plant Nanobionics and Photosynthesis

Previously several studies have provided evidence on improving photosynthetic performance through conventional or bioengineering methods. With the intervention of nanoscience, photosynthetic capacity can also be enhanced by upgrading the light absorption capacity of leaves. The use of titanium oxide nanoparticles has been hypothesized for their role in increasing photosynthesis by improving light absorbance by plant leaves. It has also been proposed that titanium oxide nanoparticles have high photocatalytic activity of anatase crystal nano-TiO_2. It protects the chloroplast from photochemical stress [31] by enhancing photosynthesis [32] through influencing electron transport chain (ETC), photophosphorylation, leaf water conductance, and transpiration rate [33]. Further, single-walled carbon nanotubes (SWCNTs) have been supplied through perfusion in the form of suspension to the leaves of *Arabidopsis thaliana* and to the chloroplast of *Spinacia oleracea.* This treatment enhances the ETC, and the shelf life of isolated chloroplast was extended by 2 h. The study revealed that the SWCNTs are able to capture solar energy in a wavelength that is weakly absorbed by chloroplast due to high electrical conductance. It has been proposed that the increase in the light absorption by chloroplast is due to the enhanced absorption in UV and NIR ranges of the spectrum of light [34].

4.5.6   Plant Genetic Engineering and Nanomaterials

The main hurdle to the release of exogenous biomolecules in plant cells is the cell wall. To conquer this boundary and accomplish genetic transformation, various methodologies for DNA delivery in the plant cell are based on *Agrobacterium* transformation and biolistic techniques. Confinements to these methodologies depend on extensive plant damage and narrow host range. The previous studies majorly on nanomaterial-based plant genetic transformation have been performed in cell cultures. The use of nontoxic, degradable, double hydroxide (LDH) clay nano-sheets or bio-clay for loading dsRNS of different plant viruses is a non-culturable approach and protects against viruses. For instance, the dsRNAs or breakdown of

RNA products not only protects against the cauliflower mosaic virus (CMV) when sprayed in tobacco leaves but also provides overall protection to newly evolved, unsprayed leaves on viral challenges 20 days after solitary spray treatment in tobacco [35]. Among culture techniques, the most successful approach is silicon carbide mediated used in different callus.

Another example of genetic transformation with the use of magnetic nanoparticles (MNPs) has been successfully reported in cotton. The MNPs form a complex with β-glucuronidase (GUS) reporter gene and penetrate with the help of magnetic force into cotton pollen grains without reducing pollen viability. The result provides the transgenic cotton plants with successfully integrated exogenous genome and is stably inherited in the offspring [36].

4.5.7 Atomically Modified Seeds

Another case study aims to alter the color of local rice varieties without the technique of genetic modification. The nanotechnology research initiative in Thailand reported this study in which rice DNA was rearranged by inserting a nitrogen atom. The researchers drilled a hole through the rice cell membrane to insert a nitrogen atom, which would stimulate the rearrangement of rice DNA. The change in color of the local rice variety from purple to green is one of the attractions of this nanoscale technique.

4.5.8 Internet of Nano-Things in Farming

Internet of Nano-Things (IoNT) means utilization of wide variety of equipment and extensive manpower for essential task. In agro-system this can be used to implement precision farming by tracking crucial data about crops and soil. Nano-sensors can collect real-time data about crop health and growth, soil moisture, quality, and usage of pesticides and insecticides. Nano-sensors can also be utilized to track the location of livestock using a centralized sytem

4.5.9 Nanobiosensors

Nanobiosensors (NBSs) have been designed to monitor plant fractions, water, and soil health in the agroecosystem. These are analytic devices that explore the physicochemical property of nanomaterials. These are tools with ultra-advanced and enhanced biosensors that provide recognition of biological elements with their chemical or physical components [38] with the help of signals produced by the transducer. This information permits plant scientists to access the needs of the crops for water and nutrient and also prediction of disease symptoms [39]. An appropriately structured system of nano-sensors would permit all kinds of improvement in crop species and most proficient agronomic management factors such as composts, pesticides, and others.

4.5.9.1 Three parts of nanobiosensors

Biologically sensitive probe: A detecting component that produces signal when interacting with the target (biomolecule) relative to the target concentration. Some of the examples of a probe or biomolecule interactions are listed below:

(a) Antibody–antigen
(b) Interactions of nucleic acid
(c) Interaction of enzymatic
(d) Interaction of microorganisms, proteins, and so on

Transducer: It is a physical part responsible for the conversion of the recognized signals into digital ones. The nanomaterials have characteristics to manage various varieties of signals such as electrochemical, visual, etc.

Data-recording unit: It includes amplification of signal and signal processor required for data storage and transfer.

Nano-networks are designed for observing plant's conditions enabling the proficient use of crop inputs as fertilizers, water, and pesticides. This information provides an important tool for plant science research with a high-resolution of crop monitoring. Further, recognition and quantification of viruses, bacteria, and fungi in plants are also applications of biosensors [40]. The treatment of bacterial spot disease in Solanaceae caused by *Xanthomonas axonopodis* pv. *vesicatoria* was accomplished using fluorescent silica nanoparticles combined with antibody [41]. Additionally, the continuous assessment of plant metabolites and hormones will provide a detailed understanding of plant biosynthetic pathways.

4.6 Constraints

Identification of gaps and obstacles: There are many issues despite considerable advances in the field of nanotechnology in farming. Some of them that require attention are as follows:

- **Designing of specific hybrid carriers:** Specific hybrid carriers should be developed to deliver the active agents included in nutrients, pesticides, fertilizers, and others for the achievement of maximum efficiency.
- The requirement of processes that can be easily scalable at the industrial level.
- Development of better methodologies/techniques to study the comparative effects of nano-formulations/nano-systems with existing commercials.
- Getting hold of better information and development techniques to tackle risk and life-cycle assessment of nanomaterials, nano-pesticides, nano-fertilizers, and their negative impacts on non-target organisms such as other plants, soil microbiota, and bees.

In addition to the aforementioned scenarios, future case studies will help to safeguard human resources and customers with respect to the food produced by the use of nanomaterials and nanoparticles. There is also a need for the development of techniques to quantify nanoparticles available in various environmental compartments [42]. The dynamics of nanoparticles, i.e., their interaction between targeted and non-targeted organisms and the occurrence of synergistic effects would not be understood by the currently available methods [43]. The novel methodological advancement will allow the assessment of newly developed nanomaterials, and further advancement in analytical methodologies would support predictive models to characterize, localize, and quantify engineered nanomaterials in the environment. This may also address the negative impact.

4.7   Key Challenges Ahead

4.7.1   Societal Effects

There have been no standard methods for the use and testing of nanotechnology or nanomaterials till now, and the products are marketed without check. So this has caused the frequent distribution of nanomaterials in the form of food or packaging. The harmful effects on humans and the environment have not been documented yet, but the infiltration of these materials into the human body is well evident. Due to ambiguous chemical property, they may cause negative effects at some point.

4.7.2   Cost and Access

A patent provides motivation and encouragement for research and investment. To support these innovative ideas, intellectual property rights (IPRs) play an important role in the globalized world. At the consumer level, IPRs increases the price and hence creates a huge gap at the economic level.

4.7.3   Environmental and Human Health Risks

The use of engineered nanoscale materials in farming, water, and food may pose a risk to individual consumption and also to the environment [44]. There are chances of migration of nanoparticles from food to the environment. The food safety experts are looking at whether legislative controls on food packaging materials are already in place or are sufficient to deal with the new properties of nanoparticles.

4.8   Conclusion and Future Prospect

Applications of nanotechnology in agriculture can prove to be a big boon in maintaining sustainable agriculture. The emergence of novel methods for enhancing productivity in agro-systems can upgrade efficiency and create cost-effective systems. Engineering of nanomaterials and their action in maintaining sustainable agriculture have changed the global agricultural scenario by innovation and fast expansion to meet the projection of global food demand. The intervention of nanosciences with agricultural sciences has ushered in a new green revolution with a decline in agro-farming risks. However, there is a vast gap in the understanding of intake capacity, acceptable edge, and ecotoxicity of various nanomaterials in the agro-environment. Another aspect of interaction between agriculture and nanotechnology is related to the valorization of the produced waste material. There is a need for deep knowledge to understand and strengthen the novel technologies. Therefore, future research is urgently required to explore the behavior and fate of changed agricultural inputs and their interaction and effects with bio-macromolecules persisting in living systems and environments.

References

1. Rastogi, A., Tripathi, D. K., Yadav, S., Chauhan, D. K., Zivcak, M., and Ghorbanpour, M. (2019). Application of silicon nanoparticles in agriculture. *Biotech*, **9**, 90–104.

2. Abedini, A., Daud, A. R., Hamid, M. A. A., Othman, N. K., and Saion, E. (2013). A review on radiation-induced nucleation and growth of colloidal metallic nanoparticles. *Nanoscale Res. Lett.*, **8**, 1–10.

3. Khan, M. R. and Rizvi, T. F. (2014). Nanotechnology: Scope and application in plant disease management. *Plant Pathol. J.*, **13**, 214–231.

4. Dwivedi, S., Saquib, Q., Al-Khedhairy, A. A., and Musarrat, J. (2016). Understanding the role of nanomaterials in agriculture. In: Singh, D. P., Singh, H. B., and Prabha, R. (Eds.), *Microbial Inoculants in Sustainable Agricultural Productivity*. Springer, New Delhi, India, pp. 271–288.

5. Kim, D. H., Gopal, J., and Sivanesan, I. (2017). Nanomaterials in plant tissue culture: The disclosed and undisclosed. *RSC Adv.*, **7**, 36492–36505.

6. Kwak, S. Y., Wong, M. H., Lew, T. T. S., Bisker, G., Lee, M. A., Kaplan, A., Dong, J., Liu, A. T., Koman, V. B., and Sinclair, R. (2017). Nanosensor technology applied to living plant systems. *Annu. Rev. Anal. Chem.*, **10**, 113–140.

7. Kurepa, J., Paunesku, T., Vogt, S., Arora, H., Rabatic, B. M., and Lu, J. (2010). Uptake and distribution of ultrasmall anatase TiO_2 alizarin red s nanoconjugates in *Arabidopsis thaliana*. *Nano Lett.*, **10**, 2296–2302.

8. Schwab, F., Zhai, G., Kern, M., Turner, A., Schnoor, J. L., and Wiesner, M. R. (2016). Barriers, pathways and processes for uptake, translocation and accumulation of nanomaterials in plants: Critical review. *Nanotoxicology*, **10**, 257–278.

9. Al-Salim, N., Barraclough, E., Burgess, E., Clothier, B., Deurer, M., and Green, S. (2011). Quantum dot transport in soil, plants, and insects. *Sci. Total Environ.*, **409**, 3237–3248.

10. Zhao, X., Meng, Z., Wang, Y., Chen, W., Sun, C., and Cui, B. (2017). Pollen magnetofection for genetic modification with magnetic nanoparticles as gene carriers. *Nat. Plants*, **3**, 956–964.

11. Lv, J., Christie, P., and Zhang, S. (2019). Uptake, translocation, and transformation of metal-based nanoparticles in plants: Recent advances and methodological challenges. *Environ. Sci. Nano*, **6**, 41–59.

12. Cunningham, F. J., Goh, N. S., Demirer, G. S., Matos, J. L., and Landry, M. P. (2018). Nanoparticle-mediated delivery towards advancing plant genetic engineering. *Trends Biotechnol.*, **36**, 882–897.

13. Etxeberria, E., Gonzalez, P., Bhattacharya, P., Sharma, P., and Ke, P. C. (2019). Determining the size exclusion for nanoparticles in citrus leaves. *Hort. Sci.*, **51**, 732–737.

14. Palocci, C., Valletta, A., Chronopoulou, L., Donati, L., Bramosanti, M., and Brasili, E. (2017). Endocytic pathways involved in PLGA nanoparticle uptake by grapevine cells and role of cell wall and membrane in size selection. *Plant Cell Rep.*, **36**, 1917–1928

15. Wang, T., Bai, J., Jiang, X., and Nienhaus, G. U. (2012). Cellular uptake of nanoparticles by membrane penetration: A study combining confocal microscopy with FTIR spectroelectrochemistry. *ACS Nano*, **6**, 1251–1259.

16. Mahakham, W., Sarmah, A. K., and Maensiri, S. (2017). Nanopriming technology for enhancing germination and starch metabolism of aged rice seeds using phytosynthesized silver nanoparticles. *Sci. Rep.*, **7**, 8263–8274.

17. Dimkpa, C. O., Hansen, T., Stewart, J., McLean, J. E., Britt, D. W., and Anderson, A. J. (2015). ZnO nanoparticles and root colonization by a beneficial pseudomonad influence essential metal responses in bean (*Phaseolus vulgaris*). *Nanotoxicology*, **9**(3), 271–278.

18. Mahajan, P., Dhoke, S. K., and Khanna, A. S. (2011). Effect of nano-ZnO particle suspension on growth of mung (*Vignaradiata*) and gram (*Cicer arietinum*) seedlings using plant agar method. *J. Nanotechnology*, **7**, 1–7.

19. Pradhan, S., Patra, P., Das, S., Chandra, S., Mitra, S., and Dey, K. K. (2013). Photochemical modulation of bio safe manganese nanoparticles on Vignaradiata: A detailed molecular biochemical, and biophysical study, *Environ. Sci. Technol.*, **47**, 13122–13131.

20. Zheng, L., Hong, F. S., Lu, S. P., and Liu, C. (2005). Effect of nano-TiO_2 on strength of naturally and growth aged seeds of spinach. *Biol. Trace Elem. Res.*, **104**, 83–91.

21. Khodakovskaya, M. V. and Biris, A. S. (2010). Method of using carbon nanotubes to affect seed germination and plant growth. U.S. Patent Application 13/509,487.

22. Delfani, M., Firouzabadi, M. B., Farrokhi, N., and Makarian, H. (2014). Some physiological responses of black-eyed pea to iron and magnesium nanofertilizers. *Commun. Soil Sci. Plant Anal.*, **45**, 530–540.

23. Li, S. X., Wang, Z. H., Miao, Y. F., and Li, S. Q. (2014). Soil organic nitrogen and its contribution to crop production. *J. Integr. Agric.*, **13**, 2061–2080.

24. Solanki, P., Bhargava, A., Chhipa, H., Jain, N., and Panwar, J. (2015). Nano-fertilizers and their smart delivery system. In: Rai, M., Ribeiro, C., Mattoso, L., and Duran, N. (Eds.), *Nanotechnologies in Food and Agriculture*. Springer, Cham Switzerland, pp. 81–101.

25. Miao, Y. F., Wang, Z. H., and Li, S. X. (2015). Relation of nitrate N accumulation in dryland soil with wheat response to N fertilizer. *Field Crops Res.*, **170**, 119–130.

26. Barrett, D. M. and Lloyd, B. (2012). Advanced preservation methods and nutrient retention in fruits and vegetables. *J. Sci. Food Agric.*, **92**(1), 7–22.

27. Ma, L., Zhang, M., Bhandari, B., and Gao, Z. (2017). Recent developments in novel shelf life extension technologies of fresh-cut fruits and vegetables. *Trends Food Sci. Technol.*, **64**, 23–38.

28. Khan, A., Khan, R. A., Salmieri, S., Tien, C. L., Riedl, B., Bouchard, J., Chauve, G., Tan, V., Kamal, M. R., and Lacroix, M. (2012). Mechanical and barrier properties of nanocrystalline cellulose reinforced chitosan based nanocomposite films. *Carbohydr. Polym.*, **90**(4), 1601–1608.

29. Cheng, S., Zhang, Y., Cha, R., Yang, J., and Jiang, X. (2016). Water-soluble nanocrystalline cellulose films with highly transparent and oxygen barrier properties. *Nanoscale*, **8**(2), 973–978.

30. Brody, A. L., Zhuang, H., and Han, J. H. (2011). Nanostructure packaging technologies. In: *Modified Atmosphere Packaging for Fresh-Cut Fruits and Vegetables*. John Wiley & Sons, Hoboken, NJ.

31. Hong, F. S., Yang, F., Ma, Z. N., Zhou, J., Liu, C., Wu, C., and Yang, P. (2005). Influences of nano-TiO_2 on the chloroplast ageing of spinach under light. *Biol. Trace Elements Res.*, **1043**, 249–260.

32. Ma, L., Liu, C., Qu, C., Yin, S., Liu, J., Gao, F., and Hong, F. (2008). Rubisco activase mRNA expression in spinach: Modulation by nanoanatase treatment. *Biol. Trace Elements Res.*, **1222**, 168–178.

33. Qi, M., Liu, Y., and Li, T. (2013). Nano-TiO_2 improve the photosynthesis of tomato leaves under mild heat stress. *Biol. Trace Elements Res.*, **156**(1–3), 323–328.

34. Marchiol, L. (2018). Nanotechnology in agriculture: New opportunities and perspectives. In: *New Visions in Plant Science*. IntechOpen, Ozge Çelik.

35. Mitter, N., Worrall, E. A., Robinson, K. E., Li, P., Jain, R., and Taochy, C. (2017). Clay nanosheets for topical delivery of RNAi for sustained protection against plant viruses. *Nat. Plants*, **3**, 16207.

36. Zhao, X., Meng, Z., Wang, Y., Chen, W., Sun, C., and Cui, B. (2017). Pollen magnetofection for genetic modification with magnetic nanoparticles as gene carriers. *Nat. Plants*, **3**, 956–964.

37. Bagal-Kestwal, D. R., Kestwal, R. M., and Chiang, B. H. (2016). Bio-based nanomaterials and their bio nanocomposites. In: Visakh, P. M. and Martinez Morlanes, M. J. (Eds.), *Nanomaterials and Nanocomposites: Zero- to Three-Dimensional Materials and Their Composites*. Wiley, pp. 255–329.

38. Khiari, R. (2017). Valorization of agricultural residues for cellulose nanofibrils production and their use in nanocomposite manufacturing. *Int. J. Polym. Sci.*, **9**, 1–10.

39. Mariano, M., El Kissi, N., and Dufresne, A. (2014). Cellulose nanocrystals and related nanocomposites: Review of some properties and challenges. *J. Polym. Sci.*, **52**(12), 791–806.

40. Duhan, J. S., Kumar, R., Kumar, N., Kaur, P., Nehra, K., and Duhan, S. (2017). Nanotechnology: The new perspective in precision agriculture. *Biotechnol. Rep.*, **15**, 11–23.

41. Yao, K. S., Li, S. J., Tzeng, K. C., Cheng, T. C., Chang, C. Y., and Chiu, C. Y. (2009). Fluorescence silica nanoprobe as a biomarker for rapid detection of plant pathogens. *Adv. Mater. Res.*, **79**, 513–516.

42. Sadik, O. A., Du, N., Kariuki, V., Okello, V., and Bushlyar, V. (2014). Current and emerging technologies for the characterization of nanomaterials. *ACS Sustain. Chem. Eng.*, **2**, 1707–1716.

43. Parisi, C., Vigani, M., and Rodriguez-Cerezo, E. (2015). Agricultural nanotechnologies: What are the current possibilities? *Nano Today*, **10**, 124–127.

44. Nakamura, J., Nakajima, N., Matsumura, K., and Hyon, S. H. (2011). In vivo cancer targeting of water soluble taxol by folic acid immobilization. *J. Nanomedic. Nanotechnol.*, **2**, 106–112.

45. Sanzari, I., Leone, A., and Ambrosone, A. (2019). Nanotechnology in plant sciences: To make a long story short. *Front. in Bioengineering and Biotechnology*, Doi: https://doi.org/10.3389/fbioe.2019.00120.

46. Shang, Y., Hasan, K., Ahammed, G. J. (2019). Application of nanotechnology in plant growth and crop protection: A review. *Molecules*, **24**, 1–23.

Multiple Choice Questions

1. The size of nanomaterials in nanometers (nm)
   (a) ≥ 10
   (b) ≥ 100
   (c) ≥ 1000
   (d) ≤ 1

2. The traditional methods of nanomaterials synthesis rely on
   (a) Oxidation reduction in liquid phase
   (a) Chemical reduction in liquid phase
   (b) Magnetic reduction in liquid phase
   (c) All of above

3. What are the techniques for nanomaterials synthesis?
   (a) Laser ablation
   (b) Chemical vapor deposition
   (c) Sonochemical
   (d) All of above

4. Route for nanoparticle entrapping:
   (a) Trichomes
   (b) Damages and wound
   (c) Leaves
   (d) All of above

5. To prevent the natural barrier, water and another materials switch from _____ to _____
   (a) Plasmodesmata to cell wall
   (b) Cell wall to vacuoles
   (c) Apoplast to symplast
   (d) Both (a) and (b)

6. Nano-priming enhances _____ enzyme activity during seed germination.

   (a) α-amylase

   (b) β-amylase

   (c) Amylopectin

   (d) Amyloplast

7. Genetic transformation in cotton is achieved by which nanoparticles?

   (a) Silicon

   (b) Silver

   (c) Magnetic

   (d) Electrical

8. An atomically modified seed is an example of

   (a) Maize

   (b) Cotton

   (c) Rice

   (d) Wheat

9. Among these, which are probe/biomolecules interactions?

   (a) Cellular interactions

   (b) Nucleic acid interactions

   (c) Magnetic interaction

   (d) Only (a) and (b)

10. A transducer

    (a) Converts biomolecules to signal

    (b) Converts the recognized signal into a digital signal

    (c) Converts probe to transducer

    (d) None of above

11. The barrier to apoplastic movement is

    (a) Plasmodesmata

    (b) Casparian strip

    (c) Cell wall

    (d) None of the above

12. _____ is the method of cell internalization.

    (a) Endocytosis

    (b) Pore formation

    (c) Membrane translocation

    (d) All of above

13. Improvement in seed germination in old rice seeds enhanced by _____ nanoparticles.

    (a) Gold

    (b) Silver

    (c) Platinum

    (d) Iron

14. Enhanced seed germination in mung bean and chickpea is due to _____ nanoparticles.

    (a) Iron oxide

    (b) Zinc oxide

    (c) Titanium oxide

    (d) All of above

15. _____treated seeds produced 73% more dry weight, three times higher photosynthetic rate, and 45% increase in chlorophyll a content.

(a) Nano-ZnO_2

(b) Nano-TiO_2

(c) Nano-FeO_2

(d) Nano-MnO_2

16. The more efficient shelf life extension was demonstrated by

(a) Nanotubes

(b) Nanogels

(c) Nanocomposites

(d) Pure polymer

17. The nanomaterial used in pesticides is an _____.

(a) Auxiliary compound

(b) Active ingredient

(c) A drug

(d) None of above

18. The _____ property of an active ingredient in water makes them difficult to use, which was overcome by nanomaterials.

(a) Permeability

(b) Solubility

(c) Insolubility

(d) Stability

19. Hurdle to the release of exogenous biomolecules in plant cells is

(a) Plasma membrane

(b) Cell wall

(c) Cell membrane

(d) Both (a) and (c)

20. The un-cultural approach for loading dsRNA is _____

(a) Bio-clay

(b) Bio-sheets

(c) Biosensors

(d) Agrobacterium transformation

21. Successful genetic transformation in cotton was achieved through

(a) Magnetic nanoparticles

(b) Electric nanoparticles

(c) UV nanoparticles

(d) Zinc nanoparticles

22. TiO_2 enhanced_____

(a) Photosynthesis

(b) Respiration

(c) Photorespiration

(d) Greenhouse gases

23. TiO_2 protects chloroplast from _____

(a) Electric nanoparticles

(b) Photochemical stress

(c) Photorespiration

(d) Photosynthesis

Answer Key

1. (b)   2. (b)   3. (d)   4. (d)   5. (c)   6. (a)   7. (c)

8. (c)   9. (d)   10. (b)   11. (b)   12. (a)   13. (b)   14. (b)

15. (b)   16. (c)   17. (a)   18. (c)   19. (b)   20. (a)   21. (a)

22. (a)   23. (b)

Short Answer Questions

1. Explain the key challenges for nanotechnology.
2. What is the need for new technology in the field of agriculture?
3. What are the hindrances in trapping nanoparticle to cell?
4. What are the applications of nanotechnology in agriculture?
5. Define in brief the role of nano-priming in seed germination.
6. What are nano-fertilizers and their role in increasing soil fertility?
7. Describe the genetic transformation through nanoparticles in cotton.
8. What are atomically modified seeds?

Long Answer Questions

1. What are nanobiosensors?
2. Describe the functioning of nanobiosensors.
3. Describe the uptake and mobilization of nanoparticles to biological system.
4. Describe in detail the type of nano-pesticides with their active ingredient, benefit, and example.
5. What are the major challenges addressed by agriculture?

Chapter 5

Nanotechnological Advances for Nutraceutical Delivery

Shaveta Sharma,[a] Puneet Sudan,[a] Vimal Arora,[b]
Manish Goswami,[b] and Chander Parkash Dora[c]

[a]*Chandigarh College of Pharmacy, CGC Campus, Landran,*
Mohali 140307, Punjab
[b]*University Institute of Pharma Sciences, Chandigarh University,*
NH-95 Chandigarh-Ludhiana Highway, Mohali 140413, Punjab
[c]*M.M. College of Pharmacy, Maharishi Markendeshwar*
(Deemed to be University), Mullana, Ambala 133207, Haryana
chanddora@gmail.com, cpdora@mmumullana.org

Nutraceuticals are super-nutrients isolated and purified from foods (fruits, vegetables, legumes, spices, etc.), which generally promote health besides their authentic nutritional values and are, therefore, used to manage various diseases/disorders. Mainly, they are categorized by their source of origin, therapeutic benefits, and chemical constitution. The prior art also suggests their long history of health benefits; however, the full potential of nutraceuticals, such as curcumin, resveratrol, quercetin, and carotenoids, has still been unexplored due to their poor physiochemical properties, i.e., low aqueous solubility, poor diffusion, low gastrointestinal stability,

Nanotechnology: Principles and Applications
Edited by Rakesh K. Sindhu, Mansi Chitkara, and Inderjeet Singh Sandhu
Copyright © 2021 Jenny Stanford Publishing Pte. Ltd.
ISBN 978-981-4877-43-5 (Hardcover), 978-1-003-12026-1 (eBook)
www.jennystanford.com

poor permeability, and low bioavailability. To enhance the efficacy of nutraceuticals, drug-delivery systems using nanotechnology (liposome, nanospheres, nanoemulsion, micelle, and nanocrystals) represent an innovative way to effectively increase their efficacy, oral bioavailability, and alleviate other barriers. Besides improving the physicochemical properties and therapeutic efficacy of nutraceuticals, developed nanoformulations also provide a targeted approach-based delivery. This chapter will give brief details of the full therapeutic potential gained with the utilization of nanotechnology in nutraceuticals.

5.1 Introduction

Since the ancient period, the most social and intelligent animal on the planet had formulated many medicines by exploiting multifarious herbal plants in the form of extracts and their phytoconstituents, which had great therapeutic potential. Dr. Stephen coined the term "Nutraceuticals" in 1989, which is made up of the two words "nutrition" and "pharmaceutical." Nutraceuticals have a long journey since a new trend in the care of companion animals emerged in the 1990s and the parallel in the human population. With the newer developments and advancements of dietary supplements, nutraceuticals also include minerals, vitamins, amino acids, and other dietary substances. "Nutraceuticals" is a term that can be explained as functional foods, having medicinal and nutritional benefits. Furthermore, these can also be called medical foods, designer foods, phytochemicals, and nutritional supplements, which are a very important part of our daily routine as "bio" yogurts (curd), healthy fortified breakfast, cereals, vitamins, herbal remedies, garlic, and soybeans. They may be a specific component of a food, such as omega-3 fish oil, which can be derived from salmon and other cold-water fish, and even genetically modified foods and supplements [1].

At present, there are no universally accepted definitions for nutraceuticals and functional foods, although there is something common between all the definitions offered by different health-oriented professional organizations. According to the *Nutrition Business Journal*, nutraceuticals can be anything that can be utilized mainly for the benefit of health. Table 5.1 encloses the types of nutraceuticals with their examples.

The major benefits of dietary active compounds and their role in human nutrition are the most significant areas of exploration. They having wide-ranging implications not only for consumers and healthcare providers but also for regulators and manufacturing industries [2]. Foods and nutrients have vital and beneficial roles in the normal performance of the human body. They not only help to maintain the health of an individual but also prove beneficial in reducing the risk of multifarious ailments and diseases [3].

Table 5.1 Types of nutraceuticals and examples

Sr. No	Types of nutraceuticals	Examples
1.	Inorganic mineral supplements	Minerals
2.	Vitamin supplements	Vitamins
3.	Digestive enzymes	Enzymes
4.	Probiotics	*Lactobacillus acidophilus*
5.	Prebiotics	Digestive enzymes
6.	Dietary fibers	Fibers
7.	Cereals and grains	Fibers
8.	Health drinks	Fruit juice
9.	Antioxidants	Vitamin C
10.	Phytochemicals	Carotenoids
11.	Herbs	Soya proteins

Although there are thousands of products of nutraceuticals available in the market, various research groups and companies are still working to improve solubility, permeability, stability, and bioavailability of nutraceuticals. In this chapter, we will focus on nanotechnology to improve the deliverability of nutraceuticals.

Nanotechnology consists of the two words "nano" and "technology." Nano refers to a nanometer match with 1 billionth of a meter (10^{-9} m). Technology refers to using tools, machines, and technologies to achieve a particular function. Nanotechnology can be explained as an expanding area of interdisciplinary research on 1–100 nm particles, with the endeavor to fashion, improve, and utilize the nanoscale structures in advanced studies [4]. Engineered nanoparticles have a considerable bang on the economy, industry, and people's lives around the globe; they are already being contained in a big variety of everyday products such as food, veterinary drugs, biocides, agriculture, soil cleaning, and water purification methods [5].

To enhance the efficacy of nutraceuticals, drug-delivery systems using nanotechnology such as liposome, nanoparticles, nanosuspension, and polymeric micelle effectively increase the solubility, therapeutic efficacy, permeability, oral bioavailability, and stability. There are many problems with the delivery of nutraceuticals to get to the target site. Poor absorption by the gastrointestinal tract (GIT) is due to the solubility problems of active pharmaceutical ingredients, e.g., vitamins and calcium supplements. Interaction of nutraceutical products with chemicals, drugs, and other nutraceuticals can affect the therapeutic activity. The increased awareness of appropriate dosage and interaction of nutraceutical products is very important in special concerns such as renal impairment and aging. The problems of absorption and solubility can be dealt with nanotechnology. By reducing the size, the surface of molecules is increased, which further increases the desired biological process. Modification of a nutraceutical molecule into a nanocarrier can overcome the barriers. Nano-delivery techniques are extensively used in the pharmaceutical approach. The low bioavailability associated with the majority of phytochemicals due to poor solubility and incomplete absorption results in diminished biological activity [6]. The use of nanotechnology will enhance the solubility, oral bioavailability, physiological performance, and thermal stability of poorly absorbed nutraceuticals. Nanotechnology can improve solubility, gastrointestinal stability, and oral bioavailability of micronutrients/nutraceuticals [7, 8]. Also in the food industry, nanotechnology plays an important role in the production of nutrients/food with better solubility, oral bioavailability, and thermal stability [9].

Nanotechnology has proven to work delicately in the field of nutrition science. It will enhance food storage and increase the shelf life of food products. It has the potential for improving health due to the increased bioavailability of the active compounds [10]. These micronutrients loaded with lipidic nanocarriers showed improved gastrointestinal stability and thus degradation safety [11]. Nutraceuticals mean any nontoxic food component used for health benefits and treatment of acute and chronic diseases such as cancer and diabetes. Curcumin, garlic acid, ginger, aloe vera, vitamin C, ellagic acid, and quercetin (Fig. 5.1) are a few examples of nutraceuticals that are used in the prevention of joint diseases, cancer, and diabetes and to maintain the level of cholesterol. Vegetables and fruits not only give nutritional value but also have therapeutic efficacy. Vitamins and minerals are used as dietary supplements but have problems in absorption.

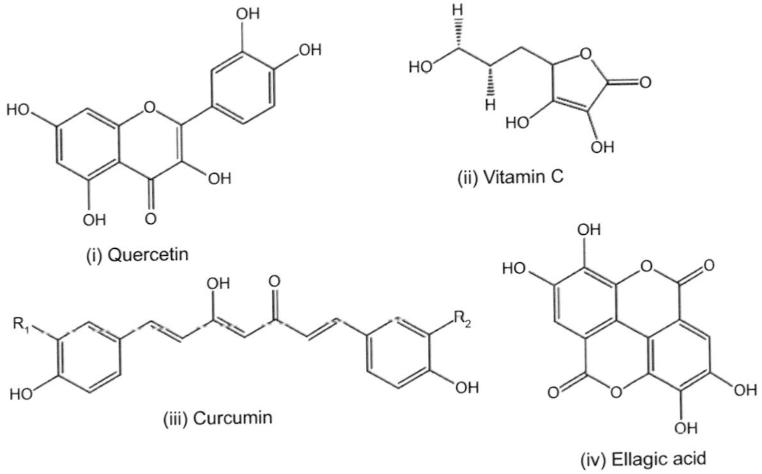

(i) Quercetin

(ii) Vitamin C

(iii) Curcumin

(iv) Ellagic acid

Figure 5.1 Chemical structure of a few examples of nutraceuticals.

5.2 Classification of Nutraceuticals

The food sources used as nutraceuticals are all natural and can be categorized as:

- Probiotics
- Dietary fiber
- Prebiotics
- Polyunsaturated fatty acids
- Antioxidant vitamin
- Polyphenol
- Spices

Further, nutraceuticals can be broadly classified into the following two groups:

1. Potential nutraceuticals
2. Established nutraceuticals

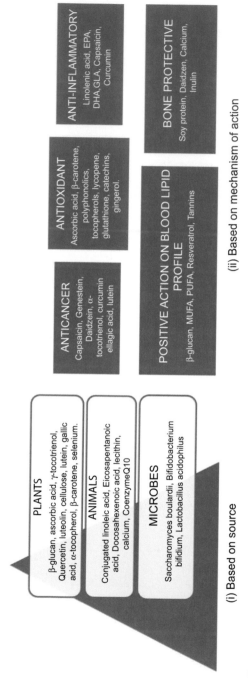

(i) Based on source

(ii) Based on mechanism of action

Figure 5.2 Classification of nutraceuticals based on (i) source and (ii) mechanism of action [12].

Potential nutraceuticals can be transitioned into an established one only after proven efficient clinical data of their health and medical benefits [12]. Moreover, nutraceuticals can also be classified based on their source and mechanism of action (Fig. 5.2).

5.3   Problems in Nutraceuticals Delivery

The prime factors behind the weak oral delivery of a nutraceutical are its target capacity and oral absorption. The poor bioavailability of nutraceuticals could be due to low serum distribution, apparent fast metabolism, and elimination due to short half-life [13].

5.3.1   Bioavailability

Bioavailability is defined as "the rate and extent of drug that reaches in the systemic circulation." Relative bioavailability measures the bioavailability of a formulation of a bioactive compared with its another formulation, whereas absolute bioavailability is calculated by the ratio of the area under the curve (AUC) of intravenous administration to the AUC of oral administration. Steps for orally administered nutraceuticals to reach systemic circulation are discussed as follows:

- **Systemic circulation:** Nutraceuticals that are absorbed can directly enter the systemic circulation. Polar compounds propagate into blood vessels and get into the liver through the hepatic portal vein. Hydrophobic compounds may be integrated into chylomicrons and can directly enter the systemic circulation, avoiding the first-pass hepatic metabolism [14].
- **Solubilization:** In the small intestine, nutraceuticals through food are solubilized. Nutraceuticals can be hydrophilic or hydrophobic. Hydrophilic nutraceuticals are readily dissolved in the aqueous solution in the GIT lumen. Hydrophobic nutraceuticals can get help from mixed micelle systems in the lumen fluid. The micelles are composed of endogenous bile salts and phospholipids. Food lipids can also lead to the development of micelles following digestion

by lipase. Triglycerides are converted into free fatty acids, monoglycerides, and diglycerides. Phospholipids are natural amphiphilic, biocompatible molecules that can easily solubilize the nutraceuticals by micellar formation.

- **Absorption:** After solubilization, nutraceuticals are absorbed by the small intestine through various epithelial routes or mainly by the lymphatic route. The primary site for nutrient and nutraceutical absorption is the small intestine, and this happens specifically through the enterocytes (absorption epithelial cells) lining the lumen of the jejunum.

5.4 Nanotechnology in Nutraceuticals

With the rising population around the globe, there is an increase in the prevalence of a variety of harmful diseases such as cancer, hypertension, diabetes, obesity, infectious diseases, pain/ inflammation, hyperlipidemia, and osteoporosis.

Nutraceuticals are beneficial and found to be useful in prophylaxis and to alleviate the aforementioned diseases, e.g., coenzyme Q-10, polyphenols, isoflavone, flavones, and carotenoids for the treatment of cytotoxicity and many other life-threatening diseases; elements such as zinc as a prophylactic treatment to sustain metabolism; calcium and magnesium for the maintenance of bone health; selenium as an antioxidant; vitamin D for the treatment of osteoporosis; vitamins E and C as antioxidant for the prevention of cancer; and vitamin B complex for the suitable performance of the nervous system. On the other hand, nutraceuticals have certain restrictions, including poor solubility, absorption and bioavailability, chemical instability, and higher first-pass metabolism, which can be overcome by nano-delivery of nutraceuticals (Fig. 5.3).

A few years back and even in the present scenario, people are trying to improve the quality of life by modifying their diet and by substituting modern medicine with alternative and natural products of herbal origin. As a result, there has been a great boom in the research and development activities related to delivery systems of nutraceuticals [15].

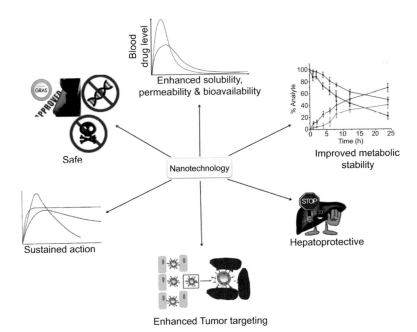

Figure 5.3 Advantages of nanotechnology-based nutraceuticals [16].

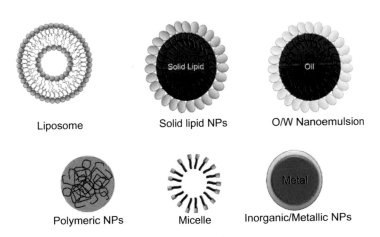

Figure 5.4 Different types of nanomaterials/nanocarriers.

The prime concern of this field is represented by the poor bioavailability of nutraceuticals, which are habitually excreted from the body without any medicinal benefit. To promote oral bioavailability, permeability, and thus bioavailability of dietary and health supplements, nanotechnology can be employed [16, 17]. The exceptional properties of nanomaterials are due to their small size, high surface-area-to-volume ratio, and higher efficiency due to higher uptake via endocytosis and other biological transport mechanisms [13–16]. Nanomaterials can be further categorized into organic and inorganic type based on their chemical composition. These nanomaterials (organic or inorganic type) can act themselves as therapeutic agents (self-assembly omega-3 fatty acids, nanosilver, and zinc-oxide-based formulations) or delivery vehicles.

The same physicochemical properties making these new materials promising for many application fields might affect human health. To prevent health risks associated with the usage of nanomaterials, it is also important to evaluate exposure and hazard [18–21]. Different nanomaterials (Fig. 5.4) significantly vary the physiochemical and biological properties of nutraceuticals, and thus their impact on human health and safety needs to be carefully watched [22].

Different formulation technologies, e.g., liposome, lipidic nanoparticles (solid lipid nanoparticles, nanostructured lipid carriers), nanoemulsion, nanosuspension, polymeric nanoparticles, lipid polymer hybrid nanoparticles, micelles, and inorganic nanoparticles, have been utilized/evaluated to improve the aforementioned limitations of nutraceuticals. Submicron emulsion, one of the subtypes of nanoparticles, increases solubility, half-life, and bioavailability of nutraceuticals with decreased gastric irritation. Self-emulsifying nano/micro drug-delivery systems have unique properties of increased palatability (by masking bioactives), stability (by preventing hydrolysis of bioactives), and lymphatic uptake of bioactives causing increased bioavailability. Lipid nanocarriers have also enhanced the permeation of neuroprotective nutraceuticals to the blood–brain barrier and thus increased bioavailability. The liposomal delivery system is beneficial to avoid sensitive bioactives from oxidation, light, and moisture. It can serve as a unique system to deliver both hydrophilic and lipophilic bioactives. Highly potent anticancer bioactives possess lower bioavailability and poor distribution. Dendrimers and inorganic nanoparticles can be utilized for targeting such bioactives and show sustained release. Polymeric nanocarriers, e.g., micelles, have also shown superior stability (protect bioactives from enzymatic hydrolysis) over lipidic nanocarriers and thus control the release of nutraceuticals in a better way. Polymeric nanoparticles derived from natural sources such as zein, β-lactoglobulin, β-casein, and nanodiamond can be used to deliver bioactives in a controlled manner. It can also protect bioactives from the harsh acidic environment and thus show lower toxicity. The benefits and innovations associated with the application of nanotechnology in the nutraceuticals-based industry are summarized in Table 5.2.

Table 5.2   Role of nanotechnology in the nutraceuticals-based industry

Sr. No	Product	Application	Significance
1.	Food enzymes	Carrier—Enzyme immobilization Protein separation—Purification	Specific coating to selectively kill pathogen
2.	Food additives	Nanocapsules, nanoparticles, anticaking agents	Improved bioavailability and absorption of nutrients and removed chemicals, pathogens from food by selective binding
3.	Food supplements	Nanosensors, nanoencapsulation Solid lipid nanoparticles	Storage enhancement, protection from degradation, and increased stability
4.	Novel food	Healthy supplements Herbal medicine	Beneficial to health
5.	Nanoparticles applied in food packing	Antimicrobial and antifungal surface coating	Detected chemicals of foodborne pathogens

5.5 Various Nanocarriers for Nutraceuticals

Many barriers obstruct nutraceutical formulations that negatively affect their effectiveness. Bioavailability, site-specific delivery, liable nature, targetability, and oral absorption are the basic causes of poor nutraceutical oral delivery. Nanosizing techniques are recently used to solve the nutraceutical delivery problems with bright results in terms of nutraceutical safety, shelf stability, solubility, enhancement of the dissolution rate, resulting in increased bioavailability [23].

5.5.1 Polymeric Micelles

Micelles are widely used to deliver nutraceuticals and have a core–shell characteristic structure in which the hydrophilic head region supports and stabilizes the hydrophobic core in the aqueous medium and increases water solubility, and the hydrophobic region carries and protects the drug. They efficiently improve the efficacy of therapeutic drugs associated with CNS diseases [24]. Micelles are formed above the critical micelle concentration (CMC) and are thermodynamically stable [14]. Beyond the CMC, individual amphiphiles immediately form micelles. Limitations among micelles are low drug-loading efficiency and low serum stability; even they are good nanocarriers for lipophilic bioactives [25]. Bioactives incorporated in micellar nanoformulation show low serum stability mainly when diluted with blood (below CMC) after systemic injection.

5.5.2 Nanoparticles

Nanoparticles-loaded nutraceuticals are formulated by adding a hydrophilic cosolvent, nutraceutical, and a hydrophobic polymer to water. Various synthetic polymers (e.g., polylactic acid (PLA), polyglycolic acid (PGA), polylactic acid glycolic acid (PLGA), polyethylene glycol (PEG), and polyglutamic acid (PGA)) and natural polymers (e.g., albumin, chitosan, heparin, and alginate) have been utilized to prepare polymeric nano-nutraceuticals [26]. Polymeric nanoparticles of curcumin were prepared with different derivatives of chitosan and sodium tripolyphosphate by the emulsification technique. This combination renders the low solubility, high crystallinity, and poor oral bioavailability of curcumin polymeric nanoparticles [27]. Polymeric nanoparticles (PNPs) show outstanding improvement in efficacy and bioavailability. PNPs are also treated as ideal candidates for the delivery of vaccines, antibiotics, and cancer therapy according to the polymer type and ability to adjust the release of drugs from PNPs [28].

5.5.3 Metal Nanoparticles

Metals such as zinc, titanium oxide, iron, and silver are prepared in the nano-size range. Metal nanoparticles are favorable due to many reasons such as relatively large surface area, good biocompatibility, catalytic activity, and superior selectivity. They are usually isolable, dispersible, and reusable catalysts. They can be applied as metal oxide such as nano-Ag, nano-ZnO, nano-Cu, and nano-TiO$_2$ and silver nanoparticles (AgNPs). Biological activity of metal nanoparticles was conducted in nutraceuticals such as garlic, ginger, and cayenne pepper. Spherical shape and an average size of 3–6 nm represented strong antibacterial activity against Gram-negative and Gram-positive bacterial strains [29].

5.5.4 Hybrid Nanoparticles

Hybrid nanoparticles (HNPs) utilize the positive outcomes of two different components. HNPs consist of both systems: lipid and polymer/organic and inorganic materials. HNPs show drug release with erosion and hydrolytic degradation when they come in contact with penetrated water/release vehicle. HNPs have a core layer protected by multiple layers of shell material. Furthermore, functionalization of these interfaces may increase the targeted uptake and may also be utilized for diagnosis setup [30]. Different types of nanohybrid particles such as curcumin gold hybrid nanoparticles have been prepared and evaluated. Curcumin has anticancer potential even with low aqueous solubility and permeability (BCS IV) and is utilized as a medicine. The formulation of various nanoparticles will resolve the limitations and disadvantages of curcumin. Encapsulated curcumin in nanoscale particles requires a low dose and increases bioavailability. Curcumin-based HNPs are more effective in therapeutics [31].

5.5.5 Targeted Delivery System

Along with higher bioavailability and lipid permeability, nanotechnology may add a successful advantage of targeting in the delivery of nutraceuticals. Active targeting occurs when the drug carrier system is bound to a cell-specific ligand or targeting carrier, while passive targeting occurs when nanoparticles reach the targeting site due to the leaky vasculature of the tumor [32]. An ideal nano-delivery carrier should be site specific, bind to targeted sites, and have a maximum load of drugs with lesser toxicity than its conventional counterpart. Here, the targeted concept of a delivery system comes into play. Targeted nano-delivery systems can be prepared by adding (chemically or physically) ligand (small molecules, antibodies, proteins, peptides, and nucleic acid aptamers) on the surface of the delivery vehicle. One best example of the targeted delivery system is the addition of folic acid to the surface of nanoparticles, which shows higher affinity toward folate receptors and higher efficacy in various types of cancers [33, 34]. Similarly, many other task-specific carbohydrates, short peptides, antibodies, and small molecules have been designed and employed [33].

5.5.6 Nanoemulsion

Nanosized emulsion includes droplets with a mean size of 20–600 nm. They prevent flocculation due to their droplet size [35]. Vitamin E encapsulation shows that nanoemulsion formulation can improve stability with an average particle size of 277 nm. Carotene is a hydrophobic nutraceutical, requiring a novel delivery system such as nanoemulsion because of its limited solubility. β-carotene is encapsulated into oil-in-water nanoemulsions via high-pressure, dual-channel microfluidization. Moreover, its solubility, suspendibility, and stability are enhanced with different emulsifiers, e.g., Quillaja saponins, lecithin, and whey protein isolate [36].

5.5.7 Nanosuspension

Nanosuspension is beneficial in the nutraceutical industry due to enhanced drug stability and improved oral bioavailability. Curcumin solubility is also increased by the phase change from crystalline to amorphous in curcumin nanosuspension [37].

5.5.8 Liposome

Liposome is made up of a lipid bilayer that surrounds the aqueous layers. Liposomes are flexible, biocompatible, and nontoxic carriers and can protect active ingredients against enzymatic degradation. Liposomes improve the solubility, stability, and bioavailability of hydrophobic drugs. However, liposomes have limitations such as poor stability, short half-life, low loading capacity, and high cost [38].

5.6 Application of Nanotechnology for Delivery of Nutraceuticals

Nanotechnology is mainly used for delivering nutraceuticals with poor water solubility. This technology overcomes the barriers associated with the delivery of nutraceuticals [39]. Polysaccharides and proteins may produce different delivery platforms for loading nutraceuticals and functional nutrients [40]. Nanoparticles of different shapes and sizes have been prepared using various methods. Different processes can be used for nanoencapsulation of bioactives and polysaccharides depending on their physical and chemical properties. Polysaccharides can cover both hydrophilic and hydrophobic compounds [41]. Curcumin is found in the rhizome of turmeric plant (*Curcuma longa*) and is widely used for antioxidant, anticancer, anti-inflammatory, and antimicrobial activities [42]. Various research groups are still working on newer approaches to improve solubility, bioavailability, and stability and better the previous formulation of curcumin to fight against cancer and inflammation. The quality and stability of nutraceuticals can also be increased by loading in a suitable nano-delivery system. Encapsulation of active food ingredients can be attained with various nanotechnological approaches using suitable polysaccharides (e.g., pectin and starch) and food grade proteins (e.g., casein, soy proteins, gelatin, and monoglycerides, etc.). Table 5.3 demonstrates nanotechnology approaches in the food industry.

Table 5.3 Different examples of nanotechnology-based nutraceuticals

Sr. No	Nano-delivery system	Nutraceutical	Remarks	Reference
1.	Anthocyanins SLNs	Palmitic acid, span 85, egg lecithin	Increased stability	[43]
2.	Bromelain nanoparticles	Katira gum	Increased anti-inflammatory activity	[44]
3.	Caffeine nanohydrogels	Lactoferrin-glycomacropeptide	Increased permeation through the skin	[45]
4.	Capsaicin polymeric micelles	Phospholipid, sodium cholate, and PVP K30	Prolonged plasma circulation with an increase in oral bioavailability	[46]
5.	Curcumin nanohydrogel nanoparticles	Hydrolyzed tetramethyl orthosilicate, chitosan, polyethylene glycol 400	Increased antimicrobial and wound healing	[47]
6.	Glycyrrhizic acid polymeric nanoparticles	Chitosan, katira gum	Increased anti-inflammatory activity	[44]
7.	Green tea extract, nanostructured lipid carriers	Cetyl palmitate, glyceryl stearate, grape seed oil, sea buckthorn oil	Increased antioxidant activity	[48]
8.	Hesperidin SLNs	Stearic acid, glyceryl behenate, oleic acid, Tween 80	Could well mask the bitter taste, after taste, and obviate poor solubility of hesperetin	[49]

Sr. No	Nano-delivery system	Nutraceutical	Remarks	Reference
9.	Lutein polymeric nanoparticles	Poly-γ-glutamic acid, chitosan	Increase solubility	[50]
10.	Melatonin polymeric nanoparticles	Poly(D,L-lactide-co-glycolide), polyvinyl alcohol	Sustained release	[51]
11.	Quercetin nanoparticles	Bovine serum albumin (BSA)	Prolonged quercetin release and improved antioxidant activity	[52]
12.	Resveratrol SLNs	Cetyl palmitate, polysorbate 60, miglyol-812	Validated for trans-resveratrol protection, stabilization, and intestinal permeability	[53]
13.	Resveratrol SNEDDS	α-tocopherol	Improved bioavailability	[54]
14.	Thymoquinone nanoparticles	Poly(styrene-b-ethylene oxide)	Enhanced antitumor activity	[55]
15.	Curcumin solid lipid nanoparticles	Tween 80	Effective oxygen scavenging activity	[56]
16.	Curcumin nanoemulsion	Tween 20	Controlled lipid digestion rate and free fatty acid adsorption	[57]

(Continued)

Table 5.3 *(Continued)*

Sr. No	Nano-delivery system	Nutraceutical	Remarks	Reference
17.	Caffeine nanoparticles	PLGA	Significant increase in the endurance of dopaminergic neurons, fiber outgrowth, and expression of tyrosine hydroxylase	[58]
18.	Caffeine SLNs	Softisan, pluronic F68	Increased permeation through the skin	[45]
19.	Reassembled casein micelle and casein nanoparticles	Omega 3-fatty acid	Protection against cardiovascular diseases	[59]
20.	β-Carotene-based nanoemulsion	Tween 20	Higher optical clarity and increased oral bioavailability	[60]
21.	Apigenin nanoparticles	PLGA	Ameliorative potentials in combating skin cancer	[61]
22.	Apigenin nanoparticles	Mesoporous silica	Improved solubility and bioavailability	[62]
23.	Thymoquinone nanoparticles	Chitosan	Increased brain targeting through nasal route	[63]
24.	Coenzyme Q10 nanoparticles	PLGA	Increased oral bioavailability and antioxidant activity	[64]
25.	Coenzyme Q10 nanoparticles	Brij 78, and/or Tween 20	Enhanced stability	[65]

Sr. No	Nano-delivery system	Nutraceutical	Remarks	Reference
26.	Coenzyme Q10 liposome and SLNs	Lipoid S100 and Compritol® 888 ATO	Enhanced topical delivery	[66]
27.	Green tea catechins nanoparticles	Chitosan	Enhanced intestinal absorption and bioavailability	[67]
28.	Genistein nanoparticles	Eudragit	Increased oral bioavailability	[68]
29.	Resveratrol SLNs	Stearic acid	Enhanced antineoplastic activities in human colorectal cancer cells	[69]
30.	5-demethylnobiletin nanoparticles	Unsaturated fatty acid	Enhanced trans-intestinal uptake	[70]
31.	Vitamin D3 micelle	Carboxymethyl chitosan–soy protein	Enhanced encapsulation and controlled release	[71]
32.	Naringenin nanoparticles	β-Lactoglobulin	Enhanced bioavailability by suppressing crystallization	[72]
33.	Naringenin nanoparticles	Chitosan	Improved nose to brain delivery	[73]
34.	Lycopene nanoparticles	Chitosan	Enhanced cellular accumulation	[74]
35.	Kaempferol nanoparticles	Polyvinylpyrrolidone K-17PF	Improved ocular delivery	[75]
36.	Lutein nanoparticles	Zein/soluble soybean polysaccharide	Enhanced in vitro bioaccessibility	[76]

5.7   Market Potential

Nutraceuticals work on root causes and take a longer time for recovery, do not have side effects, focus on prevention and wellness, do not require prescription for buying, and do not need FDA approval. In contrast, pharmaceuticals are relatively more instant in effect for severe illnesses, have side effects, focus on illness and treatment, are sold only on prescription except OTCs, and need FDA approval. As is evident, while in the initial years, between 1999 and 2002, the nutraceutical industry grew at 7% per annum. The next few years up to 2010 saw double growth at 14% per annum. The United States has been the largest nutraceutical market so far and almost fully mature. Between 2010 and 2015, it showed a compounded growth of 10% annually. The European market is also expected to grow with an annual growth of 5% by 2021. The Indian nutraceutical market expects 21% growth annually. Different pharmaceutical companies (e.g., Amway, Nutraceutical, GSK, Unilever, Infinitus, and Parry Nutraceuticals) are eagerly looking at preventive products for a certain range of ailments. Chavanprash, or herbal honey, has been a notable success. It is incorporated as a general health supplement into anything from jam to chocolate to cheesecake to capsules. With changes in lifestyle and food habits and urbanization, more people are now vulnerable to lifestyle-based diseases/disorders. This situation has provided a massive growth opportunity for nutraceuticals globally. The entire nutraceuticals category is divided into functional foods, functional beverages, and dietary supplements. Nutraceuticals have gained global importance and have become part of daily diet due to the increased risk of diseases from improper lifestyle, and people are consciously adopting preventive healthcare measures. Gradually increasing healthcare expenses have stimulated the demand for nutraceuticals. In developing countries such as Mexico, India, Brazil, and South Africa, the compounded annual growth rates are 7.3%, 12.3%, 8.9%, and 8%, respectively. Nutraceuticals have become an opportunity for the economic growth of developed as well as many developing countries that have a rich source of medicinal herbs and traditional knowledge of plants, especially India, China, and South American countries. Various types of products such as functional foods (cereals, bakery, confectionery), functional drinks (energy drinks,

Table 5.4 Marketed nutraceuticals based on nanotechnology.

Sr. No	Name of marketed product	Product description	Remarks	Company name
1.	NanoResveratrol™	Resveratrol loaded in a natural phospholipid-based nano-delivery system	Increased gut absorption	Life Enhancement, USA*
2.	Spray for Life® Multi-Vitamin	Composition of multivitamins in a unique NanoSyzed™ solution	Increased bioavailability and rate of absorption	
3.	Spray for Life® CoQ10 complex	Composition of CoQ10, Vitamin E and L-Carnitine in a combined unique NanoSyzed™ solution.	Increased bioavailability	
4.	Spray for Life® Fast Acting melatonin	Composition of melatonin, minerals and herbs in a combined unique NanoSyzed™ solution	Increased rate of absorption	NanoSynergy® Worldwide, USA**
5.	Spray for Life® Natural D3	Composition of vitamin D3 and vitamin E in a unique NanoSyzed™ solution	Increased bioavailability and rate of absorption	
6.	Spray for Life® Sweet & Slim®	Composition of minerals, amino acids, and herbs in a unique NanoSyzed™ solution	Increased rate of absorption	
7.	Spray for Life® Vitamin C+ Zinc	Composition of vitamin C + zinc in a unique NanoSyzed™ solution	Increased bioavailability and rate of absorption	
8.	Nutri-Nano™CoQ-10 3.1x Softgels	30 nm oil-in-water nanoemulsion system	Increased bioavailability	Solgar, USA***

(Continued)

Table 5.4 (*Continued*)

Sr. No	Name of marketed product	Product description	Remarks	Company name
9.	Lypo-Spheric™ B complex plus	Liposomal Nano-spheres® that encapsulate B vitamins, minerals, and cinnamon	Increased bioavailability and rate of absorption	LivOn Labs, USA@
10.	Lypo-Spheric™ Glutathione	Liposomal Nano-spheres® that encapsulate glutathione	Increased bioavailability and rate of absorption	
11.	Lypo-Spheric™ R-Alpha lipoic acid	Liposomal Nano-spheres® that encapsulate R-Alpha lipoic acid	Increased bioavailability and rate of absorption	
12.	NanoGreens 10®	Vegetable and fruit superfood delivered through SuperSorb® technology	Maximize absorption	BioPharma® Scientific LLC, USA@@
13.	nanoice™	Menthol and plant extract delivered through SuperSorb® technology	Increased absorption and relieving	
14.	Quercefit™	Quercetin loaded in a phospholipid-based Phytosome® carrier	Increased solubility and bioavailability	Indena S.p.A., Italy@@@
15.	Meriva®-The Life Guardian™	Curcumin-loaded Phytosome®	Increased solubility and bioavailability	
16.	Ubiqsome®	CoQ10 loaded in Phytosome®	Increased solubility and bioavailability	
17.	Siliphos®	Silybin-loaded Phytosome®	Increased solubility and bioavailability	

*Life Enhancement [https://www.nanotechproject.org/cpi/products/nanoresveratroltm/, accessed on 11/03/2020]
**NanoSynergy® [http://sprayforlife.com/Products/products.html, accessed on 11/03/2020]
***Solgar [https://www.solgar.co.uk/all-products/nutri-nano-coq-10-3-1x-softgels-pack-of-50/, accessed on 11/03/2020]
@ LivOn Labs [https://www.livonlabs.com/products/, accessed on 11/03/2020]
@@ BioPharma® Scientific LLC [https://www.biopharmasci.com/Default.asp, accessed on 11/03/2020]
@@@ Indena S.p.A. [https://www.indena.com/products/healthfood/, accessed on 11/03/2020]

sports drinks, fortified juices), and dietary supplements (vitamins, minerals, probiotics, omega-3 fatty acids) exist in the market. With the advent of technology, various companies want to grab a big share of the whole nutraceuticals market and develop products with newer systems, especially nano-based delivery systems. Nowadays, various brands of nutraceuticals are available based on nanotechnology. One of the popular companies based in Italy (Indena S.p.A) has developed a phytosome-technology-based nanoproduct of various plant extracts or phytoconstituents. Different commercialized nano-nutraceuticals are listed in Table 5.4 [77–80].

5.8 Future Scope

Current advancements in nanoscience and nanotechnology have resulted in novel and innovative applications in nutraceuticals. However, the utilization of nanotechnology in food-based products is still in the initial stages of development. As a novel strategy for nutraceuticals, nowadays, different leading food companies such as Unilever, Kraft, Nestle, Heinz, GlaxoSmithKline, and Hershey are working swiftly to introduce nanotechnology into commercialization [17]. Despite the slower pace of development of nano-based nutraceuticals, the future holds great potential in newer delivery platforms for nutraceuticals, developing novel packaging materials to protect nutraceuticals, novel processing technologies, and advanced analytical tools for food quality. With the use of newer technology, newer challenges can also be anticipated, and the prime focus is on overcoming the cost barrier in introducing nanotechnology in food products. Moreover, significant changes in food-based regulations are much needed to combat the toxicity associated with nanotechnology.

5.9 Conclusion

Delivery of bioactives is an essential step for showing the efficacy and therapeutic value of nutraceuticals. Among the various tools and technologies, nanotechnology has been proved to be the most efficient method in delivering bioactives/nutraceuticals. Formulations based on nanotechnology result in nanocarriers such as nanoemulsion, nanoparticles, nanosuspension, polymeric micelles, and liposomes with various physicochemical properties. Different functional attributes have opened new possibilities for the delivery of nutraceuticals, food production, and processing. Advancement in the field has created a broad potential for nanotechnology, particularly in improving solubility, bioavailability, permeability, and efficacy of nutraceuticals. Along with these advantages, targeting of bioactives has also been reported. In brief, with the advent of delivery tools for nutraceuticals, nanotechnology plays a very important role. However, challenges associated with this technology, such as toxicity, cost, and others, also need to be considered.

References

1. Jack, D. B. (1995). Keep taking the tomatoes: The exciting world of nutraceuticals, *Mol. Med. Today,* **1**, pp. 118–121.

2. Bagchi, D. (2006). Nutraceuticals and functional foods regulations in the United States and around the world, *Toxicology,* **221**, pp. 1–3.

3. Ramaa, C., Shirode, A., Mundada, A., and Kadam, V. (2006). Nutraceuticals: An emerging era in the treatment and prevention of cardiovascular diseases, *Curr. Pharm. Biotechnol.,* **7**, pp. 15–23.

4. Sun, Y. and Xia, Y. (2002). Shape-controlled synthesis of gold and silver nanoparticles, *Science,* **298**, pp. 2176–2179.

5. Smolkova, B., El Yamani, N., Collins, A. R., Gutleb, A. C., and Dusinska, M. (2015). Nanoparticles in food. Epigenetic changes induced by nanomaterials and possible impact on health, *Food Chem. Toxicol.,* **77**, pp. 64–73.

6. Sahni, J. K. (2012). Exploring delivery of nutraceuticals using nanotechnology, *Int. J. Pharm. Investig.,* **2**, pp. 53–53.

7. Chen, H., Weiss, J., and Shahidi, F. (2006). Nanotechnology in nutraceuticals and functional foods, *Food Technol. (Chicago),* **60**, pp. 30–36.

8. Chen, L., Remondetto, G. E., and Subirade, M. (2006). Food protein-based materials as nutraceutical delivery systems, *Trends Food Sci. Technol.,* **17**, pp. 272–283.

9. Semo, E., Kesselman, E., Danino, D., and Livney, Y. D. (2007). Casein micelle as a natural nano-capsular vehicle for nutraceuticals, *Food Hydrocoll.,* **21**, pp. 936–942.

10. McClements, D. J. (2013). Utilizing food effects to overcome challenges in delivery of lipophilic bioactives: Structural design of medical and functional foods, *Expert Opin. Drug Deliv.*, **10**, pp. 1621–1632.

11. Weiss, J., Decker, E. A., McClements, D. J., Kristbergsson, K., Helgason, T., and Awad, T. (2008). Solid lipid nanoparticles as delivery systems for bioactive food components, *Food Biophys.*, **3**, pp. 146–154.

12. Pandey, M., Verma, R. K., and Saraf, S. A. (2010). Nutraceuticals: New era of medicine and health, *Asian J. Pharm. Clin. Res.*, **3**, pp. 11–15.

13. Anand, P., Kunnumakkara, A. B., Newman, R. A., and Aggarwal, B. B. (2007). Bioavailability of curcumin: Problems and promises, *Mol. Pharm.*, **4**, pp. 807–818.

14. Huang, Q., Yu, H., and Ru, Q. (2010). Bioavailability and delivery of nutraceuticals using nanotechnology, *J. Food Sci.*, **75**, pp. 50–57.

15. Prabu, S. L., Suriya Prakash, T., Dinesh, K., Suresh, K., and Ragavendran, T. (2012). Nutraceuticals: A review, *Elixir Pharm.*, **46**, pp. 8372–8377.

16. Chaudhry, Q., Scotter, M., Blackburn, J., Ross, B., Boxall, A., Castle, L., Aitken, R., and Watkins, R. (2008). Applications and implications of nanotechnologies for the food sector, *Food Addit. Contam.*, **25**, pp. 241–258.

17. Weiss, J., Takhistov, P., and McClements, D. J. (2006). Functional materials in food nanotechnology, *J. Food Sci.*, **71**, pp. R107–R116.

18. Benetti, F., Bregoli, L., Olivato, I., and Sabbioni, E. (2014). Effects of metal (loid)-based nanomaterials on essential element homeostasis: The central role of nanometallomics for nanotoxicology, *Metallomics*, **6**, pp. 729–747.

19. Jones, C. F. and Grainger, D. W. (2009). In vitro assessments of nanomaterial toxicity, *Adv. Drug Deliv. Rev.*, **61**, pp. 438–456.

20. Nel, A., Xia, T., Mädler, L., and Li, N. (2006). Toxic potential of materials at the nanolevel, *Science*, **311**, pp. 622–627.

21. Shaikh, J., Ankola, D., Beniwal, V., Singh, D., and Kumar, M. R. (2009). Nanoparticle encapsulation improves oral bioavailability of curcumin by at least 9-fold when compared to curcumin administered with piperine as absorption enhancer, *Eur. J. Pharm. Sci.*, **37**, pp. 223–230.

22. Committee, E. S. (2011). Guidance on the risk assessment of the application of nanoscience and nanotechnologies in the food and feed chain, *EFSA J.*, **9**, pp. 2140.

23. Zaki, N. M. (2014). Progress and problems in nutraceuticals delivery, *J. Bioeq. Bioavail.*, **6**, pp. 75.

24. Pardridge, W. M. (2005). The blood–brain barrier: Bottleneck in brain drug development, *NeuroRx,* **2**, pp. 3–14.

25. Kim, S., Shi, Y., Kim, J. Y., Park, K., and Cheng, J. X. (2010). Overcoming the barriers in micellar drug delivery: Loading efficiency, in vivo stability, and micelle–cell interaction, *Expert Opin. Drug Deliv.,* **7**, pp. 49–62.

26. Wang, X., Wang, Y., Chen, Z. G., and Shin, D. M. (2009). Advances of cancer therapy by nanotechnology, *Cancer Res. Treat.,* **41**, pp. 1.

27. Facchi, S. P., Scariot, D. B., Bueno, P. V., Souza, P. R., Figueiredo, L. C., Follmann, H. D., Nunes, C. S., Monteiro, J. P., Bonafé, E. G., and Nakamura, C. V. (2016). Preparation and cytotoxicity of N-modified chitosan nanoparticles applied in curcumin delivery, *Int. J. Biol. Macromol.,* **87**, pp. 237–245.

28. Kayser, O., Lemke, A., and Hernandez-Trejo, N. (2005). The impact of nanobiotechnology on the development of new drug delivery systems, *Curr. Pharm. Biotechnol.,* **6**, pp. 3–5.

29. Otunola, G. A., Afolayan, A. J., Ajayi, E. O., and Odeyemi, S. W. (2017). Characterization, antibacterial and antioxidant properties of silver nanoparticles synthesized from aqueous extracts of *Allium sativum*, *Zingiber officinale*, and *Capsicum frutescens*, *Pharmacog. Mag.,* **13**, pp. S201.

30. Zheng, M., Yue, C., Ma, Y., Gong, P., Zhao, P., Zheng, C., Sheng, Z., Zhang, P., Wang, Z., and Cai, L. (2013). Single-step assembly of DOX/ICG loaded lipid–polymer nanoparticles for highly effective chemo-photothermal combination therapy, *ACS Nano,* **7**, pp. 2056–2067.

31. Deljoo, S., Rabiee, N., and Rabiee, M. (2019). Curcumin-hybrid nanoparticles in drug delivery system, *Asian J. Nanosci. Mat.,* **2**, pp. 66–91.

32. Varshosaz, J. and Farzan, M. (2015). Nanoparticles for targeted delivery of therapeutics and small interfering RNAs in hepatocellular carcinoma, *World J. Gastroenterol.,* **21**, pp. 12022.

33. Friedman, A. D., Claypool, S. E., and Liu, R. (2013). The smart targeting of nanoparticles, *Curr. Pharm. Des.,* **19**, pp. 6315–6329.

34. Liu, R., Kay, B. K., Jiang, S., and Chen, S. (2009). Nanoparticle delivery: Targeting and nonspecific binding, *MRS Bull.,* **34**, pp. 432–440.

35. Yi, J., Liu, Y., Zhang, Y., and Gao, L. (2018). Fabrication of resveratrol-loaded whey protein–dextran colloidal complex for the stabilization and delivery of β-carotene emulsions, *J. Agr. Food Chem.,* **66**, pp. 9481–9489.

36. Luo, X., Zhou, Y., Bai, L., Liu, F., Deng, Y., and McClements, D. J. (2017). Fabrication of β-carotene nanoemulsion-based delivery systems using dual-channel microfluidization: Physical and chemical stability, *J. Colloid Interf. Sci.,* **490**, pp. 328–335.

37. Shin, G. H., Li, J., Cho, J. H., Kim, J. T., and Park, H. J. (2016). Enhancement of curcumin solubility by phase change from crystalline to amorphous in Cur-TPGS nanosuspension, *J. Food Sci.,* **81**, pp. N494–N501.

38. Nam, J. H., Kim, S.-Y., and Seong, H. (2018). Investigation on physicochemical characteristics of a nanoliposome-based system for dual drug delivery, *Nanoscale Res. Lett.,* **13**, pp. 101.

39. Augustin, M. A. and Hemar, Y. (2009). Nano-and micro-structured assemblies for encapsulation of food ingredients, *Chem. Soc. Rev.,* **38**, pp. 902–912.

40. McClements, D. J. (2015). *Food Emulsions: Principles, Practices, and Techniques,* 3rd ed. (CRC Press, USA).

41. Renard, D., Robert, P., Lavenant, L., Melcion, D., Popineau, Y., Gueguen, J., Duclairoir, C., Nakache, E., Sanchez, C., and Schmitt, C. (2002). Biopolymeric colloidal carriers for encapsulation or controlled release applications, *Int. J. Pharm.,* **242**, pp. 163–166.

42. Onoue, S., Takahashi, H., Kawabata, Y., Seto, Y., Hatanaka, J., Timmermann, B., and Yamada, S. (2010). Formulation design and photochemical studies on nanocrystal solid dispersion of curcumin with improved oral bioavailability, *J. Pharm. Sci.,* **99**, pp. 1871–1881.

43. Ravanfar, R., Tamaddon, A. M., Niakousari, M., and Moein, M. R. (2016). Preservation of anthocyanins in solid lipid nanoparticles: Optimization of a microemulsion dilution method using the Placket–Burman and Box–Behnken designs, *Food Chem.,* **199**, pp. 573–580.

44. Bernela, M., Ahuja, M., and Thakur, R. (2016). Enhancement of anti-inflammatory activity of bromelain by its encapsulation in katira gum nanoparticles, *Carbohydr. Polym.,* **143**, pp. 18–24.

45. Puglia, C., Offerta, A., Tirendi, G. G., Tarico, M. S., Curreri, S., Bonina, F., and Perrotta, R. E. (2016). Design of solid lipid nanoparticles for caffeine topical administration, *Drug Deliv.,* **23**, pp. 36–40.

46. Zhu, Y., Peng, W., Zhang, J., Wang, M., Firempong, C. K., Feng, C., Liu, H., Xu, X., and Yu, J. (2014). Enhanced oral bioavailability of capsaicin in mixed polymeric micelles: Preparation, in vitro and in vivo evaluation, *J. Funct. Foods,* **8**, pp. 358–366.

47. Krausz, A. E., Adler, B. L., Cabral, V., Navati, M., Doerner, J., Charafeddine, R. A., Chandra, D., Liang, H., Gunther, L., and Clendaniel, A. (2015).

Curcumin-encapsulated nanoparticles as innovative antimicrobial and wound healing agent, *Nanomedicine,* **11**, pp. 195–206.

48. Manea, A. M., Vasile, B. S., and Meghea, A. (2014). Antioxidant and antimicrobial activities of green tea extract loaded into nanostructured lipid carriers, *Comptes Rendus Chimie,* **17**, pp. 331–341.

49. Fathi, M., Varshosaz, J., Mohebbi, M., and Shahidi, F. (2013). Hesperetin-loaded solid lipid nanoparticles and nanostructure lipid carriers for food fortification: Preparation, characterization, and modeling, *Food Biopr. Technol.,* **6**, pp. 1464–1475.

50. Lee, J. S. and Lee, H. G. (2016). Chitosan/poly-γ-glutamic acid nanoparticles improve the solubility of lutein, *Int. J. Biol Macromol.,* **85**, pp. 9–15.

51. Altındal, D. Ç. and Gümüşderelioğlu, M. (2016). Melatonin releasing PLGA micro/nanoparticles and their effect on osteosarcoma cells, *J. Microencapsul.,* **33**, pp. 53–63.

52. Antçnio, E., Khalil, N. M., and Mainardes, R. M. (2016). Bovine serum albumin nanoparticles containing quercetin: Characterization and antioxidant activity, *J. Nanosci. Nanotechnol.,* **16**, pp. 1346–1353.

53. Neves, A. R., Martins, S., Segundo, M. A., and Reis, S. (2016). Nanoscale delivery of resveratrol towards enhancement of supplements and nutraceuticals, *Nutrients,* **8**, pp. 131.

54. Jain, S., Garg, T., Kushwah, V., Thanki, K., Agrawal, A. K., and Dora, C. P. (2017). α-Tocopherol as functional excipient for resveratrol and coenzyme Q10-loaded SNEDDS for improved bioavailability and prophylaxis of breast cancer, *J. Drug Target.,* **25**, pp. 554–565.

55. Fakhoury, I., Saad, W., Gali-Muhtasib, H., and Schneider-Stock, R. (2014). Thymoquinone nanoparticle formulation and in vitro efficacy, *NSTI-Nanotech.,* **2**, pp. 367–370.

56. Jourghanian, P., Ghaffari, S., Ardjmand, M., Haghighat, S., and Mohammadnejad, M. (2016). Sustained release curcumin loaded solid lipid nanoparticles, *Adv. Pharm. Bull.,* **6**, pp. 17.

57. Joung, H. J., Choi, M. J., Kim, J. T., Park, S. H., Park, H. J., and Shin, G. H. (2016). Development of food-grade curcumin nanoemulsion and its potential application to food beverage system: Antioxidant property and in vitro digestion, *J. Food Sci.,* **81**, pp. N745–N753.

58. Singhal, N. K., Agarwal, S., Bhatnagar, P., Tiwari, M. N., Tiwari, S. K., Srivastava, G., Kumar, P., Seth, B., Patel, D. K., and Chaturvedi, R. K. (2015). Mechanism of nanotization-mediated improvement in the efficacy

of caffeine against 1-methyl-4-phenyl-1, 2, 3, 6-tetrahydropyridine-induced parkinsonism, *J. Biomed. Nanotechnol.,* **11**, pp. 2211–2222.

59. Zimet, P., Rosenberg, D., and Livney, Y. D. (2011). Re-assembled casein micelles and casein nanoparticles as nano-vehicles for ω-3 polyunsaturated fatty acids, *Food Hydrocoll.,* **25**, pp. 1270–1276.

60. Yuan, Y., Gao, Y., Zhao, J., and Mao, L. (2008). Characterization and stability evaluation of β-carotene nanoemulsions prepared by high pressure homogenization under various emulsifying conditions, *Food Res. Int.,* **41**, pp. 61–68.

61. Das, S., Das, J., Samadder, A., Paul, A., and Khuda-Bukhsh, A. R. (2013). Efficacy of PLGA-loaded apigenin nanoparticles in Benzo [a] pyrene and ultraviolet-B induced skin cancer of mice: Mitochondria mediated apoptotic signalling cascades, *Food Chem. Toxicol.,* **62**, pp. 670–680.

62. Huang, Y., Zhao, X., Zu, Y., Wang, L., Deng, Y., Wu, M., and Wang, H. (2019). Enhanced solubility and bioavailability of apigenin via preparation of solid dispersions of mesoporous silica nanoparticles, *Iran. J. Pharm. Res.,* **18**, pp. 168–182.

63. Alam, S., Khan, Z. I., Mustafa, G., Kumar, M., Islam, F., Bhatnagar, A., and Ahmad, F. J. (2012). Development and evaluation of thymoquinone-encapsulated chitosan nanoparticles for nose-to-brain targeting: A pharmacoscintigraphic study, *Int. J. Nanomed.,* **7**, pp. 5705.

64. Swarnakar, N. K., Jain, A. K., Singh, R. P., Godugu, C., Das, M., and Jain, S. (2011). Oral bioavailability, therapeutic efficacy and reactive oxygen species scavenging properties of coenzyme Q10-loaded polymeric nanoparticles, *Biomaterials,* **32**, pp. 6860–6874.

65. Hsu, C.-H., Cui, Z., Mumper, R. J., and Jay, M. (2003). Preparation and characterization of novel coenzyme Q10 nanoparticles engineered from microemulsion precursors, *AAPS PharmSciTech,* **4**, pp. 24–35.

66. Gokce, E. H., Korkmaz, E., Tuncay-Tanrıverdi, S., Dellera, E., Sandri, G., Bonferoni, M. C., and Ozer, O. (2012). A comparative evaluation of coenzyme Q10-loaded liposomes and solid lipid nanoparticles as dermal antioxidant carriers, *Int. J. Nanomed.,* **7**, pp. 5109.

67. Dube, A., Nicolazzo, J. A., and Larson, I. (2010). Chitosan nanoparticles enhance the intestinal absorption of the green tea catechins (+)-catechin and (−)-epigallocatechin gallate, *Eur. J. Pharm. Sci.,* **41**, pp. 219–225.

68. Tang, J., Xu, N., Ji, H., Liu, H., Wang, Z., and Wu, L. (2011). Eudragit nanoparticles containing genistein: Formulation, development, and bioavailability assessment, *Int. J. Nanomed.,* **6**, pp. 2429.

69. Serini, S., Cassano, R., Corsetto, P. A., Rizzo, A. M., Calviello, G., and Trombino, S. (2018). Omega-3 PUFA loaded in resveratrol-based solid lipid nanoparticles: Physicochemical properties and antineoplastic activities in human colorectal cancer cells in vitro, *Int. J. Mol. Sci.*, **19**, pp. 586.

70. Yao, M., McClements, D. J., Zhao, F., Craig, R. W., and Xiao, H. (2017). Controlling the gastrointestinal fate of nutraceutical and pharmaceutical-enriched lipid nanoparticles: From mixed micelles to chylomicrons, *NanoImpact.*, **5**, pp. 13–21.

71. Teng, Z., Luo, Y., and Wang, Q. (2013). Carboxymethyl chitosan–soy protein complex nanoparticles for the encapsulation and controlled release of vitamin D3, *Food Chem.*, **141**, pp. 524–532.

72. Shpigelman, A., Shoham, Y., Israeli-Lev, G., and Livney, Y. D. (2014). β-Lactoglobulin–naringenin complexes: Nano-vehicles for the delivery of a hydrophobic nutraceutical, *Food Hydrocoll.*, **40**, pp. 214–224.

73. Md, S., Alhakamy, N. A., Aldawsari, H. M., and Asfour, H. Z. (2019). Neuroprotective and antioxidant effect of naringenin-loaded nanoparticles for nose-to-brain delivery, *Brain Sci.*, **9**, pp. 275.

74. Dhiman, A. and Bhalla, D. (2019). Development and evaluation of lycopene loaded chitosan nanoparticles, *Curr. Nanomed.*, **9**, pp. 61–75.

75. Zhang, F., Li, R., Yan, M., Li, Q., Li, Y., and Wu, X. (2020). Ultra-small nanocomplexes based on polyvinylpyrrolidone K-17PF: A potential nanoplatform for the ocular delivery of kaempferol, *Eur. J. Pharm. Sci.*, **147**, pp. 105289.

76. Li, H., Yuan, Y., Zhu, J., Wang, T., Wang, D., and Xu, Y. (2020). Zein/soluble soybean polysaccharide composite nanoparticles for encapsulation and oral delivery of lutein, *Food Hydrocoll.*, **103**, pp. 105715.

77. McClements, D. J. and Rao, J. (2011). Food-grade nanoemulsions: Formulation, fabrication, properties, performance, biological fate, and potential toxicity, *Crit. Rev. Food Sci. Nutr.*, **51**, pp. 285–330.

78. Van de Casteele, R. (2005). Method for the delivery of a biologically active agent, United States Patent 6,861,066.

79. Van de Casteele, R. and Gerike, M. (2006). Nanofluidized B-12 composition and process for treating pernicious anemia, United States Patent 2006, 0280761 A1.

80. Bombardelli, E. (1990). Pharmaceutical and cosmetic compositions containing complexes of flavanolignans with phospholipids, United States Patent 4,895, 839.

Multiple Choice Questions

1. Nanoparticles target the rare _____ causing cells and remove them from blood.
   - (a) Tumor
   - (b) Infection
   - (c) Cold
   - (d) Fever

2. _____ is the field in which nanoparticles are used with silica-coated iron oxide.
   - (a) Magnetic applications
   - (b) Electronics
   - (c) Medical diagnosis
   - (d) Structural and mechanical materials

3. Coating nanoparticles with ceramics leads to ____ material.
   - (a) Corrosion
   - (b) Corrosion resistant
   - (c) Wear and tear
   - (d) Soft

4. The prefix nano comes from
   - (a) French word meaning billion
   - (b) Greek word meaning dwarf
   - (c) Spanish word meaning particle
   - (d) Latin word meaning invisible

5. What are the barriers in the delivery of nutraceuticals?
   - (a) Solubility
   - (b) Bioavailability
   - (c) Permeability
   - (d) All of these

6. With nanoscale distribution, _____ in matrix improves the life and performance.
   - (a) Carbide
   - (b) Tungsten
   - (c) Hydrides
   - (d) Nitrites

7. The extensively used nanoparticle as catalyst is
   - (a) Silver
   - (b) Copper
   - (c) Gold
   - (d) Cerium

8. Due to ____ tensile strength, some nanomaterials are used in aircraft.
   - (a) High
   - (b) Low
   - (c) Moderate
   - (d) None

9. Fabrics are extensively made out of nanomaterials like_____
   - (a) Carbon nanotubes
   - (b) Fullerenes
   - (c) Mega tubes
   - (d) Polymers

10. Lignans are part of which family of compounds?

    (a) Carotenoids   (b) Polyphenols

    (c) Phytosterols   (d) None of the above

11. How can one make a drug-targeted delivery system?

    (a) Ideal size of carrier   (b) Specific site ligand

    (c) Both (a) and (b)   (d) None of these

12. Piperine is a compound found in_____

    (a) Pepper   (b) Cardamom

    (c) Turmeric   (d) Cloves

13. The food sources used as nutraceuticals are all natural and can be categorized as

    (a) Dietary fiber   (b) Probiotics

    (c) Prebiotics   (d) All the above

14. Blueberries are considered functional foods because they contain

    (a) Proteins   (b) Phytochemicals

    (c) Fats   (d) None

15. Omega-3 fatty acids are naturally high in salmon, which can be classified as _____ food.

    (a) Dietary   (b) Functional

    (c) Fortified   (d) Nutraceutical

16. Select commercial nano-nutraceuticals for cardiovascular diseases.

    (a) Greenselect Phytosome®   (b) Lucoselect Phytosome®

    (c) Both (a) and (b)   (d) None of these

17. Which marketed nanoformulation is suitable for liver health?

    (a) Miritoselect®   (b) Virtiva®

    (c) Siliphos®   (d) None of these

18. What particle size of nanoformulation is required to target various cancer cells?

    (a) More than 5 μm   (b) 5 and 200–300 nm

    (c) Less than 5 nm   (d) None of these

19. The biological activity silver nanoparticles (AgNPs) was conducted in _____ nutraceuticals.

    (a) Ginger   (b) Garlic

    (c) Cayenne pepper   (d) All of these

20. Which type of nanoparticles (NPs) is treated as ideal candidates for delivery of vaccines, antibiotics, and cancer drugs?

    (a) Polymeric NPs   (b) Solid lipid NPs

    (c) Hybrid NPs   (d) All of these

21. Which of the following nanoemulsions has higher optical clarity and increased oral bioavailability?

    (a) Curcumin   (b) β-Carotene

    (c) Both (a) and (b)   (d) All of these

22. Certain traditional thermal-processing techniques affect the nutritional properties of food.

    (a) True   (b) False

23. Which of the following nanoparticles has enhanced antitumor activity?

    (a) Thymoquinone   (b) Curcumin

    (c) Quercetin   (d) All of these

24. Which nanohydrogel is used for increasing skin permeation?

    (a) Thymoquinone   (b) Caffeine

    (c) Curcumin   (d) All of these

25. Which nanoparticles are prepared for increasing anti-inflammatory activity using katira gum?

    (a) Bromelain

    (b) Glycyrrhizic acid polymeric lipid NPs

    (c) Curcumin

    (d) Both (a) and (b)

Answer Key

1. (a)   2. (c)   3. (b)   4. (b)   5. (d)   6. (b)   7. (c)

8. (a)   9. (b)   10. (b)   11. (c)   12. (a)   13. (d)   14. (b)

15. (d)   16. (b)   17. (c)   18. (b)   19. (b)   20. (a)   21. (b)

22. (a)   23. (a)   24. (b)   25. (d)

Short Answer Questions

1. Discuss the advantages of nanotechnology in delivering nutraceuticals.

2. Why nanotechnology is an important tool for cancer research?

3. How to modify physiochemical behavior of lipophilic nutraceuticals?

4. Explain the different types of nutraceuticals.

5. Write briefly about the future scope of nanotechnology in nutraceuticals.

6. Mention the role of nanocapsules for food additives?

7. Write the role of polymers in overcoming the limitations of lipophilic nutraceuticals?

8. Briefly mention the characteristics of ideal or targeted nanoparticles.

9. Explain the different barriers to highly lipophilic nutraceuticals.

Long Answer Questions

1. Explain the different commercialized nano-based nutraceuticals.

2. Describe various nanotechnology approaches used for delivery of nutraceuticals.

3. Explain the role of numerous nanocarriers to overcome the problems of nutraceuticals.

4. Describe the significance of different curcumin nanoformulations.

Chapter 6

Nanobiotechnology: Applications and Future Prospects

**Tapan Behl, Priya Nijhawan, Arun Kumar,
and Rakesh K. Sindhu**

Chitkara College of Pharmacy, Chitkara University, Punjab, India
tapanbehl31@gmail.com, tapan.behl@chitkara.edu.in

6.1 Introduction

Nanotechnology is a therapeutic tactic that comprises instruments accomplished by manifesting the physical as well as chemical activities of a compound. While biotechnology deals with the data and methods of biology to operate molecular, heritable events to generate drugs and another varied fields of product, nanobiotechnology is a combination of biotechnology and nanotechnology through which microtechnology is fused to a biological method. Atomic-grade equipment can be made by integrating biological systems to manipulate various assets of a biological system. Nanobiotechnology has the prospective to delete barriers in science and to improve present thoughts. Nanotechnology is the production of materials and devices whose least functional composition is on the nanometer

Nanotechnology: Principles and Applications
Edited by Rakesh K. Sindhu, Mansi Chitkara, and Inderjeet Singh Sandhu
Copyright © 2021 Jenny Stanford Publishing Pte. Ltd.
ISBN 978-981-4877-43-5 (Hardcover), 978-1-003-12026-1 (eBook)
www.jennystanford.com

scale [1, 2]. Mostly nanotechnology consists of developing materials and about 1 to 100 nm in size. On the other hand, biotechnology deals with metabolic events. Connotation of biotechnology and nanobiotechnology plays a significant role in producing useful tools for study. Nanotechnology is varied, extending from an allowance of conventional device physics to de novo approaches relying on molecular self-assembly, from generating innovative materials with measurement on the nanoscale to investigate whether we can control matters on the atomic scale [3].

6.2 Pros of Nanobiotechnology

The anatomical alteration can hypothetically activate the possibility for the enlargement of numerous targeted nanotechnological substances.

- Drug targeting is manifested by knowing different pathophysiological features of disorders [4, 5].
- Higher vascular permeability binds with an impaired lymphatic drainage in tumors recovering the effect of the nanosystems in the tumors by retaining [6–8].
- Nanoparticles are used to supply necessary medicines to the central nervous system [9–12].

6.3 Applications of Nanobiotechnology

Various applications of nanobiotechnology such as disease diagnosis and molecular imaging are studied. Several de novo compounds have also been investigated under clinical trials [13, 14].

6.3.1 Diagnostic Applications

Present diagnostic procedures rely on visible signs; before doctors can identify, people suffer from certain diseases. But by time those indications appear, their management has a reduced chance. Nucleic acid diagnostics plays a vital role in identifying pathogens at an early stage of illness. The present method is polymerase chain reaction (PCR), but nanotechnology provides more efficiency and economy.

6.3.2   Detection

Numerous clinical studies detect the existence of an ailment triggered by microorganisms by identifying the interaction of antibodies with the disorder-related target. Earlier assessments used to be done by incorporating antibodies in inorganic dyes and visualizing the cascades through fluorescence microscopy. This type of test limits the specificity of the detection technique. Nanobiotechnology functions by using semiconductor nanocrystals [15].

6.3.3   Individual Target Probe

With the availability of detectors, optical and colorimetric detectors are used by doctors. Nanosphere is a manufacturing unit that produces methods that permit medical professionals to optically assess the genetic constituents of biological samples. Nanogold elements embedded with a small segment of DNA form the basis of the easy-to-read tests for the presence of genetic sequence. If the sequence in the samples attaches to the complementary DNA on several nanospheres, then it forms a condensed network of visible gold balls [16].

6.3.4   Protein Chips

Proteins help in producing the biotic phenotype of mammals in healthy and diseased individuals. Therefore, proteomics is used nowadays to diagnose diseases and for pharmaceutical purpose where medicines are produced to modify signaling pathways. Protein chips are combined with protein constituents that attach to the protein consisting of specific biochemical motif [17]. Two companies presently use this application: Agilent, Inc. and NanoInk, Inc. Agilent uses non-contact ink-jet method to develop microarrays by printing oligos and whole cDNAs on glass slides at the nanoscale. NanoInk uses the dip-pen nanolithography (DPN) method to give structure on a nanoscale dimension [18].

6.3.5   Sparse Cell Detection

Sparse cells are different from adjacent cells in normal physiological conditions. They are used in the recognition and diagnosis of several genetic disorders. Although it is difficult to detect and detach these sparse cells, nanobiotechnology exhibits novel therapeutic approaches for development in this field. Researchers have developed nanosystems proficient in efficiently arranging sparse cells from blood and tissues. This method exploits the inimitable function of sparse cells and show variation in distortion and surface charges [19].

6.3.6 Nanotechnology as a Device in Imaging

Intracellular imaging is used through the labelling of target molecules with quantum-dots-like fluorescent proteins that promote the study of intracellular signaling complexes by the optical method [20, 21].

6.3.7 Therapeutic Applications

Nanotechnology has developed de novo formulations of compounds with few adverse effects (Fig. 6.1 and Table 6.1).

6.3.8 Nanobiotechnology in the Food Sector

The advancement of nanotechnology in the food and agriculture sector has resulted in the development of nanobiotechnology, which includes pesticide delivery through bioactive nanoencapsulation, biosensors to assess microorganisms and variation in food products, and edible thin films to preserve fruits [22, 23].

6.3.9 Drug Delivery

Nanoparticles are supplied to the targeted location by marketed drugs. If therapeutics bind to a nanoparticle, then it is known as the site of disorder by radio cascade. These medicines are also developed to "release" when certain substances are present. Also adverse effects of standard drugs can be prevented by decreasing the dosage needed to treat subjects [24]. Several drugs that cannot be taken orally due to their poor bioavailability can now be used in therapy with the help of nanotechnology [25, 26]. Nanoformulations provide defense to drugs that are vulnerable to degradation when bared to exciting pH [27, 28]. Nanotechnology also delivers antigens for vaccination [29–34].

6.3.10 Liposomes

A liposome consists of a lipid bilayer, which can be incorporated in gene therapy due to its ability to pass through lipid bilayers and cell membranes of the target. Recently, the use of various liposomes in local delivery has been found to be effective [35, 36]. Zhang et al. revealed extensive expression in the central nervous system of rhesus monkeys by associating nanoparticle (such as polyethylene glycol) treated liposomes to a monoclonal antibody for human insulin reporter [37]. This investigation can be used as a futuristic approach for targeted therapy [38].

6.3.11 Surfaces

In the environment, there are numerous examples of association between molecules and surfaces. For example, the contact between blood cells and the brain depends on the interaction between cells and surface features. Nanofabrication resolves the complexity of these associations by rectifying surface features with nanoscale resolutions that result in a hybrid biological method. This hybrid method is used to monitor medicines, medical devices, and implants [39].

6.3.12 Biomolecular Engineering

The cost and time involved in traditional biomolecule production restrict the accessibility of bioactive compounds. The nanoscale synthetic method offers an alternative to traditional techniques. Progressions can be attained due to the capability of carrying out chemical and biological reactions on solid layer, in comparison to traditional processes. Engene OS (Waltham, Massachusetts) is a company involved in biomolecular engineering. The company produces engineered genomic operating methods that create programmable biomolecular instruments. These biomolecular equipment have a wide range of usage as biosensors in chemical synthesis and drug discovery.

6.3.13   Biopharmaceuticals

Nanobiotechnology produces medicines for disorders that conventional drugs cannot target. Companies traditionally focus on producing medicines for the treatment of a number of ailments. A nanoscale method for drug production will be a boon to the industry, which cannot employ a large number of organic chemists to produce a number of substances. Nanobiotechnology exhibits capability to manifest targets on solid substrates by tethering them to biomembranes and controlling where and when chemical reactions take place. This advancement will decrease formulation costs.

6.3.14   Nanotechnology in Cardiovascular Treatment

Nanotechnology is presently providing promising tools for the treatment of cardiovascular diseases by exploring the diseases at the cellular level. These tools are used in diagnosis and tissue engineering [36]. Nanotechnology evaluates and describes mechanisms existing in cardiac disorders. Additionally, it is used in developing atomic-scale machines. De novo machines are used in the prevention of congestive heart failure [40].

6.3.15   Nanotechnology in Dental Care

Nanotechnology plays a crucial role in the field of dentistry. In nanodentistry, the use of nanomaterials [41–44] improves dental hygiene. A large number of people suffer from dental problems; they can take advantage of nanodentistry [45, 46]. Also, a nanodental method in tooth treatment will be developed. The benefit of nanodentistry in teeth management is also important [47].

6.3.16   Nanotechnology in Orthopedic Treatment

Nanomaterials are used as constituents in bones formation [48, 49]. Nanomaterials and nanotubes are used for the deposition of calcium-containing minerals on graft. Nanostructured materials manifest the attachment of an implant to the bone matter by improving bone cell interactions and enrich effectiveness of orthopedic implants by reducing associated complications.

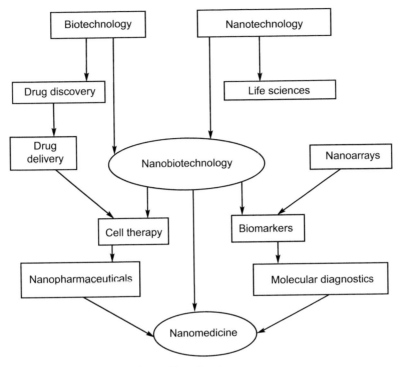

Figure 6.1 Application of nanobiotechnology.

Table 6.1 Comparison of nanobiotechnology with existing literature

Nanobiotechnology	Biotechnology
Nanobiotechnology: It deals with technology that incorporates nanomolecules into a biological system.	**Biotechnology:** It deals with nanostructures that are created for synthetic preparation.
Molecules: Nanobiotechnology uses humanmade and inorganic materials.	**Molecules:** Biotechnology uses biomolecules and organisms to produce pharmaceutical therapies.
Size: >100 nm	**Size:** 3–13 nm
Use: It is used for better electronic performance and the production and storage of energy for various applications.	**Use:** It applies itself to life sciences.

Table 6.2 Nanoresearch areas and potential applications in nanobiotechnology

Nanoresearch areas	Potential applications in nanobiotechnology
Nanofiber	• DNA analysis • DNA sequencing • Nanofiltration to obtain ultra-dense fermentation broth cell cultures
Nanoprobe	• DNA sequencing
Nanoprojectile	• Direct gene transfer
Nanobeads	• DNA vaccines • Nanovaccines
Dendrimers	• Diagnoses, treatment, and eradication of malignant tumors in small animal populations
Quantum dots	• Genetic analysis • Drug discovery • Disease diagnostics
Bucky balls	• Drug delivery
Carbon nanoparticles	• Enzyme-based biofuel production

6.4 Future Prospects of Nanobiotechnology

Nanobiotechnology implements several compounds that are useful in drug development. However, this novel process has arisen several questions as this new therapeutic approach consists of problems with toxicity [50] and its adverse effects on human subjects. Despite the existence of problems, this method has enormous scope for the future. It can play a significant role in drug delivery and gene therapy. The purpose of the present technology is target-specific drug therapy [2] and imaging applications, i.e., quantum dot technology, which involves monitoring cellular functioning in tissues [51, 52]. Constant developments in nanomedicine are implemented in the management of a number of medical disorders.

6.5   Challenges for Nanobiotechnology

The challenges are to improve equipment to evaluate contact with nanomaterials found in the environment. The exposure of mammals to nanomaterials has to be scrutinized for side effects. The challenge is more complex in the food sector. Another challenge is to create methods to identify the toxicity of nanomaterials and to assess the effects of engineered nanomaterials on humans. The next challenge is to evaluate the risk factors associated with humans. The market trial of nanobiotechnology consists of funding and patience. A number of companies have spotted a possibility of using nanotechnology for the formation of de novo compounds and the enhancement of enduring byproducts.

6.6   Future Scope

In the past few years, nanobiotechnology has been regarded as an attractive technology that has transformed the food sector. It is a tool on the nanometer scale that deals with atoms, molecules, or macromolecules having size ranging from 1 to 100 nm to produce and use materials that have novel characteristics. The twisted nanomaterials exhibit one or more external dimensions, on the scale from 1 to 100 nm, that manifest the observation and manipulation of matter at the nanoscale. These materials have unique features unlike their macroscale counterparts due to the high surface-to-volume ratio and other novel physiochemical properties. Nanobiotechnology has fetched novel industrial status, and both developed and developing countries are interested in investing more in this technology. Therefore, nanobiotechnology provides numerous prospects for the production and application of structures, materials, or systems with new properties in several fields such as agriculture, food, and medicine. The increasing consumer concerns about food quality and health benefits are forcing scientists to study ways that can elevate food quality [53].

The requirement of nanoparticle-based materials has been elevated in the food industry as many of them contain necessary nontoxic substances. Various departments, scientists, and industries have come up with novel tools, methods, and products that have application of nanobiotechnology in food science. There has been much discussion on the future prospects of nanobiotechnology. It could provide an idea for the implementation of several novel materials and devices useful in the field of medicine, electronics, biomaterials, and energy production. This approach has given rise to many problems as a novel technology, such as issues with toxicity and environmental impact of nanomaterials and their effects on worldwide economics. These worries have arisen a debate in governments on whether regulation of nanobiotechnology is warranted. Despite the presence of some issues, this technology has huge hope for the future. It has led to inventions by playing a vital role in several biomedical applications ranging from drug delivery and gene therapy to molecular imaging, biomarkers, and biosensors [54].

One of these applications in the current time is target-specific drug therapy and methods for initial diagnosis and treatment of diseases. Two types of medical applications are evolving: in clinical diagnosis and in R&D. Imaging applications, such as quantum dot technology, are already being licensed, and applications for monitoring cellular activities in tissues are coming soon. Another major type of application involves the development of highly specific and sensitive means of detecting nucleic acids and proteins. By the next 5 years, the materials being evaluated in academic and government laboratories will be commercialized. Sparse cell isolation and molecular filtration applications should, by then, make it to the market. Some of the drug-delivery systems should be commercialized or in advanced clinical trials. For example, drug-delivery systems have been developed by NanoSystems, which is evaluating the encapsulation of Taxol, a cancer drug in a nanopolymer called paclitaxel. Continuous progress in nanomedicine has opened opportunities for application in a variety of medical disciplines. Its future application as diagnostic and regenerative medicine is currently being investigated. In diagnosis, detection of diseased cells would be faster, possibly at the point of a single sick cell, while allowing diseased cells to be cured at once before they spread into and affect other parts of the body. Also individuals suffering from major traumatic injuries or impaired organ functions could benefit from the use of nanomedicine.

6.7 Conclusion

Nanobiotechnology is still in its initial phase. The field of nanobiotechnology is taking the science of impenetrably small devices closer and closer to reality. The outcome of this growth will, at some point, be so massive that it will perhaps affect nearly all sectors of science and technology. Nanobiotechnology has proposed a wide array of uses in medicine. Drug-delivery systems are only the early stages of the start of somewhat novel. Several disorders may be cured by nanotechnology in the future. Although the outlook of nanobiotechnology in drugs is high and the potential benefits are infinite, the safety of nanomedicine is not fully realized. The use of nanobiotechnology in medical therapeutics needs suitable assessment of its risk and safety factors. Applications of nanotechnology have arisen with a cumulative need of nanoparticles in several areas of food science and microbiology, food processing, food packaging, functional food development, and shelf-life extension of food. Researchers who are against the use of nanobiotechnology also agree that expansion in nanobiotechnology should continue as this sector promises great advantages, but they caution that evaluation should be done to ensure the safety of people. In the future, nanomedicine would play a vital role in the prevention and management of disorders. If everything runs smoothly, nanobiotechnology will, one day, become an inevitable part of our everyday life and will help save many lives.

References

1. Emerich, D. F. and Thanos, C. G. (2003). Nanotechnology and medicine. *Expert Opin. Biol. Ther.*, **3**: 655–663.
2. Sahoo, K. S. and Labhasetwar, V. (2003). Nanotech approaches to drug delivery and imaging. *DDT*, **8**(24): 1112–1120.
3. Nasrollahzadeh, M., et al. (2019). Applications of nanotechnology in daily life. In: *An Introduction to Green Nanotechnology*, Vol. 28, Academic Press, pp. 113–143.
4. Vasir, J. K. and Labhasetwar, V. (2005). Targeted drug delivery in cancer therapy. *Tehcnol. Cancer Res. Treat.*, **4**: 363–374.

5. Vasir, J. K., et al. (2005). Nanosystems in drug targeting: Opportunities and challenges. *Curr. Nanosci.*, **1**: 47–64.

6. Maeda, H., et al. (2000). Tumor vascular permeability and the EPR effect in macromolecular therapeutics: A review. *J. Control. Release*, **65**: 271–284.

7. Matsumura, Y. and Maeda, H. (1986). A new concept for macromolecular therapeutics in cancer chemotherapy: Mechanism of tumoritropic accumulation of proteins and the antitumor agent smancs. *Cancer Res.*, **46**: 6387–6392.

8. Allen, T. M. and Cullis, P. R. (2004). Drug delivery systems: Entering the mainstream. *Science*, **303**: 1818–1822.

9. Alyautdin, R. N., et al. (1998). Significant entry of tubocurarine into the brain of rats by adsorption to polysorbate 80-coated polybutylcyanoacrylate nanoparticles: An in situ brain perfusion study. *J. Microencapsul.*, **15**: 67–74.

10. Garcia-Garcia, E., et al. (2005). A relevant in vitro rat model for the evaluation of blood–brain barrier translocation of nanoparticles. *Cell Mol. Life Sci.*, **62**(12): 1400–1408.

11. Feng, S. S., et al. (2004). Nanoparticles of biodegradable polymers for clinical administration of paclitaxel. *Curr. Med. Chem.*, **11**: 413–424.

12. de Kozak, Y., et al. (2004). Intraocular injection of tamoxifen-loaded nanoparticles: A new treatment of experimental autoimmune uveoretinitis. *Eur. J. Immunol.*, **34**: 3702–3712.

13. Shaffer, C. (2005). Nanomedicine transforms drug delivery. *Drug Discov. Today*, **10**: 1581–1582.

14. Moghimi, S. M., et al. (2005). Nanomedicine: Current status and future prospects. *FASEB J.*, **19**: 311–330.

15. Drexler, E. K. (1992). *Nanosytems: Molecular Machinery, Manufacturing and Computation.* John Wiley & Sons, New York.

16. Nanosphere Inc: 2004, Available at http://www.nanosphere-inc.com.

17. Lee, K. B., et al. (2002). Protein nanoarrays generated by dip-pen nanolithography. *Science,* **295**: 1702–1705.

18. NanoInk Inc: 2004, Available at http://www.nanoink.net.

19. NBTC (Nano-biotechnology Center, Cornell University): 2004, Available at http://www.nbtc.cornell.edu/default.htm.

20. Lin, H. and Datar, R. H. (2006). Medical applications of nanotechnology. *Natl. Med. J. India,* **19**: 27–32.

21. Guccione, S., et al. (2004). Vascular-targeted nanoparticles for molecular imaging and therapy. *Methods Enzymol.*, **386**: 219–236.

22. Bradley, E. L., et al. (2011). Applications of nanomaterials in food packaging with a consideration of opportunities for developing countries. *Trends Food Sci. Technol.*, **22**, 603–610.

23. Bratovčić, A., et al. (2015). Application of polymer nanocomposite materials in food packaging. *Croat. J. Food Sci. Technol.*, **7**, 86–94.

24. LaVan, D. A., et al. (2002). Timeline: Moving smaller in drug discovery and delivery. *Nat. Rev. Drug Discov.*, **1**: 77–84.

25. El-Shabouri, M. H. (2002). Positively charged nanoparticles for improving the oral bioavailability of cyclosporin-A. *Int. J. Pharm.*, **249**: 101–118.

26. Hu, L., et al. (2004). Solid lipid nanoparticles (SLNs) to improve oral bioavailability of poorly soluble drugs. *J. Pharm. Pharmacol.*, **56**: 1527–1535.

27. Arangoa, M. A., et al. (2001). Gliadin nanoparticles as carriers for the oral administration of lipophilic drugs. Relationships between bioadhesion and pharmacokinetics. *Pharm. Res.*, **18**: 1521–1527.

28. Arbos, P., et al. (2004). Nanoparticles with specific bioadhesive properties to circumvent the pre-systemic degradation of fluorinated pyrimidines. *J. Control. Release,* **96**: 55–65.

29. Diwan, M., et al. (2003). Biodegradable nanoparticle mediated antigen delivery to human cord blood derived dendritic cells for induction of primary T cell responses. *J. Drug Target.*, **11**: 495–507.

30. Koping, M., et al. (2005). Nanoparticles as carriers for nasal vaccine delivery. *Expert Rev. Vaccines,* **4**: 185–196.

31. Lutsiak, M. E., et al. (2002). Analysis of poly (D,L-lactic-co-glycolic acid) nanosphere uptake by human dendritic cells and macrophages in vitro. *Pharm. Res.*, **19**: 1480–1487.

32. Yotsuyanagi, T. and Hazemoto, N. (1998). Cationic liposomes in gene delivery. *Nippon Rinsho.*, **56**: 705–712.

33. Young, L. S., et al. (2006). Viral gene therapy strategies: From basic science to clinical application. *J. Pathol.*, **208**: 299–318.

34. Davis, S. S. (1997). Biomedical applications of nanotechnology: Implications for drug targeting and gene therapy. *Trends Biotechnol.*, **15**: 217–224.

35. Hart, S. L. (2005). Lipid carriers for gene therapy. *Curr. Drug Deliv.*, **2**: 423–428.

36. Ewert, K., et al. (2005). Lipoplex structures and their distinct cellular pathways. *Adv. Genet.*, **53**: 119–155.

37. Zhang, Y., et al. (2003). Organ specific gene expression in the rhesus monkey eye following intravenous nonviral gene transfer. *Mol. Vis.*, **9**: 465–472.

38. Gupta, A., et al. (2016). Nanoemulsions: Formation, properties and applications. *Soft Matter.*, **12**, 2826–2841.

39. Elan Corporation, PLC: 2004, Available at http://www.elan.com.

40. Wickline, S. A., et al. (2006). Applications of nanotechnology to atherosclerosis, thrombosis, and vascular biology. *Arterioscler Thromb Vasc Biol.*, **26**: 435–441.

41. Panyam, J. and Labhasetwar, V. (2003). Biodegradable nanoparticles for drug and gene delivery to cells and tissue. *Adv. Drug Deliv. Rev.*, **55**: 329–347.

42. West, J. L. and Halas, N. J. (2000). Applications of nanotechnology to biotechnology commentary. *Curr. Opin. Biotechnol.*, **11**: 215–217.

43. Shi, H., et al. (1999). Template-imprinted nanostructured surfaces for protein recognition. *Nature,* **398**: 593–597.

44. Sims, M. R. (1999). Brackets, epitopes and flash memory cards: A futuristic view of clinical orthodontics. *Aust. Orthod. J.*, **15**: 260–268.

45. Slavkin, H. C. (1999). Entering the era of molecular dentistry. *J. Am. Dent. Assoc.*, **130**: 413–417.

46. Ure, D. and Harris, J. (2003). Nanotechnology in dentistry: Reduction to practice. *Dent. Update*, **30**: 10–15.

47. Fartash, B., et al. (1996). Rehabilitation of mandibular edentulism by single crystal sapphire implants and overdentures: 3–12 year results in 86 patients. A dual center international study. *Clin. Oral Implants Res.*, **7**: 220–229.

48. Shellhart, W. C. and Oesterle, L. J. (1999). Uprighting molars without extrusion. *J. Am. Dent. Assoc.*, **130**: 381–385.

49. Webster, T. J., et al. (2004). Nanobiotechnology: Carbon nanofibres as improved neural and orthopedic implants. *Nanotechnology,* **15**: 48–54.

50. Price, R. L., et al. (2003). Selective bone cell adhesion on formulations containing carbon nanofibers. *Biomaterials,* **24**: 1877–1887.

51. Buzea, C., et al. (2007). Nanomaterials and nanoparticles: *Sources Toxicity Biointerphases,* **2**: MR17.

52. Milunovich, S. and Roy, J. (2001). The next small thing: An introduction to nanotechnology. *Merrill Lynch Report,* September 4.

53. Hamad-Schifferli, K., et al. (2002). Remote electronic control of DNA hybridization through inductive coupling to an attached metal nanocrystal antenna. *Nature,* **415**: 152–155.

54. Dasgupta, N., et al. (2017). Applications of nanotechnology in agriculture and water quality management. *Environ. Chem. Lett.,* **15**(4): 591–605.

Multiple Choice Questions

1. Which one of these statements is true?
    (a) Nanospheres form a network of visible gold balls
    (b) Nanospheres form a network of visible red balls
    (c) Nanospheres form a network of visible silicon balls
    (d) Nanospheres form a network of visible aluminum balls

2. Nanobiotechnology is a combination of
    (a) Biology and nanotechnology
    (b) Biotechnology and nanotechnology
    (c) Nanobiotechnology and nanospheres
    (d) Nanotechnology and nanospheres

3. Nanotechnology produce materials with size range_____
    (a) 1 to 100 nanometers   (b) 100 to 500 nanometers
    (c) 0 to 100 nanometers   (d) 0 to 50 nanometers

4. Nanoparticles are used to supply _____
    (a) Signals to CNS   (b) Medicines to CNS
    (c) Both (a) and (b)   (d) None of the above

5. Nanobiotechnology function by using _____
    (a) Semiconductor nanocrystal   (b) Magnetic detectors
    (c) Nanoparticles   (d) Nanospheres

6. Which industry is using protein chip application?
    (a) Agilent industry   (b) Nanoink industry
    (c) Both (a) and (b)   (d) Regenta industry

7. Which method is used for the production of protein chip?
    (a) Dip-pen nanolithography method
    (b) Metal oligo method

(c) Simer alpha method

(d) Visible gold method

8. Nanotechnology is used to deliver_____ for vaccination

   (a) Antisera   (b) Antibody

   (c) Antigens   (d) Both (a) and (b)

9. Nanobiotechnology help in the management of

   (a) Cardiovascular disorders

   (b) Dental problem

   (c) Bone and joint disorder

   (d) All of the above

10. Which method is used in imaging application of nanobiotechnology?

   (a) Quantum dot method   (b) Regenital method

   (c) Oligio method   (d) Quadratic dot method

11. Genetic disorder is diagnosis by which method

   (a) Optical method   (b) Sparse cell detection method

   (c) Nanotubes method   (d) Quantum dot method

12. Nanobiotechnology is used as a device in _____ imaging

   (a) Intracellular   (b) Intercellular

   (c) Extracellular   (d) Both (b) and (c)

13. Engene OS (Waltham, Massachusetts) is an industry in the area of _____

   (a) Biopharmaceuticals   (b) Drug delivery

   (c) Protein chips   (d) Biomolecular engineering

14. Name the method used to develop microarrays by printing oligos and whole cDNAs onto glass slides at the nanoscale

   (a) Non-contact ink-jet method

   (b) Quantum dot method

   (c) Optical method

   (d) Microarray method

15. What is the full form of DNP?

   (a) Dip-pen nanotechnology method

   (b) Dot-pin nanobiotechnology method

   (c) Dip-pen nanolithography method

   (d) Dip-point nanospheres method

Answer Key

1. (a)   2. (a)   3. (a)   4. (b)   5. (a)   6. (a)   7. (a)
8. (c)   9. (d)   10. (a)   11. (d)   12. (a)   13. (d)   14. (a)
15. (a)

Short Answer Questions

1. Describe any five applications of nanobiotechnology in detail.
2. Enlist the challenges associated with nanobiotechnology.
3. Write a note on the following:
   (a) Sparse cell detection.
   (b) Diagnostic applications.
   (c) Biomolecular engineering.
4. Describe the application of nanobiotechnology associated with the field of pharmacy.
5. Describe individual target probe and protein chips in detail.

Long Answer Questions

1. Enumerate the application of nanobiotechnology.
2. Descriptive role on future prospects and challenges of nanobiotechnology.
3. Write a brief note on nanobiotechnology.
4. Enumerate pros of nanobiotechnology and drug delivery in detail.

Chapter 7

Nanocomposites: Preparation, Characterization, and Applications

Anurag Sangwan,[a] Parth Malik,[b] Rachna Gupta,[c] Rakesh Kumar Ameta,[b] and Tapan K. Mukherjee[d]

[a]*UGC-HRDC, Guru Jambheshwar University of Science and Technology, Hisar, Haryana, India*
[b]*School of Chemical Sciences, Central University of Gujarat, Gandhinagar, India*
[c]*Department of Biotechnology, Visva-Bharati, Santiniketan (Bolpur), West Bengal, India*
[d]*Department of Internal Medicine, University of Utah, Salt Lake City, Utah, USA*
tapan400@gmail.com

Increasing dependence on traditional materials has enforced the need for replacing them with more capable alternatives, promising stronger resilience, and strength to be fitted in dynamic and robust applications. These requirements have necessitated the controlled mixing of more than one kind of materials, which can provide the proportionate improvements in performance attributes. Thereby, composites have rightly emerged as the amicable substitutes wherein faster response, greater stress-bearing capability, and

Nanotechnology: Principles and Applications
Edited by Rakesh K. Sindhu, Mansi Chitkara, and Inderjeet Singh Sandhu
Copyright © 2021 Jenny Stanford Publishing Pte. Ltd.
ISBN 978-981-4877-43-5 (Hardcover), 978-1-003-12026-1 (eBook)
www.jennystanford.com

longer lasting times could be attained. The best instances include development of stronger materials for sophisticated areas like those of defense and corrosion, where either there is a prompt requirement, or we are already spending too much to counter. The advent of nanomaterials has further strengthened the performance attributes of these materials, leading to improved mechanical strength, load bearing, and tolerance of environmental stresses. The shear stress of carbon nanotubes (CNTs) is no more a fascination to discuss or illustrate as also the flexibility of fullerenes and nano-metallic solid structures. Nanocomposites basically involve the reinforcements of nanomaterials (may be one or more) to a bulk material, which is presumed to play the role equivalent to that of an adsorbent matrix. These entities have been extensively evaluated for their mechanical strength, buckling capability, corrosion tolerance, extreme order conductivity variations, thermal stress tolerance, and light-reflecting characteristics. The research is in swift progress for still finer development of stronger materials. With these insights, this article discusses the development of nanocomposites alongside their performance ascertaining characterizations and breakthrough applications of recent origin.

7.1   Introduction

Development of better and more resilient materials has been a persistent requirement of humankind, more so with the present-day dynamic lifestyle and mounting needs of automation and faster communication. Since the sustainable natural resources supporting the life on earth are finite, it has emerged as a liability to conserve them and ensure their judicious use, preventing any unchecked exploitation and wastage. Owing to this, a persistent thought of global economy has rallied behind generation of clean energy with reduced waste creation and recycling of nutrients, wherever possible. Composites are intentionally created entities in this regard, made with the intent to co-propagate the performance-optimizing aspects of two or more individual objects [1–3]. For instance, one material has good electrical conduction ability but gets corroded easily. Thereby, commercial use of such a material creates economic pressure for its regular generation. However, if this material is combined with another substance capable of resisting corrosion, it could provide a better product having enhanced electrical conductance as well as the ability of reasonably sufficient usage. Similarly, several electronics industries require materials with high dielectric constant, flexibility, simplicity in development, and mechanical as well as textural strength. A single entity with all such attributes is significantly tough to be located and unearthed. Thereby, ceramic materials (having high dielectric constant and exhibiting brittle texture even at high temperature) impart a better suitability, particularly with their ability to withstand high temperature. On the contrary, the polymeric materials have easier processing attributes but have low dielectric constants. Composites comprising micron-level ferroelectric ceramics as liquid crystal (LC) filler or as thermoplastic polymer matrices do not have suitable processing attributes. Hence, it is difficult to modify them into thin uniform films, required in multipurpose microelectronics utilities. Under these circumstances, nanocomposites (NCs) provide amiable remedies wherein, a much greater useful combination of material properties could be exercised through nanometer-scale mixing. Thus, composites address such natural limitations of individual material entities.

Nanocomposites are a further specific class of composites having a nanomaterial as fillers dispersed in a bulk matrix. Typical composition resembles a coexistence of matrix and filler, where the fillers could be particles, fibers, or fragments that are generally surrounded and bound together as distinct entities using the matrix [4, 5]. The nomenclature "nanocomposites" spans significant diversity of materials, including three-dimensional (3D) metal matrix materials to two-dimensional (2D) lamellar composites, unidimensional nanowires (NWs) to non-dimensional (0D) core–shell configurations, collectively substantiating diversified mixed and layered materials [6, 7]. The obvious curiosity after knowing what NCs are remains the need to know about the fundamental essence of NCs with their intent. It is not surprising, perhaps, that nanomaterials have been the subject of intense interest not only in academic research but also from the point of view of (scale-up driven) industrial applications. The reason behind this is that nanomaterials simply have tremendously high surface areas for exploring and improving their interaction properties, which consequently results in significant distinctions of physical, chemical, optical, and magnetic properties [8, 9]. As a consequence, the properties of NCs depend not only on the behavior/stimulus specific response of their individual constituents but also on their respective inclusion extent, the interaction-driven morphology, and interfacial characteristics. Studies have even yielded NCs that are thousand times tougher and stronger than their bulk component. With reference to their applications of a better material behavior, NCs constitute a combination of organic and inorganic materials, wherein the inorganic phase could be 3D framework systems (such as zeolites), 2D layered substances (such as clays, metal oxides, metal phosphates), and even 1D rod-like structures. Thereby, the essence of making NCs revolves around accomplishing the nonexistent qualitative attributes, similar to light weight reinforced mechanical adhesives, noncorrosive and long-lasting battery cathodes, better shock- and stress-bearing capabilities, higher mechanical strength, buckling capability, and several others.

It might be yet another surprise wherein the terminologies organic and inorganic phases have been used when so many other criteria of defining materials are known and established. This is because of the fact that inorganic layered materials prevail in several distinct configurations, bearing significantly accessible well-defined, symmetric intra-lamellar space [10, 11]. This particular property rightly describes the utility of layered materials as matrices for polymers, laying the foundation of hybrid NCs. Lamellar NCs signify a rare possibility wherein interfacial interactions among the constituent phases are maximized. As a result, concurrent variations in the matrix and fuller compositions and their mutual interactions could provide NCs with a broad range of properties. This is best illustrated by the distinct constitutional makeup of intercalated and exfoliated lamellar NCs, with the former comprising alternate assembly of polymer chains having inorganic layers being in a fixed compositional stoichiometry, together contributing to a systematically even count polymer layers in the intra-lamellar space.

Contrary to this, in exfoliated NCs, the relative extent of polymer chains intervening matrix layers varies with interlayer separation being in the order of 100 angstroms. While intercalated NCs find their usage driven by their electronic and charge transport characteristics, their exfoliated counterparts are preferred for superior mechanical properties. Thus, the most generalized definition of NC could be a several-phase-constituted solid material, with at least one nanoscale dimension. It is perhaps under the control of our will to control the number of nanoscale dimensions, usually optimized with reference to specific applications or requirements in consideration. Typical nanomaterials used in making NCs include nanoparticles (NPs), nanofibers, and nanoclays. In one line, the essence of making NCs is their inherent ability to provide varied material properties (chiefly being mechanical and physical) without changing the wholesome chemical compositions [12, 13].

7.2 Formation Rationale and Variations of Nanocomposites

Based on the fundamental nature of the material of the basal matrix, three main variations of NCs are recognized: ceramic, polymer, and metal matrices. It is pertinent to know here that generation of NCs is strictly with respect to better application fulfilment, and there are some localized aspects only, where NCs fare better than their bulk counterparts. Some of the notable challenges in NC fabrication include the difficulty in homogeneous distribution of the filler material, wherein the structure–property relationship needs to be understood better. Thereby, there is an urgent need to develop simpler mechanisms of filler (exclusively NPs) material exfoliation and dispersion. Besides this, the expenditure incurred in making NCs is also significantly high compared to those of normal composites.

Thus, though NCs promise better wear-and-tear resistance alongside higher load-bearing capability, the specific formation mechanisms necessitate the requirement of sophisticated processing, thereby increasing the incurred expenditure. The stage is optimum for a discussion of different types of NCs, being distinguished into metal, ceramic, and polymer, on the basis of specific matrix regime (Fig. 7.1).

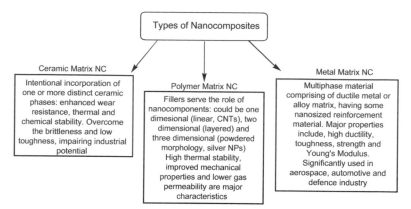

Figure 7.1 Different classes of nanocomposites with their characteristic features. Requirement of bulk matrix and filler ingredient as the nanomaterial contributor is common to all.

7.2.1 Ceramic Matrix Nanocomposites

The ceramic matrix nanocomposites (CMNCs) are materials made after the intentional inclusion of one or more different ceramic phases, especially with reference to hydrophilic resistance, thermal and chemical stability enhancement. The rationale of making CMNCs is attributed to their brittleness and low toughness of ceramic materials, limiting their industrial applications. Examples of CMNCs include matrix wherein energy-distributing agents (fiber, platelets, or particles) are selected for moderating the brittleness and enhancing fracture durability [14, 15].

7.2.1.1 Processing methods

Several methods have been optimized for CMNC synthesis, and latest among which are single-source precursor methodologies, decided through hybrid precursor melt spinning, subsequently being subjected to curing and pyrolysis.

Some previously known methods of this regime are conventional powder, polymer precursor route, spray pyrolysis, and vapor deposition methodologies (chemical and pulse vapor depositions) [16–19]. A brief glance of some of the NCs using these methods can be found in Table 7.1 along with the salient advantages and performance sensitive cautions.

Table 7.1 Different CMNC processing methods with their advantages and limitations

Method	System	Advantages	Limitations
Powder procedure	Al_2O_3/SiC	Easier in approach	Low formation pace and phase dispersion, agglomeration
Polymer precursor method	Al_2O_3/SiC, SiN/SiC	Possibility of finer particle preparation, better reinforcement dispersion	Agglomeration and ultrafine particle dispersion (causing phase segregation)
Sol–gel procedure	SiO_2/Ni, ZnO/Co, TiO_2/Fe_2O_3, La_2O_3/TiO_2, Al_2O_3/SiO_2, $NdAlO_3/Al_2O_3$	Simple, versatile, chemical homogeneity	Compact, low void formation compared to mixing method

7.2.1.2 Structural overview

A typical CMNC comprises a matrix containing nanosized reinforced components, which could be particles, whiskers, fibers, or even nanotubes. High brittleness and easy fracturing of ceramic material necessitate the incorporation of ductile metal phase within matrix, thereby improving durability as well as mechanical properties (hardness and fracture toughness). These improvements are the outcomes of more controlled interaction between different phases, matrices, and reinforcements. Understanding the structure–property relationship of CMNCs requires a thorough knowledge of the inter-relationship between the surface area and the volume of reinforcement materials. The variation of matrix and filer inclusion extents of CMNCs is a decisive factor to confer several improved properties, of which fracture strength toughness and high-temperature resistance are major. Table 7.2 presents a few examples of CMNCs, with the characteristically improved material properties, corresponding to the specified matrix and reinforcement combination.

Table 7.2 Some characteristically improved properties corresponding to different CMNC matrix and reinforcement constituents

Matrix/Reinforcement	Improved properties
Si_3N_4/SiC	Strength and durability
Al_2O_3/SiC	Strength and durability
B_4C/TiB_2	Strength and durability
$Al_2O_3/NdAlO_3$	Photoluminescence

CMNCs are highly durable industrial materials, owing to their high-temperature resistance, significant mechanical strength, and chemical inertness. Bioceramic materials have no undesirable effect on humans and are, therefore, highly preferred for designing prosthetics. On being freely left, CMNCs remain unaltered for a reasonably long time. The recyclability of these materials is a positive sign for their usage, promising a long-lasting life along with reduced carbon emission and cost effectiveness [20].

7.2.2 Metal Matrix Nanocomposite

Metal matrix nanocomposites (MMNCs) represent multiphased materials comprising a ductile metal or alloy matrix, having the provision of implanting nanoscale reinforcement. Salient MMNCs traits include high ductility, toughness, strength, and Young's modulus, making them useful in aerospace and automotive industries.

7.2.2.1 Processing and preparation

Metal matrices for the preparation of MMNCs include Al, Mg, Pb, Sn, W, and Fe, while the reinforcing entities (the nanomaterial constituents) vary from metal powders such as silica, clays (as crystalline grade). More common reinforcements include low-particle size and well-versed interaction-chemistry-bearing layered silicates and clays.

Among the preparation methods, there exist chemical and physical methods, making use of bottom-up and top-down principles. The physical methods include spray pyrolysis, liquid metal infiltration, rapid solidification, plasma and chemical vapor depositions, and electrodeposition.

Table 7.3   Select compositions of MMNCs with their fabrication method

Method	System
Spray pyrolysis	W/Cu, Fe/MgO
Liquid filtration	Pb/Cu, W/Cu-Nb/Cu, Pb/Fe-Nb/Fe, Al-C60
Sol–gel	Fe/SiO$_2$, Ag/Au, Au/Fe/Au
Rapid solidification	Al/X/Zr (X= Si, Cu, Ni), AL/Pb, Fe alloy
Ultrasonication	Al/SiC

Contrary to this, chemical methods include colloidal assembly and sol–gel processing. Some of the relatively less used methods for making MMNCs are melt falling drop quenching and one-step preparation of carbon dot and Au nanoparticle consisting NCs [21–24]. Table 7.3 summarizes the different methods to make MMNCs, along with the potential advantages and cautions.

7.2.3   Polymer Matrix Nanocomposites

Polymer matrix nanocomposites (PMNCs) consist of a polymer material as matrix, whereas the fillers can be different nanomaterials such as 1D (carbon nanotubes, CNTs), 2D (layered materials, montmorillonite), and 3D powdered textures (e.g., AgNPs) [25]. Interactions between the polymer matrix and the nanoscale reinforcement on a molecular scale are accountable to the attractive forces operating between the two. Consequently, the inclusion of a small extent of nanofiller (having dimensions below 100 nm from that of the matrix) brings about a change in the material properties. Typical methods for making PMNCs are same as that for composites, vis-à-vis in situ intercalative polymerization, solvent approach, and mixing melted polymer matrix. The major characteristics of PMNCs include significant inertness toward varying temperatures, better mechanical characteristics (high abrasion resistance), and meagre gas porosity (higher barrier capacity) [26].

7.2.3.1   Processing approaches

Several mechanisms are known for the preparation of PMNCs, more popular of which include solution-driven intercalation, in situ intercalative polymerization, melt intercalation, and template synthesis (sol–gel methodology) [27]. The first major method, the intercalative polymerization approach, involves the formation of polymer in the midst of two intercalated sheets. The working relies on layered silicate swelling within liquid monomer, thereby initiating the polymerization (through the availability of heat/radiation or by the diffusion of a suitable initiator) [28]. Initially, this method was applied for producing nylon-montmorillonite nanocomposite, but after some time, it became a common mechanism for making several thermoplastics. This method promises a convenient method to make thermoset-clay NCs [29]. A few demerits of this method include slow reaction rate, dependence of clay exfoliation through clay swelling, diffusion rate of monomers in the layered segments, and incomplete polymerization-driven oligomer formation [20].

In the melt-intercalation approach, there is no requirement of solvent and polymer matrix, and the typical makeup comprises a molten layered silicate. Conventional methods such as injection molding or extrusion make use of high-temperature thermoplastic polymer mechanical mixing with organophillic clay [30]. In this stage, polymer chains get exfoliated to generate NCs. The method is easily the best choice for making thermoplastic NCs and can be a significant remedy in the event of unsuccessful in situ polymerization or due to the inappropriate polymer selection. Crucial hurdles of this method include its restrictive suitability to polyolefins, accounting for a majority of used polymers [20].

In the sol–gel method, the formation of clay minerals occurs within the polymer matrix through the application of an aqueous or semi-solid solution comprising polymer and silicate as constituent units. Processing begins by nucleation, subsequent to which, the polymer inclusion facilitates the development of inorganic host crystals, while being enclosed in between the layers. This method is generally preferred for making double-layered hydroxide-based NCs and is much less suited for layered silicates [28]. Two major concerns of the process are high temperature requirement for clay mineral synthesis (that often decomposes the polymers) and the aggregation tendency during the silicate growth. Table 7.4 provides different methods for making PMNCs along with their advantages and cautions.

Table 7.4   An overview of PMNC preparation methods along with potential merits and demerits

Method	Major use	Advantages	Disadvantages
In situ interactive polymerization	Manufacture of nylon-montmorillonite NCs and thermoplastics	Tethering effect enabling the linkage	Slow reaction rate, dependence of clay exfoliation
Melt interaction	In making thermoplastic NCs	Replacement method for in situ polymerization	Limited application to polyolefins
Template preparation (sol–gel technology)	In making double-layered hydroxide-based NCs	Aids in the dispersion of silicate layers in a single-step process	High temperature for making clay minerals, often involves polymer decomposition and aggregation during silicate growth

7.2.3.2   Properties of PMNCs

The native characteristics of PMNCs project their usefulness in being widely used across cutting-edge applications. Compared with traditional materials, these composites possess greater specific strength and specific modulus, damage tolerance, flame retardant and damping properties. Glass fiber presents a befitting example of a high specific strength and Young's modulus material, having high density with lower specific modulus of glass fiber resin matrix [31]. In terms of stress tolerance, NCs exhibit matrix cracking (in place of fracture causing crack propagation), fiber pull-out, interfacial structure breaking, and fiber breaking. An unfortunate event of minute fiber fracturing results in load transfer through the matrix on intact fibers, a generalized frequent observation. On being substituted by an NC in a shorter time limit, there is seldom any loss of bearing capacity despite the observations of defects and cracks [31]. In terms of flame retardation, the search in earlier studies revealed that addition of even 5% nanosized clay particles could induce a 63% lowering of nylon-6 flammability. Herein, recent studies have shown a boost in this flame-reading ability through enabling molecular-level clay dispersion. This phenomenon is manifested through enhancement of barrier properties by the clay generated "tortuous path," delaying gas movement through the matrix (Fig. 7.2) [32]. Similar to the above attributes, NCs impart the ability of vibration absorption through the incorporation of a damping material, enabling a cessation of vibrations in a shorter time [31].

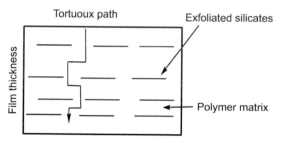

Figure 7.2 Nelson's path model for NCs conferred barrier enhancement [32].

The stage is now perfectly augmented to switch over to the characterization techniques relevant for analyzing NCs, since these are specially designed materials after keeping the intended applications in consideration. Based on the exclusive results of characterization, different NCs are used for their specified applications. Three different levels of NCs prevail: optical, crystallographic, and spectroscopic. Each mode of characterization provides implicit information about the fabricated NCs, although analysis from crystal structure study is the first and basic level of study. This is because the implicit confirmation of a material nature (from its chemical composition point of view) is inferred from the database of its particular geometry.

7.3 Characterizations of Nanocomposites

The characterizations of NCs have increased the interest of researchers with theoretical and experimental techniques due to their versatile applicability. Here a wide range of methodologies are practiced as inspecting techniques for NCs by influencing their physical and chemical behaviors, biological interactions, optical characteristics, and generation cost. Hence, a thorough and in-depth characterization of NCs is required to achieve an efficient accommodation in their nanoscale features to realize their manifested benefits. Characterization methods for NCs practiced significantly for NC characterization over the last decade include scanning electron microscopy (SEM), transmission electron microscopy (TEM), optical microscopy (OM), and atomic force microscopy (AFM). These techniques have emerged significantly crucial to investigate and correlate the structure, size, and morphology of NCs with their intended applications [33–36]. Most extensively, SEM has been used for NC structural characterization despite having a lower resolution compared to AFM or TEM [37–41]. It is difficult to obtain accurate and detailed knowledge regarding the constitutional distribution within the matrix of a PMNC. Though advancements for better resolution prevail, in the form of SEM and field emission gun SEM (FESEM), all of these operate at precisely low potential differences, rendering it infeasible to screen organic entities without conductive coating. Apart from SEM, AFM and TEM are also used for the structural prediction of NCs, particularly cellulose- or chitin-like nanomaterials distribution and dispersion [37–39, 42, 43]. The structure of PMNCs is conventionally illustrated through combined TEM and wide-angle X-ray diffraction (WAXD) [38]. The text ahead chronologically describes the microscopic, crystallographic, and spectroscopic techniques practiced for NC characterization.

7.3.1 Optical Characterization

Optical microscopy is significant in characterizing the macroscopic structures when the matrix is transparent and the sample is extremely thin. However, it is not suitable for characterization of CNTs within NCs. For instance, polarized optical microscopy (OM) is used for assessing the crystallization behavior of CNT-modified and unmodified carbon nanocomposites (CNCs) along with silylated CNCs, where CNC without surface modification gets aggregated [44–46].

7.3.1.1   Scanning electron microscopy

SEM is easily the most rigorous microscopic methods for surface characterization, having 1–5 nm ranged resolution, so that the analyzed materials have a large field depth, helpful in producing 3D images. The working principle comprises an electron gun through which electrons are accelerated across the lenses that focus the beam over a very little size. These electrons penetrate the sample being analyzed to about 1 μm depth, generating image-forming signals. Among all the generated signals, three most important are backscattered electrons (BEs), secondary electrons (SEs), and X-rays. The BEs comprise elastically scattered electrons conferring the specimen atomic-number-dependent compositional contrast. These electrons have high energy and come from 1 μm or more specimen depth. SEs comprise low-energy electrons, emerging from the specimen surface, and are mostly used for topography imaging [47]. The PMNCs where both matrices and reinforcing fillers are nonconductive mostly constitute low-atomic-number elements, used for topography visualization.

7.3.1.2   Transmission electron microscopy

This microscopic characterization tool is characterized by the transmission of high-energy electrons through an ultrathin specimen domain. The image is generated on account of electron emission after the incident electron beam hits the specimen [48]. The electrons are emitted from an electron filament, having two or more condenser lenses beneath it, for de-magnifying the electrons emitted by the gun as well as controlling its diameter, while they hit the specimen (held inside the objective positioned below the condenser lens). After the objective lens, there are two lenses, namely, intermediate and projector, producing a real and magnified image on the fluorescent imaging screen or film. The contrast of TEM image is due to electron scattering, where the imaging mode is described by the bright field (BF), having an objective opening inserted to form the images by direct unscattered electrons. Thicker or high-density regions in the specimen generally scatter strongly and appear darker because of the obstructed flow by the objective aperture. The image region with no specimen is bright in BF. The diffraction, mass thickness, and phase contrast are three basic mechanisms contributing toward image formation [45, 48]. The NCs having low-atomic-number element only induce a weak electron scattering, resulting in below par image contrast. For these materials, the mass thickness contrast mechanism can be applied using thin sample staining involving heavy metal enabled scanning of interesting features. For instance, uranyl acetate is an appropriate staining agent for CNTs to induce better contrast [44]. Thereby, extensive characteristics of nanoparticle-reinforced polymer composites are affected by inter-matrix nanoparticle phase distribution. The corresponding capability to ascertain the nanoparticle surface distribution or being rooted in a polymer matrix is crucial, in the course of relating the resulting composite attributes to characteristic enhancement mechanism. TEM is, therefore, a universally acclaimed technique for determining NCs thought thickness, making it arguably the most suitable microscopy subject to homogeneous CNT distribution and dispersion [38].

7.3.1.3 Atomic force microscopy

AFM is a scanning probe module that inspects an NC sample using nanometer resolution probe, having 10 nm as typical radius of curvature. The monitoring of probe and sample surface interaction response is responsible for producing the sample topography [49–51]. Two operating modes, namely, contact and noncontact, are used where the measurements could be made under vacuum, vapor, and liquid environments. The contact mode provisions the AFM cantilever tip scan the specimen surface, maintaining a constant force between the tip and sample. While ascertaining the impacts and counter-impacts of this probe–specimen interaction, the typical probe vibration frequency aids in maintaining a constant force over the sample in noncontact/tapping mode. Advantages of this mode are lower lateral forces contrary to contact mechanism, making it feasible to acquire phase image like additional contrast channels. The sub-nanometer height and nanometer-scale lateral resolutions are permitted for reference inspection of cellulose NCs, propelling in the concomitant surface roughness [52, 53]. In the noncontact mode, the phase imagining regulates the amplitude of probe vibration and nanoparticles distribution within the polymer matrix. It also implies recording cantilever oscillation phase lag, relative to cantilever's piezo driver's transmitted signal. The phase lag is much sensitive to adhesion variations, viscoelasticity and thus manifests contrast among the different material components of NCs. Additionally, phase imaging highlights edges, typically remaining unaffected by significant amplitude differences, inferring clear observation of fine features, like those of grain edges. Such phase imaging characteristics make AFM a useful technique for ascertaining typical nanoparticle distribution within polymer composites [42, 43, 54]. It is essential to scan multiple sample sections through varied NC thickness as conventionally AFM is used to screen the nanoparticles scattering through the composite thickness. The imaging for topography and phase provides information about the typical CNC alignment within NCs where the degree of CNC alignment is qualitatively described via a variety of image analysis methods [55–57]. Typically, raw AFM image depicts the post-processing long CNC axis, where the angle of the CNC long axis with respect to geometry or CNC alignment is measured. Such working explains the 2D CNC alignment, supposing that all CNCs recline along the AFM imaging plane. For additional degrees of freedom along the CNC orientation, a method for 3D alignment of short fiber composites involving elliptical cross section on a polished surface could be appropriate [58–60].

7.3.2 Crystal Structure Inspection

The WAXD is used to ascertain the CNC alignment within the composites where distinctions in diffracted X-ray strength are measured corresponding to each specified diffraction plane as a function of orientation along with composite geometry [56, 61–63]. The diffracting X-rays with reference to primary X-ray beam direction, like monochromatic X-ray, $\gamma = 0.1541$ nm from CuKα radiation, generated at a fixed accelerating potential difference and current, are measured by WAXD. Diffracting X-rays appear merely in specific circumstances, depending on the internal solid crystalline morphology. For instance, the 2θ diffraction peaks at ~14.5°, 16.6°, 20.4°, 22.7°, and 34.4° are obtained for cellulose, corresponding to plane miller indices [58, 64]. The extent of CNC arrangement can be estimated using 2D WAXD diffraction patterns, providing diffraction intensities for (5–50)° 2θ and Azimuthal angle ϕ (360°) rotation about one NC axis. The resulting 2D diffraction patterns depict the Debye–Scherrer rings, inferring Azimuthal intensity dispersal for each 2θ diffraction peak [59]. On linking the diffraction patterns with the axis of the to-be-scanned composite sample, one can retrieve further details pertaining to the inter-composite CNC alignment [65]. For CNC, diffraction patterns are obtained corresponding to accommodate a composite specimen so that the alignment remains along the axis [59, 65].

7.3.3 Spectroscopic Characterization

Owing to the mounting interest in the structural chemistry of NCs, their molecular characterization has become necessary for understanding their characteristics as well as new material development. Spectroscopic techniques provide significant details about polymers, fillers, and composites, wherein variations in fluorescent intensity, nuclear magnetic spin frequency matching, infrared and Raman frequencies elucidate the filler dispersion extent across the host matrix, the extent of polymer and filler particle interaction along with the polymer chain dynamics at the polymer–filler interface.

7.3.3.1 Fluorescence spectroscopy

Analysis by fluorescence spectroscopy requires amalgamation in the presence of a fluorescent probe, employed at an extremely dilute concentration such that the sample bulk is not unduly attenuated. The probe is selected for its sensitivity to ascertain variations in the immediate surroundings, through varied emission features. Typical investigation of the luminescence attributes of a small probe embedded within a polymer matrix deciphers key polymer science prospects, like counting dynamics of polymer chains via excimer fluorescence, phase separation and polymer mixing ability, transport phenomenon or polymer deterioration [66, 67]. Application of fluorescence attributes of polymer composites extensively intercepts a characteristic photophysical process such as energy transfer or fluorescence suppression. The interface properties and dispersion in PMNCs are investigated by the Förster resonance energy transfer (FRET) mechanism [68]. This mechanism explains energy transfer from a donor fluorophore (having excited electronic state) to an acceptor via non-radiative dipole–dipole coupling. It needs an overlap between donor's fluorescence and acceptor's absorption spectra. Furthermore, typical FRET competence depends on the intermolecular separation, should be ideally within 1–10 nm. FRET is applied to reveal nanoscale features happening at the polymer–filler system interface, depicting the labelled phases with appropriate donor–acceptor chromophores [68]. An attractive application of fluorescence spectroscopy remains the stress-softening process investigation in filled elastomers (Mullins effect), inferred by moderate decrement in the load subsequent to a material's stretching after the first load release [69].

7.3.3.2   Solid-state NMR spectroscopy

The solid-state nuclear magnetic resonance (NMR) allows the screening of polymer–filler interfaces on account of mutual NMR spectral sensitivity with the relaxation parameters to native as well as segmental molecular activities (of polymer chains) [70]. The interactions of polymer–filler typically contribute to an adsorption layer formation, wherein the chain motilities are far too limited than the bulk structure, distinguishing the workable polymer nature in the interfacial region. In solid-state H^+ NMR using spin-echo technique, the transverse magnetization relaxation function is also studied (for instance in poly(dimethylsiloxane)) [70].

7.3.3.3   Infrared and Raman spectroscopy

Infrared ray (IR) absorption and Raman scattering are the spectroscopic protocols extensively employed to collect vibrational-features-determined information about the polymeric systems [71, 72]. The IR radiation originates from the direct interaction between the frequencies of the IR incident ray with a specified vibrational mode. On the other hand, the Raman effect is basically an inelastic light scattering, happening subsequent to a molecule's irradiation with monochromatic/laser illumination. Raman scattering allows scrutiny of thick polymeric samples but not very thin films that remain fit to be analyzed using IR spectroscopy, owing to a robust absorption of IR radiations by the polymer functional groups. The analysis of such spectra not only provides organic and inorganic phase interaction estimate but also respective intercalation and exfoliation polymer states. Such composites contain layered silicates, filler dispersion/functionalization, and varying-orientation-bearing polymer chains/anisometric particles. A combination of AFM and IR techniques to retrieve IR spectra with a nanoscale spatial resolution is also used to screen the NCs [73]. Such an analysis mode lowers the extent of normal IR spectroscopy, just like the diffraction-imposed spatial resolution limits. It is used for characterizing the nanoscale aspects of polymer NCs, within the cross and interdisciplinary domains of conventional and material sciences [74].

7.4 Material Significance of Nanocomposites in Inter- and Cross-Disciplinary Domains

7.4.1 Aerospace Sector

Numerous composite materials are involved in building the aerospace structures such as equipment inclusions, aircraft internal parts, coatings, cockpit, crew gear, heat-contracting tubing, space accommodating mirrors, housings, nozzles, and solar array substrates. Apart from their physicochemical inertness, light weight, low operating cost, and fire resistance, these materials also have a few disadvantages that restrict their use. First, their higher electrical resistance hinders an involvement in electromagnetic shielding, lightning strike defense, antennas, and circuits [75]. Then, a lower thermal conductivity enhances the pressure on discrete structures depending on electrical heaters. Last, they are less resistant to moisture absorption, easily influenced by environmental clues, and degraded with time [76, 77]. Therefore, to overcome such limitations, researchers have developed some NCs based on matrix and filler used. The PMNCs and CMNCs having thermal, chemical, electrical, and mechanical properties have fulfilled the aerospace requirements with nanofiller inclusions such as nanoclay, alumina,

CNTs, TiO_2, PZT (lead zirconate titanate), and others. For instance, polyamide/clay nanocomposite exhibited enhancement in heat resistance, Young's modulus, strength, biodegradability besides reduced flammability and gas permeability, making them suitable for aerospace applications [78, 79]. Moreover, higher reduction in gas permeability in case of polyamide/clay nanocomposite enhances their usefulness for floor patterning in the cargo part of airliners as well as manipulation of lightweight and strong cryogenic gas storage tanks [80]. Similarly, montmorillonite- and DOPO-based phosphorus tetraglycidyl NCs have reduced the manifold necessities in aircraft design [81, 82], whereas polyurethane-consisting CZT (cadmium zinc telluride) offers lightening-strike resistance, electromagnetic shielding, and radar-absorbing abilities to aircraft coating [75, 83]. Furthermore, CZT also imparts a stronger tensile strength with enhanced properties to the ceramic matrix than conventional carbon, which only provides toughness [84].

7.4.2 Automotive Industry

Till today, 80% of polymer/clay NCs are used in automotive applications. Owing to their extraordinary properties such as stiffness, thermo-mechanical resistance, and low power consumption, they are widely used in car part industry. In addition, they have the ability to be painted together with other car parts and undergo similar actions as metallic materials in vehicle manipulation. Such response of these materials makes them superior to the previously used polypropylene and glass fillers in car parts. For example, several entities within a car, such as timing belt, handles, rearview mirror, engine cover, bumper, gas tank, and others, are made using nylon (polyamide) as NCs by companies such as Bayer, Honeywell Polymer, Toyota Motors, RTP Company, UBE, and Unitika. Moreover, footboards of Safari and Astro cars are made up of polyolefin NCs such as polyethylene and polypropylene, produced by General Motors [85].

7.4.3 Food Packaging

NCs also play a vital role in the food packaging industry by extending the shelf life of products based on their restricting activities, constitution, and relative interactions with food particles. Their antimicrobial activities are more effective than micro- or macroscale materials due to large aspect ratios and high surface reactivity [86]. Therefore, these properties can be helpful in preventing damage of the food product through one-to-one access between active NC ingredients and products, as well as via obstructing the O contact besides ethylene and water vapor that provide protective environment.

7.4.3.1 Direct contact of food with active components

Here the active components are nanofillers such as NPs, clay, and others, which possess antimicrobial activities and enhance the food's shelf life, quality, and safety through inhibiting microbial growth as well as manipulating the biotic and abiotic factors. The antimicrobial efficacy of NCs highly depends on various features such as size, agglomeration tendency, dispersion, and nanofiller–surface–matrix interactions. Several efforts have been made to enhance the food product's shelf life using NC packaging materials. For instance, packaging films consisting of ethylene vinyl acetate, functionalized nanoclay, and brewery waste extract exhibited antibacterial activities toward *Escherichia coli* and *Staphylococcus aureus*, extending the healthy existence durations of beef and meat products [87]. Nanocomposite films of polyethylene organically formed montmorillonite, having necessary oil mixtures of both carvacrol and thymol, have emerged effective in minimizing the effects of *Botrytis cinerea* on strawberry [88]. Likewise, incorporation of Ag-montmorillonite into packaging materials prolonged the shelf life of fresh fruit salad [89]. Another study exhibited significant extension in the shelf life of cold storage carrots using similar material in alginate-based films [90]. Similarly, nisin-loaded soy lecithin nanoliposomes, inserted into the HPMC-formed packaging films, showed inhibited growth of *Listeria monocytogenes*.

7.4.3.2 Gas barrier packaging materials

Nanocomposite packaging films bring about changes in the internal or external environment through oxygen impermeability attributes. Since oxygen within a food package decreases the duration of stable existence of food materials via involvement of several activities such as microbial growth, enzyme activation, and oxidative reactions. Furthermore, this oxygen also generates ethylene, which further aggravates respiration in fruits and vegetables. Thereby, to prevent such a loss of food shelf life and concurrent food product quality, oxygen should be scavenged or removed from the package to a designated level needed to ensure healthy existence of a given food product. Apart from oxygen, other volatile compounds released from vegetables and fruits during storage should also be restricted to enhance the products' shelf life. However, improved gas barrier properties of NCs provided multilayered NCs, allowing firm packaging such as thermoformed containers. These containers provide a modified atmosphere for packaged fresh foods that rely on respiration rate, film permeability, gas transport, weight and surface area, free head space occupancy, and atmospheric structure within containers.

Many studies have reported the scavenging of residual oxygen within the package. For example, the synergistic effect of altered atmosphere in platters having Ag–cellulose absorbents together results in microbial growth inhibition and delayed senescence in processed melon [91]. Likewise, cellulose pads infused with Ag NCs retard the growth of bacteria, yeast, and molds in kiwi fruit [92]. Similar to oxygen scavenging, ethylene is also reported to be scavenged using TiO_2 NPs deposited on polypropylene films [93]. The addition of nanofiller within the polymer matrix improves several barrier features for NCs, thereby enclosing gases and water vapor. This inclusion of nanofillers confers an improved packaging quality compared to native polymer matrix. For example, gas barrier attributes of low-density polyethylene (LDPE) got enhanced even more than seven counts subsequent to montmorillonite incorporation [94]. Similarly, the inclusion of cellulose into a poly (ε-caprolactone) matrix reduced the water vapor permeability on increasing nanofiller concentration [95]. Likewise, in ethylene-co-vinyl alcohol (EVOH)/poly(lactic acid) (PLA), LDPE, polypropylene (PP), and high-density polyethylene constituted thin layers, the oxygen permeability lowered on nanocaly addition [93, 96–98]. A few commercial-grade NCs utilized for packaging are compiled in Table 7.5.

Table 7.5 Few commercial-grade NCs used for food packaging

S. No.	Nanocomposites (commercial name)	Uses	References
1.	Nylon-6 NCs	PET multilayer bottles for beverages and foods	[99]
2.	Imperm®, a nanoclay	Beer and carbonated drink bottles	[100]
3.	Durethan® KU2-2601	Coating juice bottles	[101]
4.	AEGIS™ OX	Beer bottles	[101]
5.	Biomaster® Ag based NCs	Sustained Ag⁺ release assuring food packaging	[101, 102]
6.	Agion® Ag based NCs	Sustained Ag⁺ release assuring food packaging	[101, 102]

7.4.4 Biomedical Applications

For the last 50 years, ceramic NCs have received great interest in the field of biomedicine for their unique structural and functional properties that allow withstanding soaring loading rates, thermal energy, physical stress, and chemical etching too unembellished for metal surface. Thereby, bone and dental implant applications are the most challenging for ceramics. But their brittleness restricts the potential applications to few, apart from defects induced during mechanical loading, which result in disastrous failure. Thus, to improve the activities of ceramic matrix in the biomedical domain, nanostructured ceramics are being explored to overcome extensive structural and biological complications. A nanophase ceramic composite can be produced using nanocrystalline materials or NCs having a large area-to-volume ratio, allowing significant modulation of physicochemical properties. The CNCs with improvements in strength, stability, stiffness, and creep resistance have enabled significant advances in medical science such as development of orthopedic, dental implants, tissue engineering, and drug-delivery systems for cancer treatment.

7.4.4.1 Orthopedic implants

Alumina and zirconia ceramics are crystalline in nature where atoms are bound together with strong ionic and covalent bonds. Owing to such chemical linkages, these materials have high strength, hardness, and elastic modulus, making them ideal for use in orthopedic implants. Moreover, their biocompatibility, chemically inertness, and resistance against corrosion are also fruitful in the case of lifetime implantation [103]. However, their intrinsic brittleness limits their usage as orthopedic implants. Therefore, this issue can be solved via potential improvements in the implant techniques and materials quality. Recently, numerous zirconia–alumina NCs have been reported with excellent control toward aging, resulting from their toughness and mechanical strength [104, 105]. In zirconia–alumina NCs, two types of composites can be synthesized: either zirconia-toughened alumina (ZTA) where zirconia particles are impregnated into the alumina matrix or alumina-toughened zirconia (ATZ) having alumina particles in the zirconia matrix. In the ATZ system, composites have high fracture toughness, whereas the ZTA system has composites with strong stiffness and low fracture toughness [106, 107]. Both these composites are commercialized under ZTA Biolox® Delta and ATZ BIO HIP® terminologies. In contrast to ATZ, the ZTA system is widely used in biomedical applications due the alumina matrix monodispersion of zirconia, subsequently getting transformed into a stable monoclinic phase under mechanical loading [104]. Another study developed Ce-TZP/Al_2O_3 NCs exhibiting high aging resistance, wear resistance, and biocompatibility [107–109]. Besides these NCs, Garmendia et al. established YTZP (yttria-stabilized zirconia)-CNT NCs via adding CNT into a zirconia matrix that showed an improvement in the mechanical, physical, and chemical properties.

7.4.4.2 Dental implants

CNCs find significant utility in the development of dental implants as they provide both strength and toughness that are necessary for a long-span fixed partial prostheses and implant supports. For dental implants, YTZP is an important ceramic having exceptional mechanical properties and biodegradability. But with zirconia ceramics, aging is still an issue as the implants are kept in humid, acidic, or basic oral environment. There are some other features that require detailed investigation such as the influence of thermally aided manipulation on microstructure and strength, attachment as cover for porcelains, cement annealing, visible light transparency linked with aesthetic refurbishment, X-ray denseness, and experimental prevalence feasibilities. Many features of zirconia–alumina NCs present befitting opportunities for further optimization and betterment of these properties. For example, Philipp et al. investigated the veneered ceria stabilized zirconia–alumina NCs for making dental implants after its functioning for a year [110]. While Nevarez-Rascon et al. observed the properties of ATZ and ZTA NCs such as size, density, hardness, and fracture toughness, which were in good agreement with international standards thereof [105]. Therefore, ATZ and ZTA composites combination proved to be ideal than their pure oxides for dental purpose. Besides this, Kong et al. attempted the enhancement of zirconia–alumina matrix biocompatibility toward load-bearing applications via adding bioactive $Ca_3(PO_4)_2$ like HAP and TCP, which in turn improved the bone regeneration and osteointegration [111].

7.4.4.3 Tissue engineering

Tissue engineering has achieved a remarkable success in tissue repair and corrective surgeries aimed at regeneration. The fundamental science of tissue engineering makes use of a highly porous scaffold that acts as a supportive material for cells seeding in vitro where cells adhere, proliferate, and differentiate. This scaffold permits an invasion of blood vessels that provide strength as well as nutrients to the cells. Moreover, it should be biodegradable with controlled degradation rate and must vanish as soon as new cells regenerated. Several composites are being explored to fulfil these features and ensure self-sustainable functioning within the physiological boundaries. Recently, various ceramics such as hydroxyapatite and tri- and biphasic $Ca_3(PO_4)_2$ have been studied in bone tissue engineering [112]. But owing to their brittleness, their use is localized and far too restricted for load-bearing applications [113]. Thus, polymer–ceramic nanocomposites have been developed exhibiting similar inorganic–organic composition and mechanical properties of natural bone via adding $Ca_3(PO_4)_2$ [114]. Like ceramic, polymeric scaffolds of poly(lactic acid) (PLA), poly(glycolic acid) (PGA), poly(lactic acid-co-glycolic acid) (PLGA), and poly(ε-caprolactone) (PCL) are being used for bone tissue engineering. However, their hydrophobic, biological inertness, and general weak texture have restricted their utilization in bone regeneration. With an intent to overcome these issues, Wei et al. prepared nanohydroxyapatite (nHAp)/PLLA composite on adding nanohydroxyapatite into PPLA scaffolds, which revealed increased protein absorption and thus ultimately improved cell adhesion [115]. Similarly, Wang et al. proposed nanohydroxyapatite/polycaprolactone (nHAp/PCL) scaffolds for improved bone marrow stromal cell growth compared to PCL scaffold [116]. Another investigation assessed the cellular proximity and compatibility of mesenchymal stem cells, with micro HAp and nano HApPCL composites. Their results indicated a higher cell proliferation for NCs having a large surface-area-to-volume ratio [117]. Apart from this, Hong et al. prepared a bioactive glass ceramic nanoparticle and PLLA NC gibbet. Such novel scaffold has high modulus (8.0 MPa), enabling paramount mineralization for imitation of physiological environment [118].

7.4.4.4 Drug-delivery system for cancer treatment

The development of targeted drug-delivery system has reduced the harmful effects related with the treatment of cancer, in particular resulting from lack of anticancer drug specificity. However, an extraordinary ability of NCs for targeting and controlling anticancer drug has attracted great attention among researchers toward the designing of nanoscale composites for cancer therapy. Therefore, such advancement in NCs has enhanced the versatility of drug-delivery systems via conjugating with drugs to treat specific tumor.

Many NCs have and are being used for delivery of anticancer drugs, having enabled their increased concentrations at a specific site and finally improving the efficiency. For instance, Rajan et al. established chitosan-based polyoxalate carriers for cisplatin delivery in MCF-7 cells that released the drug in a controlled manner [119]. These NCs having oxalic acid, succinic acid, tartaric acid, and citric acid were conjugated with ethylene glycol, subsequent to being examined as targeted and controlled drug-delivery vehicles for cancer treatment. Similarly, Dhanavel et al. prepared chitosan/palladium-5% nanocomposite for 5-flurouracil and curcumin co-delivery to arrest acquired drug resistance by cancer cells. Such

NCs with dual drug binding exhibited better anticancer effect on HT-29 cells than single drug [120]. In another study, nanocomposite tagged with hyaluronidase favored drug diffusion from the carrier to cells as due to degradation activity of this enzyme. The chitosan-hyaluronidase-5-flurouracil polyethylene glycol-gelatin (CS-HYL-5-Fu-PEG-G) combined association displayed an improved 5-Fu bioavailability as well as controlled release [121]. In another significant effort, Rasoulzadeh et al. encapsulated doxorubicin (DOX) in carboxymethyl cellulose containing graphene oxide (CMC/GO) hydrogel where CMC works in a pH-sensitive mode, mediating drug release at necessary pH. The prepared nanocomposite was used against human colon cancer cell line (SW480) and exhibited significant inhibition [122]. Similarly, gelatin-based magnetic hydrogel NCs were used as nanocarriers for DOX [123]. Moreover, curcumin-loaded PMMA-PEG/ZnO was examined toward gastric tumor by Dhivya et al. Their results indicated that such NCs are beneficial for loading hydrophobic drugs, which consequently resulted in improved bioavailability [124]. Quite often, for nucleus-specific delivery of anticancer drugs, functionalized reduced graphene oxide NCs are used. In one study, such NCs were conjugated with DOX that subsequently provided poly-L-lysine-functionalized graphene oxide that transferred the drugs precisely within the nucleus of HER2 breast cancer cells [125]. Apart from this, clay-based NCs also played a vital role in targeting cancer cell. Various types of clay are identified for constructing NCs. Halloysite (Hal) obtained from aluminum silicate is used to prepare nanocomposite hydrogel comprising Hal-sodium hyaluronate/polyhydroxyethyl methacrylate for colon cancer therapy using 5-Fu [126]. Likewise, palygorskite (Pal), a hydrated magnetic aluminum silicate used for nanocomposite synthesis was explored toward cancer treatment [127].

7.4.5 Electronic Applications

NCs are widely employed in chemical sensors, electroluminescent devices, batteries, memory-based gadgets, photovoltaic cells, supercapacitors, light-emitting diodes, and others [128]. PNCs conjugated with nanofillers have been explored for sensing purposes through multiple mechanisms. In this case, nanofillers include CNTs,

Au, Ni, Cu, Pt, metal oxide, and others [129]. Moreover, polymer-dependent solar cells are capable to prepare low-cost large flexible panels, for example CdS NPs embedded in a polymer matrix are used to make solar cells [130]. Nevertheless, graphene also takes part in electronic applications and increases the life span of electronic devices, which in turn reduces the cost [131]. In addition, graphene/polyaniline (PANI) NCs are used to detect ammonia, which is more sensitive for nanocomposite than PANI alone [132]. Similarly, such NCs are also used for H_2 gas detection with 16.57% higher detection efficacy than PANI (9.38%) and graphene alone (0.83%) [133].

Therefore, from the above-mentioned various applications of nanocomposites, we can conclude that mixing of two or more nanoformed fillers have tremendously enhanced the properties than particle alone. Table 7.6 summarizes the still remaining prominent utilities of different types of NCs.

Table 7.6 List of applications for different kinds of nanocomposites

S. No.	Matrix	Nanofiller	Application	References
1.	Epoxy	Clay	Anticorrosion	[134]
2.	Pullulan	Clay	Oxygen barrier	[135]
3.	Polyvinyl alcohol	ZnO	Biosensor	[136]
4.	Ni-P	PTFE	Antibiofouling	[137]
5.	Polyvinylidene fluoride	Au	Dielectric	[138]
6.	PANI	Nanocellulose	Electrochromic	[139]

7.5 Future Directions and Inspirations

Composites have, thus, rightly emerged as the most suitable materials to meet the ever-increasing demand for faster and multifunctional materials. There is, no doubt, a compulsive need to improve the wear-and-tear attributes alongside prolonging the shelf life of available options. With nanocomposites, it has become possible to accomplish multiple needs via varying incorporation of nanomaterials. The advances in the stress response and long-lasting functional abilities of NCs are substantially attributed to the existing richness of nanomaterials, which are responsible for conferring one-, two-, and three-dimensional attributes to the intact products. In developed countries, NCs have not only replaced most of the plastic gadgets but also iron and iron-comprising other materials constituted entities have been revolutionized through better anti-corrosive abilities. The inclusion of graphene, CNTs, carbon fibers, and the derivatives thereof has easily over-seeded their expected outcomes, and there is still plentiful to be enquired and investigated. While metal- and ceramics-based NCs are more suitable in daily use requirements, those of polymer NCs are exclusively befitting for strategic applications. A perfect example illustrating this is the use of Kevlar in the tires of US President's official vehicle, owing to its remarkable strength and toughness. Similarly, CNTs and fullerenes have been used to make the air bags of Toyota's automobiles, enabling quicker stimulus-driven expansion and contraction. Remarkable strength and mechanical properties of NCs have propelled them as rightful candidates toward making stronger ballistic support, missiles, roads, and several other automotive sector products. The clutches of two-wheelers, window-locking system of four-wheelers, and sensing-based devices (such as barcodes, nanoscalpels, nanotweezers, and nanofibers) have witnessed the advances of nanomaterial wonders. These gadgets have not only reduced the expenditure on raw materials and care with respect to safeguarding from wear-and-tear damages but are also better in monitoring, quicker in response, and much easier to preserve. The crucial factor of mechanical strength is known to impair the working life of several materials; in this context, NCs have emerged as highly suitable replacements through improved stress response manifested by the included nanomaterials. Not limited to industrial and strategic domains, the versatility of NCs has blessed the biomedical diagnostics with a much-needed boost. Implants of nanomaterials combined with biomaterials (termed bionanocomposite) are being used in the corrective replacement of dislocated and damaged bones. NCs have been proposed in surface inactive forms, wherein the delivery of drugs has become much controlled and precise, ensuring the timing and quantity of delivered drug in line with the requirement. The quest for bettering the existing or proven attributes still continues, although it is a big task to facilitate an efficient propagation of laboratory science and fundamentals to the normal lives of citizens. Only an integrated approach with the intent to understand the scientific rationale of various nanomaterial inclusions (with respect to their shape and size dependent activities) can revolutionize the science of NCs and improve the understanding of stronger and more resilient materials design.

7.6 Conclusion

We have presented the scientific rationale of composites as the modern-day improved materials with manifested abilities of improved performance through increased load-bearing capability along with reduced wear and tear. The formative strategies of different grades of composites propel them as robust materials having the potential of delivering as per the requirements and specific applications. The requirement arena comprises from household devices to industrial appliances and instruments along with stronger ballistic materials and more resilient gadgets. Incorporation of nanomaterials (having high aspect ratios and relatively finer constitutional distribution) confers a new realm to performance attributes of composites. One fine illustration of this strategy is the inclusion of multiwalled CNTs in the coal tar pitch matrices using a molding press. The nanomaterials as filler significantly strengthen the intermolecular forces across the entire structure alongside conferring stronger stress- and shock-withstanding abilities. Similarly, the inclusion of different nanoparticles confers stronger catalytic and adsorption performance to the bulk materials. Computational techniques can serve as incentives in this regard, through prediction of distribution and interaction behavior of nanomaterials using relative stability assessment before manufacturing. In short, nanocomposites are and have swiftly emerged as the most amicable solutions to progressively increasing requirements of better and stronger materials, enabling volumes to save energy, raw materials and reduce the intended labor needs.

References

1. McEvoy, M. A. and Correll, N. (2015). Materials that couple sensing, actuation, computation, and communication, *Science*, **347**, pp. 1261689 (1–8).

2. *History of Composite Materials.* Mar-Bal Incorporated. Accessed: Feb. 26, 2020. [Online]. Available at: https://www.mar-bal.com/language/en/applications/history-of-composites/

3. Elhajjar, R. S., Valeria, L., and Muliana, A. (2017). *Smart Composites: Mechanics and Design (Composite Materials)*, 1st ed. (CRC Press).

4. Fakirov, S. (2015). Composite materials: Is the use of proper definitions important? *Materials Today*, **18**, pp. 528–529.

5. Kamigaito, O. (1991). What can be improved by nanometer composites? *Journal of Japan Society of Powder and Powder Metallurgy*, **38**, pp. 315–321 in Kelly, A. (1994). *Concise Encyclopedia of Composites Materials* (Elsevier Ltd.).

6. Madkour, L. H. (2019). Classification of nanostructured materials. In: *Nanoelectronic Materials (Advanced Structured Materials 116)* (Springer, Cham, Switzerland), pp. 269–307.

7. Visakh, P. M. and María, J. M. M. (2016). *Nanomaterials and Nanocomposites: Zero- to Three-Dimensional Materials and Their Composites* (Wiley VCH).

8. Tetiana, A., Dontsova, S. V. N., and Ihor, M. A. (2019). Metal oxide nanomaterials and nanocomposites of ecological purpose, *Journal of Nanomaterials*, **5942194**, pp. 1–31.

9. Njuguna, J., Ansari, F., Sachse, S., Zhu, H., and Rodriguez, V. M. (2014). Nanomaterials, nanofillers, and nanocomposites: Types and properties. In: *Health and Environmental Safety of Nanomaterials* (Woodhead Publishing Ltd.), pp. 3–27.

10. Kanatzidis, M. G., Wu, C. G., Marcy, H. O., and Kannewurf, C. R. (1989). Conductive-polymer bronzes. Intercalated polyaniline in vanadium oxide xerogels, *Journal of the American Chemical Society*, **111**, pp. 4139–4141.

11. Liu, Y. J., DeGroot, D. C., Schindler, J. L., Kannewurf, C. R., and Kanatzidis, M. G. (1991). Intercalation of poly(ethylene oxide) in vanadium pentoxide (V_2O_5) xerogel, *Chemistry of Materials*, **3**, pp. 992–994.

12. Paul, D. R. and Robeson, L. M. (2008). Polymer nanotechnology: Nanocomposites, *Polymer*, **49**, pp. 3187–3204.

13. Liu, Y. J., Schindler, J. L., DeGroot, D. C., Kannewurf, C. R., Hirpo, W., and Kanatzidis, M. G. (1996). Synthesis, structure and reactions of poly (ethylene-oxide)/V_2O_5 intercalative nanocomposites, *Chemistry of Materials*, **8**, pp. 525–534.

14. Lange, F. F. (1973). Effect of microstructure on strength of Si_3N_4-SiC composite system, *Journal of the American Ceramic Society*, **56**, pp. 445–450.

15. Becher, P. F. (1991). Microstructural design of toughened ceramics. *Journal of the American Ceramic Society*, **74**, pp. 255–269.

16. Ghasali, E., Yazdani-rad, R., Asadian, K., and Ebadzadeh, T. (2017). Production of Al-SiC-TiC hybrid composites using pure and 1056 aluminium powders prepared through microwave and conventional heating methods, *Journal of Alloys and Compounds*, **690**, pp. 512–518.

17. Yan, X., Sahimi, M., and Tsotsis, T. T. (2017). Fabrication of high-surface area nanoporous SiOC ceramics using pre-ceramic polymer precursors and a sacrificial template: Precursor effects, *Microporous and Mesoporous Materials*, **241**, pp. 338–345.

18. He, J., Gao, Y., Wang, Y., Fang, J., and An, L. (2017). Synthesis of ZrB_2-SiC nanocomposite powder via polymeric precursor route, *Ceramics International*, **43**, pp. 1602–1607.

19. Brooke, R., Cottis, P., Talemi, P., Fabretto, M., Murphy, P., and Evans, D. (2017). Recent advances in the synthesis of conducting polymers from the vapour phase, *Progress in Materials Science*, **86**, pp. 127–146.

20. Lee, H. S., Choi, M. Y., Anandhan, S., Baik, D. H., and Seo, S. W. (2004). Microphase structure and physical properties of polyurethane/organoclay nanocomposites, *ACS PMSE Preprints*, **91**, 638.

21. Kobayashi, T. (2016). Applied environmental materials science for sustainability. In: *Advances in Environmental Engineering and Green Technologies*, 1st ed. (IGI Global).

22. Kashinath, L., Namratha, K., and Byrappa, K. (2017). Sol-gel assisted hydrothermal synthesis and characterization of hybrid ZnS-RGO nanocomposite for efficient photodegradation of dyes, *Journal of Alloys and Compounds*, **695**, pp. 799–809.

23. Dermenci, K. B., Genc, B., Ebin, B., Olmez-Hanci, T., and Gürmen, S. (2014). Photocatalytic studies of Ag/ZnO nanocomposite particles produced via ultrasonic spray pyrolysis method, *Journal of Alloys and Compounds*, **586**, pp. 267–273.

24. Ren, Q., Su, H., Zhang, J., Ma, W., Yao, B., Liu, L., and Fu, H. (2016). Rapid eutectic growth of $Al_2O_3/Er_3Al_5O_{12}$ nanocomposite prepared by a new

method: Melt falling-drop quenching, *Scripta Materialia*, **125**, pp. 39–43.

25. Ogasawara, T., Ishida, Y., Ishikawa, T., and Yokota, R. (2004). Characterization of multi-walled carbon nanotube/phenylethynyl terminated polyimide composites, *Composites Part A: Applied Science and Manufacturing*, **35**, pp. 67–74.

26. Alexandre, M. and Dubois, P. (2000). Polymer-layered silicate nanocomposites: Preparation, properties and uses of a new class of materials, *Materials Science and Engineering: R: Reports*, **28**, pp. 1–63.

27. Rehab, A. and Salahuddin, N. (2005). Nanocomposite materials based on polyurethane intercalated into montmorillonite clay, *Materials Science and Engineering: A*, **399**, pp. 368–376.

28. Hussain, F., Hojjati, M., Okamoto, M., and Gorga, R. E. (2006). Review article: Polymer-matrix nanocomposites, processing, manufacturing, and application: An overview, *Journal of Composite Materials*, **40**, pp. 1511–1575.

29. Anandhan, S. and Bandyopadhyay, S. (2011). Polymer nanocomposites: From synthesis to applications. In: *Nanocomposites and Polymers with Analytical Methods* (Cuppoletti, J., Ed.) (InTech), pp. 3–28.

30. Haraguchi, K. (2011). Synthesis and properties of soft nanocomposite materials with novel organic/inorganic network structures, *Polymer Journal*, **43**, pp. 223–241.

31. Wang, R. M., Zheng, S. R., and Zheng, Y. (2011). *Polymer Matrix Composites and Technology* (Woodhead Publishing Ltd.).

32. Bai, H. and Ho, W. S. W. (2009). New sulfonated polybenzimidazole (SPBI) copolymer-based proton-exchange membranes for fuel cells, *Journal of the Taiwan Institute of Chemical Engineers*, **40**, pp. 260–267.

33. Bondeson, D., Mathew, A., and Oksman, K. (2006). Optimization of the isolation of nanocrystals from microcrystalline cellulose by acid hydrolysis, *Cellulose*, **13**, pp. 171–180.

34. Saito, T., Kimura, S., Nishiyama, Y., and Isogai, A. (2007). Cellulose nanofibers prepared by tempo-mediated oxidation of native cellulose, *Biomacromolecules*, **8**, pp. 2485–2491.

35. Lahiji, R. R., Xu, X., Reifenberger, R., Raman, A., Rudie, A., and Moon, R. J. (2010). Atomic force microscopy characterization of cellulose nanocrystals, *Langmuir*, **26**, pp. 4480–4488.

36. Gong, G., Mathew, A. P., and Oksman, K. (2011). Strong aqueous gels of cellulose nanofibers and nanowhiskers isolated from softwood flour, *TAPPI Journal*, **10**, pp. 7–14.

37. Petersson, L., Kvien, I., and Oksman, K. (2007). Structure and thermal properties of poly(lactic acid)/cellulose whiskers nanocomposite materials, *Composites Science and Technology*, **67**, pp. 2535–2544.

38. Bondeson, D. and Oksman, K. (2007). Dispersion and characteristics of surfactant modified cellulose whiskers nanocomposites, *Composite Interfaces*, **14**, pp. 617–630.

39. Kvien, I., Tanem, B. S., and Oksman, K. (2005). Characterization of cellulose whiskers and their nanocomposites by atomic force and electron microscopy, *Biomacromolecules*, **6**, pp. 3160–3165.

40. Siqueira, G., Mathew, A. P., and Oksman, K. (2011). Processing of cellulose nanowhiskers/cellulose acetate butyrate nanocomposites using sol-gel process to facilitate dispersion, *Composites Science and Technology*, **71**, pp. 1886–1892.

41. Svagan, A. J., Jensen, P., Dvinskikh, S. V., Furó, I., and Berglund, L. A. (2010). Towards tailored hierarchical structures in cellulose nanocomposite biofoams prepared by freezing/freeze-drying, *Journal of Materials Chemistry*, **20**, pp. 6646–6654.

42. Goetz, L., Mathew, A., Oksman, K., Gatenholm, P., and Ragauskas, A. J. (2009). A novel nanocomposite film prepared from crosslinked cellulosic whiskers, *Carbohydrate Polymers*, **75**, pp. 85–89.

43. Etang Ayuk, J., Mathew, A. P., and Oksman, K. (2009). The effect of plasticizer and cellulose nanowhisker content on the dispersion and properties of cellulose acetate butyrate nanocomposites, *Journal of Applied Polymer Science*, **114**, pp. 2723–2730.

44. Kvien, I. and Oksman, K. (2008). Microscopic examination of cellulose whiskers and their nanocomposites. In: *Characterization of Lignocellulosic Materials* (Blackwell Publishing, Oxford, UK), pp. 342–356.

45. Sawyer, L. C. and Grubb, D. T. (1987). *Polymer Microscopy*, 1st ed. (Chapman and Hall, London).

46. Pei, A., Zhou, Q., and Berglund, L. A. (2010). Functionalized cellulose nanocrystals as biobased nucleation agents in poly(l-lactide) (PLLA)-crystallization and mechanical property effects, *Composites Science and Technology*, **70**, pp. 815–821.

47. Goldstein, J., Newbury, D. E., Joy, D. C., Lyman, C. E., Echlin, P., Lifshin, E., Sawyer, L., and Michael, J. R. (2003). *Scanning Electron Microscopy and X-Ray Microanalysis*, 3rd ed. (Springer, USA).

48. Goodhew, P. J., Humphreys, J., and Beanland, R. (2001). *Electron Microscopy and Analysis* (Taylor & Francis, London).

49. Braga, P. C. and Ricci, D. (2004). *Atomic Force Microscopy: Biomedical Methods and Applications* (*Methods in Molecular Biology 242*) (Humana Press).

50. Eaton, P. and West, P. (2010). *Atomic Force Microscopy* (Oxford University Press, USA).

51. Bandyopadhyay, S., Samudrala, S. K., Bhowmick, A. K., and Gupta, S. K. (2007). Applications of atomic force microscope (AFM) in the field of nanomaterials and nanocomposites. In: *Functional Nanostructures: Processing, Characterization and Applications* (Seal, S., Ed.) (Springer), pp. 504–568.

52. Brown, E. E. and Laborie, M.-P. G. (2007). Bioengineering bacterial cellulose/poly(ethylene oxide) nanocomposites, *Biomacromolecules*, **8**, pp. 3074–3081.

53. Aulin, C., Ahola, S., Josefsson, P., Nishino, T., Hirose, Y., Österberg, M., and Wågberg, L. (2009). Nanoscale cellulose films with different crystallinities and mesostructures: Their surface properties and interaction with water, *Langmuir*, **25**, pp. 7675–7685.

54. Ten, E., Turtle, J., Bahr, D., Jiang, L., and Wolcott, M. (2010). Thermal and mechanical properties of poly(3-hydroxybutyrate-co-3-hydroxyvalerate)/cellulose nanowhiskers composites, *Polymer*, **51**, pp. 2652–2660.

55. Cranston, E. D. and Gray, D. G. (2006). Formation of cellulose-based electrostatic layer-by-layer films in a magnetic field, *Science and Technology of Advanced Materials*, **7**, pp. 319–321.

56. Gindl, W., Emsenhuber, G., Maier, G., and Keckes, J. (2009). Cellulose in never-dried gel oriented by an ac electric field, *Biomacromolecules*, **10**, pp. 1315–1318.

57. Hoeger, I., Rojas, O. J., Efimenko, K., Velev, O. D., and Kelley, S. S. (2011). Ultrathin film coatings of aligned cellulose nanocrystals from a convective-shear assembly system and their surface mechanical properties, *Soft Matter*, **7**, pp. 1957–1967.

58. Thygesen, A., Oddershede, J., Lilholt, H., Thomsen, A. B., and Ståhl, K. (2005). On the determination of crystallinity and cellulose content in plant fibres, *Cellulose*, **12**, pp. 563–576.

59. Sugiyama, J., Vuong, R., and Chanzy, H. (1991). Electron diffraction study on the two crystalline phases occurring in native cellulose from an algal cell wall, *Macromolecules*, **24**, pp. 4168–4175.

60. Lee, Y. H., Lee, S. W., Youn, J. R., Chung, K., and Kang, T. J. (2002). Characterization of fiber orientation in short fiber reinforced

composites with an image processing technique, *Materials Research Innovations*, **6**, pp. 65–72.

61. Yoshiharu, N., Shigenori, K., Masahisa, W., and Takeshi, O. (1997). Cellulose microcrystal film of high uniaxial orientation, *Macromolecules*, **30**, pp. 6395–6397.

62. Bohn, A., Fink, H.-P., Ganster, J., and Pinnow, M. (2000). X-ray texture investigations of bacterial cellulose, *Macromolecular Chemistry and Physics*, **201**, pp. 1913–1921.

63. Gindl, W. and Keckes, J. (2006). Drawing of self-reinforced cellulose films, *Journal of Applied Polymer Science*, **103**, pp. 2703–2708.

64. Wada, M., Sugiyama, J., and Okano, T. (1993). Native celluloses on the basis of two crystalline phase ($I\alpha/I\beta$) system, *Journal of Applied Polymer Science*, **49**, pp. 1491–1496.

65. Iwamoto, S., Isogai, A., and Iwata, T. (2011). Structure and mechanical properties of wet-spun fibers made from natural cellulose nanofibers, *Biomacromolecules*, **12**, pp. 831–836.

66. Bokobza, L. (1990). Investigation of local dynamics of polymer chains in the bulk by the excimer fluorescence technique, *Progress in Polymer Science*, **15**, pp. 337–360.

67. George, G. A. (1985). Characterization of solid polymers by luminescence techniques, *Pure and Applied Chemistry*, **57**, pp. 945–954.

68. Zammarano, M., Maupin, P. H., Sung, L.-P., Gilman, J. W., McCarthy, E. D., Kim, Y. S., and Fox, D. M. (2011). Revealing the interface in polymer nanocomposites, *ACS Nano*, **5**, pp. 3391–3399.

69. Clough, J. M., Creton, C., Craig, S. L., and Sijbesma, R. P. (2016). Covalent bond scission in the Mullins effect of a filled elastomer: Real-time visualization with mechanoluminescence, *Advanced Functional Materials*, **26**, pp. 9063–9074.

70. Dewimille, L., Bresson, B., and Bokobza, L. (2005). Synthesis, structure and morphology of poly(dimethylsiloxane) networks filled with in situ generated silica particles, *Polymer*, **46**, pp. 4135–4143.

71. Cole, K. C. (2008). Use of infrared spectroscopy to characterize clay intercalation and exfoliation in polymer nanocomposites, *Macromolecules*, **41**, pp. 834–843.

72. Zhang, X., Bhuvana, S., and Loo, L. S. (2011). Characterization of layered silicate dispersion in polymer nanocomposites using Fourier transform infrared spectroscopy, *Journal of Applied Polymer Science*, **125**, E175–E180.

73. Dazzi, A. and Prater, C. B. (2016). AFM-IR: Technology and applications in nanoscale infrared spectroscopy and chemical imaging, *Chemical Reviews*, **117**, pp. 5146–5173.

74. Marcott, C., Lo, M., Dillon, E., Kjoller, K., and Prater, C. (2015). Interface analysis of composites using AFM-based nanoscale IR and mechanical spectroscopy, *Microscopy Today*, **23**, pp. 38–45.

75. Peng, H. X. (2011). Polyurethane nanocomposite coatings for aeronautical applications. In: *Multifunctional Polymer Nanocomposites* (Leng, J. and Lau, A. K. T., Eds.) (CRC Press, Boca Raton), pp. 337–387.

76. Mahieux, C. A. (2006). *Environmental Degradation in Industrial Composites* (Elsevier, Chicago).

77. Martin, R. (2008). *Ageing of Composites* (Woodhead Publishing, Cambridge).

78. Bharadwaj, R. K. (2001). Modeling the barrier properties of polymer layered silicate nanocomposites, *Macromolecules*, **34**, pp. 9189–9192.

79. Ray, S. S., Yamada, K., Okamoto, M., and Ueda, K. (2002). Polylactide-layered silicate nanocomposite: A novel biodegradable material, *Nano Letters*, **2**, pp. 1093–1096.

80. Gilman, J. W., Kashiwagi, T., and Lichtenhan, J. D. (1977). Nanocomposites: A revolutionary new flame retardant approach, *SAMPE Journal*, **33**, pp. 40–46.

81. Kojima, Y., Usuki, A., Kawasumi, M., Okada, A., Fukushima, Y., Kurauchi, T., and Kamigaito, O. (1993). Mechanical properties of nylon 6-clay hybrid, *Journal of Material Research*, **8**, pp. 1185–1189.

82. Meenakshi, K. S., Sudhan, E. P. J., Kumar, S. A., and Umapathy, M. J. (2011). Development and characterization of novel DOPO based phosphorus tetraglycidyl epoxy nanocomposites for aerospace applications, *Progress in Organic Coatings*, **72**, pp. 402–409.

83. Park, J. G., Louis, J., Cheng, Q., Bao, J., Smithyman, J., Liang, R., Wang, B., Zhang, C., Brooks, J. S., Kramer, L., Fanchasis, P., and Dorough, D. (2009). Electromagnetic interference shielding properties of carbon nanotube buckypaper composites, *Nanotechnology*, **20**, pp. 1–7.

84. Inam, F., Bhat, B. R., Vo, T., and Daoush, W. M. (2014). Structural health monitoring capabilities in ceramic-carbon nanocomposites, *Ceramics International*, **40**, pp. 3793–3798.

85. Leaversuch, R. (2001). Nanocomposites broaden roles in automotive, barrier packaging, *Plastics Technology*, pp. 47–64.

86. Damm, C., Munstedt, H., and Rosch, A. (2008). The antimicrobial efficacy of polyamide 6/silver-nano- and microcomposites, *Materials Chemistry and Physics*, **108**, pp. 61–66.

87. Barbosa-Pereira, L., Angulo, I., Lagarón, J. M., Paseiro-Losada, P., and Cruz, J. M. (2014). Development of new active packaging films containing bioactive nanocomposites, *Innovative Food Science and Emerging Technologies*, **26**, pp. 310–318.

88. Campos-Requena, V. H., Rivas, B. L., Pérez, M. A., Figueroa, C. R., and Sanfuentes, E. A. (2015). The synergistic antimicrobial effect of carvacrol and thymol in clay/polymer nanocomposite films over strawberry gray mold, *LWT-Food Science and Technology*, **64**, pp. 390–396.

89. Costa, C., Conte, A., Buonocore, G. G., and Del Nobile, M. A. (2011). Antimicrobial silver montmorillonite nanoparticles to prolong the shelf life of fresh fruit salad, *International Journal of Food Microbiology*, **148**, pp. 164–167.

90. Costa, C., Conte, A., Buonocore, G. G., Lavorgna, M., and Del Nobile, M. A. (2012). Calcium-alginate coating loaded with silver-montmorillonite nanoparticles to prolong the shelf-life of fresh-cut carrots, *Food Research International*, **48**, pp. 164–169.

91. Fernández, A., Picouet, P., and Lloret, E. (2010). Cellulose-silver nanoparticle hybrid materials to control spoilage-related microflora in absorbent pads located in trays of fresh-cut melon, *International Journal of Food Microbiology*, **142**, pp. 222–228.

92. Lloret, E., Picouet, P., and Fernández, A. (2012). Matrix effects on the antimicrobial capacity of silver-based nanocomposite absorbing materials, *LWT-Food Science and Technology*, **49**, pp. 333–338.

93. Manikantan, M. R. and Varadharaju, N. (2011). Preparation and properties of polypropylene-based nanocomposite films for food packaging, *Packaging Technology and Science*, **24**, pp. 191–209.

94. Xie, L., Lv, X. Y., Han, Z. J., Ci, J. H., Fang, C. Q., and Ren, P. G. (2012). Preparation and performance of high-barrier low density polyethylene/ organic montmorillonite nanocomposite, *Polymer-Plastics Technology and Engineering*, **51**, pp. 1251–1257.

95. Follain, N., Belbekhouche, S., Bras, J., Siqueira, G., Marais, S., and Dufresne, A. (2013). Water transport properties of bio-nanocomposites reinforced by *Luffa cylindrica* cellulose nanocrystals, *Journal of Membrane Science*, **427**, pp. 218–229.

96. Lagarón, J. M., Cabedo, L., Cava, D., Feijoo, J. L., Gavara, R., and Gimenez, E. (2005). Improving packaged food quality and safety. Part 2: Nanocomposites, *Food Additives and Contaminants*, **22**, pp. 994–998.

97. Arunvisut, S., Phummanee, S., and Somwangthanaroj, A. (2007). Effect of clay on mechanical and gas barrier properties of blown film LDPE/clay nanocomposites, *Journal of Applied Polymer Science*, **106**, pp. 2210–2217.

98. Horst, M. F., Quinzani, L. M., and Failla, M. D. (2012). Rheological and barrier properties of nanocomposites of HDPE and exfoliated montmorillonite, *Journal of Thermoplastic Composite Materials*, **27**, pp. 106–125.

99. Zeng, Q. H., Yu, A. B., Lu, G. Q., and Paul, D. R. (2005). Clay-based polymer nanocomposites: Research and commercial development, *Journal of Nanoscience and Nanotechnology*, **5**, pp. 1574–1592.

100. Anadao, P. (2012). Polymer/clay nanocomposites: Concepts, researches, applications and trends for the future. In: *Nanocomposites: New Trends and Developments* (Ebrahimi, F., Ed.) (InTech), pp. 1–16.

101. Bumbudsanpharoke, N. and Ko, S. (2015). Nano-food packaging: An overview of market, migration research, and safety regulations, *Journal of Food Science*, **80**, R910–R923.

102. Vanderroost, M., Ragaert, P., Devlieghere, F., and Meulenaer, B. D. (2014). Intelligent food packaging: The next generation, *Trends in Food Science and Technology*, **39**, pp. 47–62.

103. Rahaman, M. and Yao, A. (2007). Ceramics for prosthetic hip and knee joint replacement, *Journal of the American Ceramic Society*, **90**, pp. 1965–1988.

104. Menezes, R. R. and Kiminami, R. H. G. A. (2008). Microwave sintering of alumina–zirconia nanocomposites, *Journal of Materials Processing Technology*, **203**, pp. 513–517.

105. Nevarez-Rascon, A., Aguilar-Elguezabal, A., Orrantia, E., and Bocanegra-Bernal, M. H. (2009). On the wide range of mechanical properties of ZTA and ATZ based dental ceramic composites by varying the Al_2O_3 and ZrO_2 content, *The International Journal of Refractory Metals and Hard Materials*, **27**, pp. 962–970.

106. De Aza, A. H., Chevalier, J., Fantozzi, G., Schehl, M., and Torrecillas, R. (2002). Crack growth resistance of alumina, zirconia and zirconia toughened alumina ceramics for joint prosthesis, *Biomaterials*, **23**, pp. 937–945.

107. Nawa, M., Nakamoto, S., Sekino, T., and Niihara, K. (1998). Tough and strong Ce-TZP/alumina nanocomposites doped with titania, *Ceramics International*, **24**, pp. 497–506.

108. Uchida, M., Kim, H. M., Kokubo, T., Nawa, M., Asano, T., Tanaka, K., and Nakamura, T. (2002). Apatite forming ability of a zirconia/alumina nano-composite induced by chemicals treatment, *Journal of Biomedical Materials Research*, **60**, pp. 277–282.

109. Tanaka, K., Tamura, J., Kawanabe, K., Nawa, M., Oka, M., Uchida, M., Kokubo, T., and Nakamura, T. (2002). Ce-TZP/Al$_2$O$_3$ nanocomposites as a bearing material in total joint replacement, *Journal of Biomedical Materials Research*, **63**, pp. 262–270.

110. Philipp, A., Fischer, J., Hammerle, C. H., and Sailer, I. (2010). Novel ceria-stabilized tetragonal zirconia/alumina nanocomposite as framework material for posterior fixed dental prosthesis: Preliminary results of a prospective case series at 1 year of function, *Quintessence International*, **41**, pp. 313–319.

111. Kong, Y. M., Bae, C. J., Lee, S. H., Kim, H. W., and Kim, H. E. (2005). Improvement in biocompatibility of ZrO$_2$-Al$_2$O$_3$ nano-composite by addition of HA, *Biomaterials*, **26**, pp. 509–517.

112. Vallet-Regi, M. (2010). Evolution of bioceramics within the field of biomaterials, *Comptes Rendus Chimie*, **13**, pp. 174–185.

113. Paul, W. and Sharma, C. P. (2006). Nanoceramic matrices: Biomedical applications, *American Journal of Biochemistry and Biotechnology*, **2**, pp. 41–48.

114. Rho, J. Y., Kuhn-Spearing, L., and Zioupos, P. (1998). Mechanical properties and the hierarchical structure of bone, *Medical Engineering and Physics*, **20**, pp. 92–112.

115. Wel, G. and Ma, P. X. (2004). Structure and properties of nano-hydroxyapatite/polymer composite scaffolds for bone tissue engineering, *Biomaterials*, **25**, pp. 4749–4757.

116. Wang, Y., Liu, L., and Guo, S. (2010). Characterization of biodegradable and cytocompatible nano-hydroxyapatite/polycaprolactone porous scaffolds in degradation in vitro, *Polymer Degradation and Stability*, **95**, pp. 207–213.

117. Nejati, E., Firouzdor, V., Eslaminejad, M. B., and Bagheri, F. (2009). Needle-like nano hydroxyapatite/poly(L-lactide acid) composite scaffold for bone tissue engineering application, *Materials Science and Engineering C*, **29**, pp. 942–949.

118. Hong, Z., Reis, R. L., and Mano, J. F. (2008). Preparation and in vitro characterization of scaffolds of poly(L-lactic acid) containing bioactive glass ceramic nanoparticles, *Acta Biomaterials*, **4**, pp. 1297–1306.

119. Rajan, M., Murugan, M., Ponnamma, D., Sadasivuni, K. K., and Munusamy, M. A. (2016). Polycarboxylic acids functionalized chitosan nanocarriers for controlled and targeted anticancer drug delivery, *Biomedicine and Pharmacotherapy*, **83**, pp. 201–211.

120. Dhanavel, S., Nivethaa, E. A. K., Narayanan, V., and Stephen, A. (2017). In vitro cytotoxicity study of dual drug loaded chitosan/palladium nanocomposite towards HT-29 cancer cells, *Material Science and Engineering C*, **75**, pp. 1399–1410.

121. Rajan, M., Raj, V., Al-Arfaj, A. A., and Murugan, A. M. (2013). Hyaluronidase enzyme core-5-fluorouracil-loaded chitosan-PEG-gelatin polymer nanocomposites as targeted and controlled drug delivery vehicles, *International Journal of Pharmaceutics*, **453**, pp. 514–522.

122. Rasoulzadeh, M. and Namazi, H. (2017). Carboxymethyl cellulose/graphene oxide bionanocomposite hydrogel beads as anticancer drug carrier agent, *Carbohydrate Polymers*, **168**, pp. 320–326.

123. Reddy, N. N., Varaprasad, K., Ravindra, S., Reddy, G. S., Reddy, K. M. S., Reddy, K. M., and Mohana Raju, K. (2011). Evaluation of blood compatibility and drug release studies of gelatin based magnetic hydrogel nanocomposites, *Colloids and Surfaces A*, **385**, pp. 20–23.

124. Dhivya, R., Ranjani, J., Bowen, P. K., Rajendhran, J., Mayandi, J., and Annaraj, J. (2017). Biocompatible curcumin loaded PMMA-PEG/ZnO nanocomposite induce apoptosis and cytotoxicity in human gastric cancer cells, *Materials Science and Engineering C*, **80**, pp. 59–68.

125. Zheng, X. T., Ma, X. Q., and Li, C. M. (2016). Highly efficient nuclear delivery of anti-cancer drugs using a bio-functionalized reduced graphene oxide, *Journal of Colloid and Interface Science*, **467**, pp. 35–42.

126. Rao, K. M., Nagappan, S., Seo, D. J., and Ha, C. S. (2014). pH sensitive halloysite-sodium hyaluronate/poly (hydroxyethyl methacrylate) nanocomposites for colon cancer drug delivery, *Applied Clay Science*, **97**, pp. 33–42.

127. Han, S., Liu, F., Wu, J., Zhang, Y., Xie, Y., and Wu, W. (2014). Targeting of fluorescent palygorskite polyethyleneimine nanocomposite to cancer cells, *Applied Clay Science*, **101**, pp. 567–573.

128. Baibarac, M. and Gomez-Romero, P. (2006). Nanocomposites based on conducting polymers and carbon nanotubes: From fancy materials to functional applications, *Journal of Nanoscience and Nanotechnology*, **6**, pp. 1–14.

129. Sandler, J. K. W., Kirk, J. E., Kinloch, I. A., Shaffer, M. S. P., and Windle, A. H. (2003). Ultra-low electrical percolation threshold in carbon-nanotube-epoxy composites, *Polymer*, **44**, pp. 5893–5899.

130. Kothurkar, N. K. (2004). Solid state, transparent, cadmium sulfide-polymer nanocomposites, University of Florida.

131. Omar, G., Salim, M. A., Mizah, B. R., Kamarolzaman, A. A., and Nadlene, R. (2019). Electronic applications of functionalized graphene nanocomposites. In: *Functionalized Graphene Nanocomposites and Their Derivatives*, 1st ed (Elsevier), pp. 245–263.

132. Wu, Z., Chen, X., Zhu, S., Zhou, Z., Yao, Y., Quan, W., and Liu, B. (2013). Enhanced sensitivity of ammonia sensor using graphene/polyaniline nanocomposite. *Sensors and Actuators B Chemical*, **178**, pp. 485–493.

133. Al-Mashat, L., Shin, K., Kalantar-Zadeh, K., Plessis, J., Han, S., Kojima, R., Kaner, R., Li, D., Gou, X., Ippolito, S., and Wlodarski, W. (2010). Graphene/polyaniline nanocomposite for hydrogen sensing, *The Journal of Physical Chemistry C*, **114**, pp. 16168–16173.

134. Shi, X., Nguyen, T. A., Suo, Z., Liu, Y., and Avci, R. (2009). Effect of nanoparticles on the anticorrosion and mechanical properties of epoxy coating, *Surface & Coatings Technology*, **204**, pp. 237–245.

135. Introzzi, L., Blomfeldt, T. O., Trabattoni, S., Tavazzi, S., Santo, N., Schiraldi, A., Piergiovanni, L., and Farris, S. (2012). Ultrasound-assisted pullulan/montmorillonite bionanocomposite coating with high oxygen barrier properties, *Langmuir*, **28**, pp. 11206–11214.

136. Habouti, S., Kunstmann-Olsen, C., Hoyland, J. D., Rubahn, H. G., and Es-Souni, M. (2014). In situ ZnO-PVA nanocomposite coated microfluidic chips for biosensing, *Applied Physics A: Materials Science and Processing*, **115**, pp. 645–649.

137. Liu C. and Zhao, Q. (2011). Influence of surface-energy components of Ni-P-TiO_2-PTFE nanocomposite coatings on bacterial adhesion, *Langmuir*, **27**, pp. 9512–9519.

138. Toor, A., So, H., and Pisano, A. P. (2017). Improved dielectric properties of polyvinylidene fluoride nanocomposite embedded with poly(vinylpyrrolidone)-coated gold nanoparticles, *ACS Applied Materials & Interfaces*, **9**, pp. 6369–6375.

139. Zhou, Y., Zhang, H., and Qian, B. (2007). Friction and wear properties of the co-deposited Ni-SiC nanocomposite coating, *Applied Surface Science*, **253**, pp. 8335–8339.

Multiple Choice Questions

1. Al alloys for engine/automobile parts are reinforced to increase their
   - (a) Strength
   - (b) Wear resistance
   - (c) Elastic modulus
   - (d) Density

2. Composite materials are classified based on
   - (a) Type of matrix
   - (b) Size and shape of reinforcement
   - (c) Both (a) and (b)
   - (d) None

3. Nanocomposites have barrier properties for
   - (a) Oxygen
   - (b) Water vapor
   - (c) Ethylene
   - (d) All of these

4. Which type of nanocomposites are mostly used for orthopedic implants?
   - (a) Polymer nanocomposite
   - (b) Graphene nanocomposite
   - (c) Ceramic nanocomposite
   - (d) None of these

5. The major load carrier in dispersion-strengthened composites is
   - (a) Matrix
   - (b) Fiber
   - (c) Combination of (a) and (b)
   - (d) Cannot be defined

6. Which co-drugs are loaded together on chitosan/palladium-5% nanocomposite by Dhanavel et al.?
   - (a) 5-Fluorouracil/graphene
   - (b) ZnO/curcumin
   - (c) 5-Fluorouracil/curcumin
   - (d) None of these

7. The factor rendering ceramics unsuitable for load bearing is
   - (a) Brittleness
   - (b) Instability
   - (c) Density
   - (d) All of these

8. Thermodynamically, formation of nanocomposites is
   - (a) Endothermic
   - (b) Exothermic
   - (c) Both (a) and (b)
   - (d) None of these

9. The way to determine colloidal stability using calorimetry in a nanoparticle system is

    (a) UV/Vis spectroscopy

    (b) FT-IR spectroscopy

    (c) Electrokinetic measurements

    (d) None of these

10. The technique used for ascertaining CNT crystallization is

    (a) Raman spectroscopy

    (b) Non-polarized optical microscopy

    (c) Solid-state NMR spectroscopy

    (d) Polarized optical microscopy Raman spectroscopy

11. The spatial resolution in IR limits imposed by diffraction is lowered by

    (a) Combination of AFM and IR techniques

    (b) Combination of SEM and IR techniques

    (c) Combination of TEM and IR techniques

    (d) Combination of XRD and IR techniques

12. The interfacial dispersion in polymer nanocomposites is investigated by

    (a) Förster resonance energy transfer (FRET)

    (b) Scanning electron microscopy (SEM)

    (c) Fourier-transform infrared spectroscopy (FT-IR)

    (d) Fluorescence quenching measurement

13. The nanocomposites having maximum interactions among the constituent phases are

    (a) Polymer matrix composites   (b) Layered nanocomposites

    (c) Lamellar nanocomposites   (d) Nanoclays

14. Which of the following is the most used method for making CMNC?

    (a) Polymer precursor method   (b) Powder procedure

    (c) Ball milling   (d) Sol–gel process

15. The properties of metal matrix nanocomposites making them useful for aerospace and automotive industries are

    (a) High ductility

    (b) Low brittleness

    (c) High toughness and strength

    (d) High toughness, strength, and Young's modulus

16. Which of the following is not a type of nanocomposites?

    (a) Ceramic matrix nanocomposites

    (b) Linear matrix nanocomposites

    (c) Metal matrix nanocomposites

    (d) Polymer matrix nanocomposites

17. Which of the following is a type of nanocomposite?

    (a) Ceramic matrix nanocomposites

    (b) Metal matrix nanocomposites

    (c) Polymer matrix nanocomposites

    (d) All of the above

18. Which of the following is the processing method of ceramic matrix nanocomposites?

    (a) Spray pyrolysis   (b) Polymer precursor

    (c) Melt interaction   (d) Ultrasonication

19. Which of the following is the processing method of metal matrix nanocomposites?

    (a) Polymer precursor   (b) Melt interaction

    (c) Powder procedure   (d) Spray pyrolysis

20. Which of the following is the processing method of polymer matrix nanocomposites?

    (a) Sol–gel   (b) Melt interaction

    (c) Ultrasonication   (d) Polymer precursor

21. _____ is the type of spectroscopic characterization.

    (a) Impedance

    (b) Nuclear magnetic resonance

    (c) Scanning electron microscopy

    (d) None of the above

22. Which one of the following is a commercial-grade nanocomposite used for coating juice bottles?

    (a) Imperm®   (b) Biomaster®

    (c) Durethan® KU2-2601   (d) Agion®

23. Imperm®, a commercial-grade nanocomposite, is used for _____

    (a) Beer and carbonated drink bottles

    (b) Coating juice bottles

    (c) Beer bottles

    (d) None of the above

24. Advantage/s of nanocomposites in food packaging is/are _____

    (a) Enhanced conductivity, mechanical strength, and stability

    (b) Gas barrier properties

    (c) Lighter and biodegradable

    (d) All of the above

25. Pullulan-based nanocomposites are used for _____

    (a) Anticorrosion   (b) Biosensor

    (c) Oxygen barrier   (d) Dielectric

Answer Key

1. (c)   2. (c)   3. (d)   4. (c)   5. (a)   6. (c)   7. (a)
8. (b)   9. (c)   10. (d)   11. (a)   12. (a)   13. (c)   14. (d)
15. (d)   16. (b)   17. (d)   18. (b)   19. (d)   20. (b)   21. (b)
22. (c)   23. (a)   24. (d)   25. (c)

Short Answer Questions

1. What are the limitations for composite materials?
2. What should be the major characteristics of nanocomposites used in the aerospace industry?
3. Can ceramic nanocomposites be considered a special kind of polymer nanocomposites? If not, then please cite suitable justification(s).
4. Illustrate the mechanism of nanocomposite fabrication with improved inflammable prospects.
5. How the requirements of nanocomposites for biomedical purpose are distinct from those of aerospace?
6. How CNTs and fullerenes can be utilized to improve the performance of nanocomposites intended for drug delivery?
7. Outline the salient attributes of zirconia–alumina nanocomposites making them favored for designing orthopedic implants.
8. What is the basic difference between Young's modulus and elastic modulus? Cite suitable application-oriented examples.
9. How nanocomposites can be used to prolong the food material shelf life and natural essence?

Long Answer Questions

1. What are the applications for nanocomposites?

2. Differentiate between polymer and ceramic nanocomposites with respect to their applications?

3. Explain the application-specific method selection for polymer nanocomposites.

4. Elucidate the significance of atomic force microscopy over other microscopic techniques for nanocomposite development.

5. Explain the salient attributes of nanocomposites to be used for tissue engineering keeping the risk of in vivo toxicity and cross-reaction in mind?

Chapter 8

Nanobiosensors and Their Applications

**Ankit Kumar Singh, Agnidipta Das, and
Pradeep Kumar**
*Department of Pharmaceutical Sciences and Natural Products,
Central University of Punjab, Bathinda 151001, India*
pradeepyadav27@gmail.com

8.1   Introduction

A biosensor is an analytical sensor system related to a biological sensor that converts a biological response into an electrical signal through a transducer for assessing the concentration of substances and other biologically relevant parameters. By using biological interactions, these parameters are used for the estimation of a material and then assessing these interactions into a readable form by using transducer and electromechanical interpretation. A biosensor is also defined as "a chemical sensing device in which a biologically derived recognition is coupled to a transducer, to allow the quantitative development of some complex biochemical parameter." The main function of a biosensor is to detect biologically active materials such as proteins, antibodies, enzymes, organelles, and immunological molecules biochemically. A bioreceptor is a

Nanotechnology: Principles and Applications
Edited by Rakesh K. Sindhu, Mansi Chitkara, and Inderjeet Singh Sandhu
Copyright © 2021 Jenny Stanford Publishing Pte. Ltd.
ISBN 978-981-4877-43-5 (Hardcover), 978-1-003-12026-1 (eBook)
www.jennystanford.com

component of a biosensor used for biologically sensitive material, which acts as a template for the material to be detected [1]. Auxiliary enzymes or co-reactants co-immobilized with the analyte are the second-generation biosensors, which are used to improve the analytical quality and to simplify the performance. Enzyme-linked immunosorbent assay (ELISA) is also a second-generation biosensor. Third-generation biosensors such as biomolecules get involved in biosensing materials such as surface plasmon resonance (SPR) biosensors. Lastly, the micro-, nano-, or bio-nanoelectromechanical systems are the fourth-generation biosensors [2].

The history of biosensors dates back to as early as 1906 when M. Cremer demonstrated in a glass membrane. Between 1909 and 1922, Griffin and Nelson first demonstrated immobilization of the enzyme invertase on aluminum hydroxide and charcoal. The first true biosensor was discovered by Leland C. Clark, who he is known as the "father of biosensors" in 1956 for oxygen detection. The first working nanosensor was designed in 1999 by the Georgia Institute of Technology [3]. Nanobiosensors are sensors made up of nanomaterials. Nanomaterials are a unique gift of nanotechnology to humankind for various prospective purposes. Shape, size, surface characteristics, and inner structure are important parameters of nanomaterials. The size restrictions of nanomaterials make them complex, since their constituent atoms are located near the surface. Physicochemical properties differ greatly from the same materials on a large scale, and they have unique physical and chemical properties. In nanobiosensors technology, they play a very efficient role in the sensing mechanism. Nanobiosensors measure a biochemical or biological event by using any electronic, optical, or magnetic probe. These are hybrid biosensors that are produced by utilizing latest nanotechnological advancements and advanced electronic technologies such as fabrication technology.

Nanoelectromechanical systems (NEMS) are devices that integrate nanomaterials with electrical and mechanical systems, which are active in their electrical transduction mechanisms. Nanotubes, nanowires, nanorods, nanoparticles, and thin films made up of nanocrystalline matter are some of the widely used nanomaterials [4]. Nanobiosensors are a revolution in the sensor technology enabling the rapid analysis of multiple samples at desired time and place. Nanobiosensors show high performance in selectivity, biocompatibility, nontoxicity, reversibility, rapid response, and the sensitivity of determination by utilizing nanomaterials to introduce lots of brand new signal transduction technologies, which have been used recently [3].

8.2 Principle and Working

The working principle of nanobiosensors based on the binding of biologically sensitive element with nanostructured transducer, which modulates the physiochemical binding signal. Obtained biological response is measured in the form of electronic response. The working principle of a biosensor is depicted in Fig. 8.1.

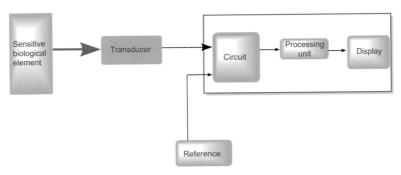

Figure 8.1   Working principle of a biosensor.

A biosensor is mainly divided into the following three sections:

1. **Sensor or bioreceptor:** It is biochemical in nature and generated by the specific interactions between bioanalytes and bioreceptors. It is an immobilized sensitive biological material (e.g., enzyme, antibody tissue, DNA probe, microorganisms, organelles, enzymes, antibodies, nucleic acids, cell receptors, etc.). Although antibodies and oligonucleotides are widely employed, enzymes are by far the most commonly used biosensing elements in biosensors. The desired biological material (usually a specific enzyme) is immobilized by conventional methods. This immobilized biological material is in intimate contact with the transducer.

2. **Transducer:** A transducer is electrical in nature. It is used to convert a biochemical signal, which is a nonelectronic signal, from the reaction of analyte with the bioreceptor into an

electronic signal. A transducer can convert product-linked changes into electrical signals, which can be amplified and measured. For the development of biosensors, electrochemical transducers are often used because these systems offer some advantages such as low cost, simple design, or small dimensions. The electrical signals from the transducer have high-frequency signal components of an apparently random nature.

3. **Detector:** A detector system is associated with electronics, which comprises a signal-conditioning circuit (amplifier), processor, and display unit. It receives the electrical signal from the transducer component and amplifies it adequately to allow proper reading and study of the corresponding response. In addition, it may be based on gravimetric, calorimetric, or optical detection [5].

The detection techniques in biosensor technology are basically of the following three types:

1. **Electrochemical and electrical detection:** Such biosensors mainly base the measurement of binding event on currents or voltages shift. The principle of electrical biosensors is related to electroanalytical techniques such as voltammetry, amperometry, coulometry, and impedance measurement.

2. **Optical detection:** The optical detection technique is related to chemiluminescence and bioluminescence. The term chemiluminescence is related to the generation of energy in the form of light or produces an electronically excited species, which emits a photon in order to reach the ground state during the reaction of compounds such as peroxide (which emits energy during a chemical reaction). The bioluminescence technique is related to the production of light by a living organism, which has been reported using luciferin in the synthetic compound. For the measurement of surface activity, the SPR technique is used. The SPR technique measures the adsorption of material on a planar metal surface.

3. **Mass-based detection:** In biochemical entities, mass-based detection is a mechanical detection technique. It is achieved by the use of micro-scaled or nano-scaled cantilever sensors. They are tiny plates, typically 0.2–1 μm thick, 20–100 μm wide, and 100–500 μm long. The sensitivity of a cantilever sensor

is superior to that of traditional quartz crystal microbalance (QCM) and surface acoustic wave (SAW) transducers. When the cantilever is excited, it starts to resonate at its resonant frequency, which is compared with biomolecules attached to its surface. The difference between the two resonant frequencies determines the mass change on the surface. Their sensing application is used in the detection of temperature, heat changes, humidity infrared light and UV radiation, and blood glucose monitoring [6].

8.3 Nanobiosensors: Types and Functionalization

Nanobiosensors are made from nanomaterials such as nanoparticles, nanotubes, quantum dots, or other biological nanomaterials. In the recent years, a wide variety of nanoparticles are available with different properties such as small size, high speed, smaller distance for electron to travel, lower power, and lower voltages. They are classified on the basis of nanomaterial-based biosensors [6]. The classification of nanobiosensors is depicted in Fig. 8.2.

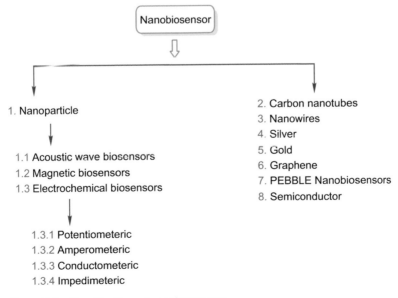

Figure 8.2 Classification of nanobiosensors.

8.3.1 Nanoparticle-Based Biosensors

A nanoparticle is a particle of size from 1 to <100 nm. Nanoparticles exhibit several unique properties that can be applied to develop chemical sensors and biosensors possessing desirable features such as enhanced sensitivity and lower detection limits. Novel properties that distinguish nanoparticles from the bulk material typically originate from their vast surface area showing a dominance of the surface properties over bulk properties. Nanoparticle-based biosensors can be divided into three subtypes: acoustic wave biosensors, magnetic biosensors, and electrochemical biosensors [7].

8.3.1.1 Acoustic wave biosensors

Acoustic wave sensors measure the changes in acoustic wave or mechanical waves as a detection mechanism to obtain medical, biochemical, and biophysical information about the analyte of interest. Acoustic wave sensors have the potential of being extremely sensitive, since they can detect both mechanical and electrical property changes that include changes in mass, elasticity, viscosity, dielectric (optical) properties, and conductivity (electronic, ionic, and thermal). This is because the acoustic wave that probes the medium of interest has both mechanical displacements and an electric field [8]. Acoustic wave biosensors have improved sensitivity, accuracy and are more precise in their detection [9]. The classification of acoustic waves nanobiosensors is depicted in Fig. 8.3.

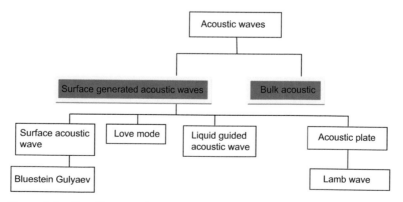

Figure 8.3   Classification of acoustic waves nanobiosensors.

Acoustic waves are mainly two types: bulk acoustic waves (BAWs) and surface-generated acoustic waves (SGAWs). The BAWs travel in the bulk of the material and only interact with the external environment at the opposite surfaces of the material, and SGAWs travel directly along or near the surface. SAWs are more appropriate for sensing applications [8]. An SAW is depicted in Fig. 8.4.

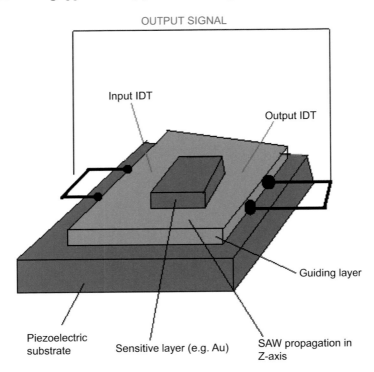

Figure 8.4 Surface accosting wave biosensors.

8.3.1.2 Magnetic biosensors

Magnetic particles are used in biosensors that bind to target molecules in a biological assay. Magnetic nanoparticles in a biosensor system provide promising alternatives to fluorescent labels. Nano-sized magnetic nanoparticles demonstrate super paramagnetic activity due to different magnetic behaviors relative to their bulk material because of the reduced number of magnetic domains. Magnetic nanoparticles concentrate the analyte before detection so that the receptor unit can be easily mixed with the analyte solution and interacted specifically with the target. The agglomerated nanoparticles can also be separated after applying an external magnetic field. For *Escherichia coli* detection, an ultra-high sensitive magneto-resistant biosensor was developed [10]. For the measurement of weak signals, superconducting quantum interference devices (SQUIDs) have been used [11]. These SQUIDs are mostly ferrite-based materials, used either individually or in combined form. These biosensors are very useful in biomedical applications. These have been used for a variety of testing purposes that demand extreme sensitivity, including engineering, medical, and geological equipment because they measure changes in a magnetic field with such sensitivity. The magnetic materials are appropriate for several analytical applications because the magnetic compounds constituted of iron coupled with other transition metals have different properties. With the incorporation of magnetic nanoparticles, the conventionally used biodetection devices have further become more sensitive and powerful [12].

8.3.1.3 Electrochemical biosensors

Normally, electrochemical biosensors are based on the enzymatic catalysis of a reaction that generates or consumes electrons. Electrochemical biosensors contain three electrodes: reference electrode, working electrode, and counter electrode.

Electrochemical nanosensors consist of sensor molecules that facilitate or analyze the biochemical reactions, medical diagnosis, and biological monitoring of diseases with the help of improved electrical means, which are physically adhered to the probe surface. A strong and specific interaction of the probe with the target analyte initiates a measurable electrochemical signal. These devices are mostly based on metallic nanoparticles [13]. Electrochemical biosensors are highly sensitivity and capable of unique identification of the response. The measurement of electrical properties from biological systems is usually electrochemical in nature, whereby a bioelectrochemical component works as the main transduction element. Electrochemical detection techniques are predominantly used in enzymes. This is mainly due to biocatalytic activity and specific binding capabilities [14]. Metallic nanoparticles help in achieving immobilization of the reactants; hence chemical reactions between the biomolecules can be easily carried out using them. Electrochemical biosensors combine the sensitivity of electroanalytical methods with the inherent bioselectivity of the biological component. Conjugated gold nanoparticles have been used in the designing of electrochemical biosensors for the identification of glucose, xanthine, and hydrogen peroxide [15]. Electrochemical biosensors have been in use for almost 50 years and seem to possess great promise for the future [16]. A nanomaterial-based electrochemical biosensor and its parts are depicted in Fig. 8.5.

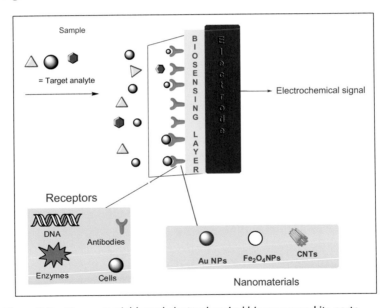

Figure 8.5 Nanomaterial-based electrochemical biosensor and its parts.

Electrochemical biosensors can further classified into four types depending on their working type: potentiometric, amperometeric, conductimetric, and impedimetric.

Potentiometric electrochemical biosensors: Garry Rechnitz developed the potentiometric electrodes [17]. Potentiometric sensors can be used to determine the analyte concentration by measuring the potential difference between the working and reference electrodes at different analyte concentrations under the conditions of no current flow. This potential difference is used to determine the quantity of analyte. An ion-selective electrode and pH meter are examples of potentiometric electrochemical biosensors. Potentiometric sensors have been developed for pathogen detection [18]. A potentiometric electrochemical biosensor is depicted in Fig. 8.6.

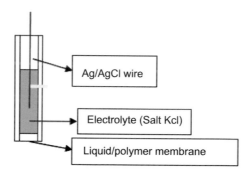

Figure 8.6 Potentiometric electrochemical biosensors.

Amperometeric electrochemical biosensors: The principle of an amperometeric transducer sensor is based on measuring the direct current generated by the enzymatic or bioaffinity reaction at the electrode surface, at a constant working potential with respect to the reference electrode. This type of sensor is one of the promising biosensing tools for routine analysis of toxins because it shows simplicity in fabrication, electrode disposability, and additional selectivity since the redox potential used in analysis is typically characteristic of the analytes [19]. The activity of recognition element varies before and after interaction with a target molecule. The product must be electroactive and undergo a redox process. The current is the rate of electrons transferred and is proportional to the analyte concentration [20].

Conductimetric electrochemical biosensors: Conductimetric methods are fundamentally nonselective. Only with the advent of modified selectivity surface and much improved instrumentation, these have become more feasible methods for sensor designing. Conductimetric methods are attractive because they have low cost and simplicity and no reference electrodes are needed. Improved instrumentation has contributed to the rapid and easy determination of analytes based only on the measurement of conductivity. The most predominant materials used in these sensors will be examined first. They are used mostly as gas sensors due to their conductivity changes following surface chemisorption [21]. The conductimetric measuring method can be used in enzyme catalysis to determine substance concentration and enzyme activity [22].

Impedimetric electrochemical biosensors: Impedimetric sensors mainly perform two types of measurements: electrical impedance (EI) and electrochemical impedance (ECI). EI is a two-electrode system—the working electrode (WE) and the counter electrode (CE)—and performs measurement by using a lock-in amplifier and function generator or frequency response analyzer (FRA) with a pair of electrodes, which may be of equal or different areas. ECI is a three-electrode system, in which measurement is performed using similar instruments along with a frontend potentiostat [23]. Impedimetric biosensors have been used successfully for the growth monitoring of microorganisms due to the production of conductive metabolites. Potentiometric and amperometeric biosensors are more frequently used than impedimetric biosensors. An impedance assay is used to monitor the hybridization of DNA fragments, which are intensified by polymerase chain reaction. An impedance biosensor is used for determining the ethanol level in some alcoholic beverages by using immobilized yeast. Assays of the pathogenic fungus *Ichthyophonus hofery* and *Erwinia carotovora* rot Malthus 2000 (Malthus Instruments, Crawley, UK) were also carried out by using an impedance-based commercial device. False positive tests are the biggest drawback of impedance biosensors due to electrolytes from the sample [24].

8.3.2 Nanotube-Based Sensors (Carbon Nanotubes)

Carbon nanotubes (CNTs), also called buck tubes, were first discovered by Japanese electron microscopist Sumio Iijima in 1991. Depending on their number of walls, CNTs are designated as single-walled nanotubes (SWNTs) or multi-walled nanotubes (MWNTs) [25].

For ultrasensitive and ultrafast biosensing systems, CNT biosensors are recognized as building blocker [26]. CNTs have been used as fabrication materials for biosensors. The functionalization of CNTs allows the tubes to be biocompatible and soluble, since any desired chemical species can be easily bonded to them. CNTs can be employed as an electrode as well as a transducer material in biosensors. Single-walled carbon nanotubes (SWCNTs) exhibit extreme sensitivity to their surrounding environment. Pristine CNTs were formed in the form of chemical-sensitive field-effect transistors (FETs) and were capable of detecting biomolecules [27]. CNTs are used for enhancing the electrical signal detection due to smoother electron transfer flow characteristics [28]. Functions of CNTs are depicted in Fig. 8.7.

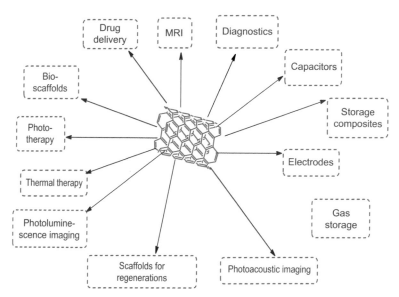

Figure 8.7 Functions of carbon nanotubes.

8.3.2.1 Functionalization of CNTs

Generally CNTs are exceptionally resistant to wetting. So they hardly disperse or dissolve in water and in organic media, but with the functionalization, their solubility, dispersive ability, manipulation, and processability are enhanced [29].

Chemical functionalization: Chemical functionalization can be described depending on the covalent bonding between functional moieties and carbon scaffold of CNTs at the tube ends or sidewalls. This sidewall functionalization is due to the alteration of sp^2 hybridization state to sp^3 hybridization and a concurrent p-conjugation system loss on graphene layer. A highly reactive chemical entity such as fluorine greatly helps to achieve this type of functionalization, and the fluorine can be removed by applying anhydrous hydrazine. The strength of C–F bonds at the fluorinated sites of CNTs is lower than the bonds of alkyl fluorides. Thereby, these sites proved to be efficient for further substitutions of functional groups such as alkyl, amino, and hydroxyl groups for functionalization. Along with fluorination of sidewalls, cycloaddition reactions such as Diels–Alder reaction, chlorination, bromination, carbene and nitrene addition are also useful tools of chemical functionalization [30]. Another method of functionalization of CNT is immobilization of proteins at the target site of nanotubes only at their end using nanocrystals of Au to protect them as mask at the sidewalls and the immobilized proteins absorb on the surface of Au nanocrystals [31].

Physical functionalization: Physical functionalization can fix valuable functional groups on the CNT surface through the covalent method, but this method of functionalization has many limitations. Physical functionalization can also be obtained via the noncovalent method also through polymer wrapping process, achieved by π–π stacking and van der Waals interactions between aromatic-ring-containing polymer chains and CNTs. In this process, a CNT suspension is formed and polymers such as polystyrene or poly(phenylene vinylene) are wrapped around the CNTs providing a CNT-super molecular complex [30].

8.3.3 Nanowire-Based Sensors

The nanowire made from nanostructure is best suited for biomedical sensors. They have high sensitivity, uniformity, reproducibility, and scalability with relatively simple fabrication process. Nanowire FETs are a type of nanowire sensors that have originated from the standard planar FETs, which consist of a gate, source, drain, and the body.

Silicon and silica nanowires are most widely used due to their high compatibility with the standard complementary metal-oxide semiconductor (CMOS) technology. The biomolecules have charges; the accumulation of biomolecules on the isolation layer does not alter the conductivity of nanowires since the original silicon and silica surface is not sensitive to biomolecules. Therefore, it is necessary to make the surface functionalized and bound by receptors for sensing the specific charged species (e.g., DNA, RNA, viruses, etc.) by functionalizing the nanowire FETs [32]. Nanowires are one-dimensional cylindrical nanostructure like CNTs, having lengths and diameter from a few micrometers to centimeters within the nano range. They have excellent electron transport properties. Many different types of nanowires exist such as superconducting metallic, semiconducting. Sensor-based nanowires are very few in number. For biochemical detection, semiconductor-based silicon nanowires doped with boron are used. Detection and isolation of streptavidin molecules from a mixture have been carried out using silicon nanowires coated with biotin. The small size and capability of these nanowires make them ideal candidates for biodetection of pathogens and a wide range of biological and chemical properties. Nanowires are better than nanotubes in two ways:

1. Their design can be modified during synthesis to control operational parameters.
2. They offer more opportunities for functionalization due to the existence of compatible materials on their surfaces [4].

8.3.3.1 Functionalization of Si/metal nanowires

Salinization is widely used for silicon nanowire (SiNW) surface deformation. Before surface deformation, the SiNW surface is subjected to vapor plasma treatment. The plasma cleans the nanowire surface and creates a hydrophilic surface by forming a hydroxyl-terminating silicon oxide surface. The hydrated layer is activated using organosilane, a reactive group. The target biomaterial is covalently bonded using one among various methods. First, aldehydes are removed from the SiNW by reacting with 3-(trimethoxysilyl) propyl aldehyde. Next, the deformed nanowire completes a covalent bond between the biomaterial and SiNW by reacting with the biomaterial in the presence of sodium cyanoborane. In another method, the SiNW fixes the biomaterial deformed by acrylic phosphoramide after reacting with mercapto propyl trimethoxy silane in Ar atmosphere. This technology was used for the functionalization of a DNA probe deformed with acrylic phosphoramide at SiNW and 5N-end [33].

8.3.3.2 Functionalization of conducting polymer nanowire

A conducting polymer (CP) nanowire is a new material used in chemical sensing and biosensing. The advantage of a CP nanowire is that an unrefined reactant can be manufactured easily through well-known chemical or electrochemical processes. Their conductivity can be increased by changing the ratio of the dopant or the monomer/dopant. A CP nanowire can be functionalized before/after synthesis using methods that differ from those used for Si or metal nanowires. The main advantage is that SiNW and CNT devices requiring functionalization and arrangement after synthesis bond with functional biomolecules within the CP nanowire at once and the synthesis progresses with internal electrical contact. Most researches on the functionalization of a CP are based on thin films. However, such researches can also be applied to CP nanowires. Various functional groups can be immobilized by a CP parent. Such a process may enable the development of a material that induces a combination of features because of the CP and its functional group. CP functionalization can be performed at three stages: before polymerization, during polymerization, and after polymerization. The functionalization of a CP nanowire before polymerization initially involves the covalent bond of a specific group of a monomer. Then, the preparation of the functionalized polymer progresses. For example, it is possible to replace the process of attaching hydrogen to nitrogen in pyrrole with a specific group. It is possible to apply such a method to cases in which a specific group has a stable feature during polymerization. Among various processes for polymerizing CP nanowires, functionalization is one in which specific anions are bonded by static electricity during electric polymerization. Such a method can realize functionalization because anions doped into the polymer parent are irreversibly caught. Functionalization after the polymerization of a CP nanowire is a process in which a proper functional group within a polymer makes a covalent stable bond with other functional groups of specific molecules. A good example of the same is the post-polymerization functionalization of poly(N-substituted pyrrole) thin film [33].

8.3.4   Silver

One of the most widely used metal nanoparticles are silver nanoparticles (AgNPs), which have received much interest in biological detection. AgNPs have attractive physicochemical properties, so they are frequently useful in electrochemical biosensors. AgNPs have been used extensively in antibacterial, antifungal, and other biomedical applications, the health-care industry, and in food storage, environment, and household utensils. Also it has been demonstrated that hydrophobic silver–gold (Ag–Au) composite nanoparticles are used in biosensing because they show strong adsorption and good electrical conducting properties [34].

8.3.5   Gold

Gold nanoparticles are 1–100 nm in size and available at 1 nM to 5 nM concentration. Their dispersion in water is known as colloidal gold. Gold nanoparticles exhibit strong optical absorption and scattering extending from the visible to near IR spectral range, which is known as the SPR band. It has the following properties:

- Suitable for protein and oligonucleotide adsorption and surface modification.
- Suitable for clinical and point-of-care applications.
- Excellent for gold particle aggregation due to high monodispersity.
- Consistency, purity, and reactive groups on uniform nanoparticles provide consistent and repeatable coating and binding.

8.3.5.1   Uses of gold nanoparticle biosensors

- Used in a variety of sensors such as colorimetric and surface-enhanced Raman spectroscopy for the detection of proteins, pollutants, and other molecules.

- Used to detect biomarkers in the diagnosis of cancer, heart disease, and infectious diseases.
- Commonly used in lateral flow immunoassays.
- Gold nanoparticles designed to detect proteins within cells, using just laser light, could enable simple and highly sensitive monitoring tools for blood clots and other disorders [35].

8.3.6 Graphene

Graphene has honeycomb structure with sp^2-bonded carbon. For the last few years, graphene-based nanomaterials have been the focus of attention because they have high mechanical strength, elasticity, thermal conductivity, and demonstration of room-temperature quantum Hall effect. Graphene is produced by chemical vapor deposition (CVD) growth, mechanical exfoliation of graphite, or exfoliation of graphite oxide [36]. Graphene has an intrinsically high electron transfer ability and high surface-to-volume ratio. It makes them used as a perfect material for fabricating biosensors. Graphene is superior in the detection of small biomolecules such as NADH than CNTs [34].

8.3.7 PEBBLE Nanobiosensors

Probes encapsulated by biologically localized embedding (PEBBLE) nanobiosensors consist of sensor molecules entrapped in a chemically inert matrix by a microemulsion polymerization process that produces spherical sensors in the size range of 20 to 200 nm. Various sensor molecules can be entrapped, including those that detect optical change in pH or Ca^{2+} ions or can detect the fluorescence. These nanosensors are capable of monitoring real-time inter- and intracellular imaging of ions and molecules, while at the same time they are also insensitive to interference from proteins and show great reversibility and stability to leaching and photobleaching. In human plasma, they demonstrate a robust oxygen-sensing capability, little affected by light scattering and auto fluorescence [37].

8.3.8 Semiconductor

Semiconductors are promising materials used for construction of biosensors. Two major types of devices are based on semiconductors: light addressable potentiometric sensors (LAPSs) and quantum dots. LAPS is a semiconductor-based chemical sensor, in which a measuring site is determined by the illumination on the sensing surface. It is applied in the material sciences, biology, chemistry, and medicine. The quantum dots were discovered in the 1980s by Alexey Ekimov. Quantum dots are specific nanoparticles containing semiconductor material having spherical shape. The semiconductive properties can be acquired due to final size confinement, and even graphite can serve as a material for the construction of dots. Quantum dots can be linked to antibodies as label in a method resembling standard ELISA or used as a reagent in the determination of antioxidants where the antioxidant protects quantum dots from quenching by a pro-oxidative reagent [38].

8.3.8.1 Functionalization of colloidal quantum dots

Colloidal quantum dots are functionalized through the following methods: ligand exchange at the quantum dots surface, covalent conjugation of DNA to the quantum dots surface ligands, and DNA functionalization on core quantum dots synthesis in aqueous solution. To apply colloidal quantum dots to biological applications, it is necessary to disperse them into water through the chemical substitution of the surface and not an organic solvent. This is realized by replacing trioctylphosphine oxide (TOPO) with a ligand that can easily disperse, such as mercaptoacetic acid, mercaptopropionic acid, dihydrolipoic acid (DHLA), DL-cysteine, as well as thiolated acid [33].

8.4 Selection and Optimization of Nanomaterials for Nanobiosensor Technology

Sensors are made from various materials depending on the requirement, purpose, and use. The general requirements for selection and optimization of nanomaterials depend on high sensitivity, fast response, and good selectivity. The "sensitivity" term is related to test or assay or diagnostic and how an analyte is detected by a sensor. The output signal of the sensor to a change in a property of the sensor is called internal sensitivity. A sensor with a large internal sensitivity is able to pick out a minute change more easily than one with a small internal sensitivity due to bound analyte in a bulk sensor property [39]. The term specificity is related uniquely to a particular subject and ability to distinguish between analyte and any "other" material. In comparison to sensitivity, specificity may be difficult to measure and confirm. In the environment containing a high concentration of other materials, the specificity of a sensor becomes particularly important when trying to detect an analyte at low concentration. Many of which may bind nonspecifically to the sensor and thus produce an anomalous signal. In the context of nanobiosensors, dynamic range is another important term, which is related to the range over which the sensor is able to accurately produce an output signal indicative of the analyte quantity [40]. For large-scale applications, low material cost and easy fabrication are also required. Super small size and large surface-to-volume ratio improved the sensitivity of the sensors as compared to traditional materials. For example, TiO_2 nanomaterials can be prepared in large scale at mild temperature, which facilitates low-cost fabrication. Their transduction principle and system simulation functions are used for sensor applications [41].

The selectivity term is the ability of the nanobiosensor or device to differentiate the target biomolecules from other similar molecules, which is the major challenge in solving complex real-world samples. The resolution of the nanobiosensors is related to detecting the smallest change in target biomolecule concentration and quantify it. Biological affinity probes are a limiting factor for a nanobiosensor rather than its electrical detection mechanism. In biological probes, selectivity and efficiency could be enhanced with improved understanding of the immobilization chemistry and minimizing nonspecific binding issues [6]. "Nanofabrication" is the production of nanomaterials such as nanotubes, nanowires, and nanoparticles, which consists of two vital operations: design and manufacturing of nanoscale adhesive surfaces via the technology of integrated circuits and the engineering of nanomaterial surfaces through the process of micromachining. Nanoelectromechanical systems (NEMS) have highly sensitive electromechanical properties. NEMS are devices having both electrical and mechanical functions. They can achieve mass resolution, and small fluidic mechanical devices can exhibit fast response times. [4].

8.5   Applications of Nanobiosensors

Nanobiosensors have currently become increasingly important in the world. They have various biological, clinical, diagnostic, and other miscellaneous applications such as industrial manufacturing, aerospace, ocean exploration, environmental protection, resources investigation, medical diagnosis, and bioengineering. Nanosensors are also used for the detection of gases, chemical and biochemical variables, as well as physical variables and electromagnetic radiation.

8.5.1   Biological, Biomedical, and Diagnostic Applications

The development of nanotechnological tools and devices can easily characterize the minute responses of individual cells and biomolecules in the micro, nano, and pico range when subjected to mechanical, electrical, magnetic, or biochemical stimuli. The design of substrates based on these technologies offers new ways to probe cellular responses to changes in their physical environment. In fact, this allows the interactions of cells and their surrounding matrix to be carried out in well-controlled conditions. Cell membranes exhibit complexity, which ranges from their molecular contents to their emergent mechanical properties and dynamic spatial organization. These compositional and geometrical organizations of membrane components are known to influence cell functions such as signal transduction [42].

The bioreceptor is combined with a suitable transducer, which produces a signal after interaction with the target molecule. The presence of the biological element makes the biosensor systems extremely specific and highly sensitive, giving an upper edge over the conventional methods. Enzymes are the most used biocatalytic elements, enabling the detection of analytes in various ways since enzymatic reactions are followed by the consumption or production of various species. Nanobiosensors have been used for the detection of serum antigens, carcinogens, and biochemicals associated with many metabolic disorders. Nanobiosensors help in the early detection of various diseases and disorders such as diabetes, cancer, allergic responses, urinary tract infections, and HIV-AIDS before the disease is diagnosed on the basis of serum analysis [7]. At present, DNA-based nanobiosensors are used to provide cost-effective, simple, fast, sensitive, and specific detection of some genetic, cancer, and infectious diseases [43]. ELISA, nucleic acid hybridization techniques, electrochemical DNA detection, and glucose monitoring are also used in this technique [44].

In nanobiosensor assay, antibodies are labeled with an enzyme or nanoparticle in the presence of a substrate responsible for generating the signal. Since the amount of labeled antibodies is correlated to the amount of antigen (analyte), the signal produced is used to quantify the antigen. Different types of assays are depicted in Fig. 8.8.

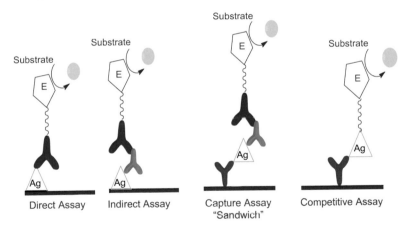

Figure 8.8 Direct, indirect, capture, and competitive assays.

Direct assay: In the direct assay method, a detection antibody against the antigen is used and the antigen is directly absorbed on the biosensing surface where it is performed.

Indirect assay: In the indirect assay method, a primary antibody is used as an intermediate between the analyte and labeled antibody. In this, the antigen is absorbed on the biosensing platform substrate.

Capture assay "sandwich": In this method, a capturing antibody is pre-immobilized on the substrate surface and a sandwich with the analyte is formed between the capturing antibody and the detecting antibody. A labeled antibody against the detection antibody is used to detect the analyte.

Competitive assay: In this method, an artificial analyte or analogous analyte is taken and labeled externally. The sample analyte competes with these labeled pseudoanalytes to capture the antibody. The native analyte in the sample displaces the labeled analyte from the capturing antibodies and washing away the labeled antigen [45].

In diagnostic technologies, nanobiosensors also facilitate the development of non-polymerase chain reaction for the development of target-specific and molecular-based therapies. Nanobiotechnology-based biosensors play an important role in treatment, diagnosis, and development of personalized medicine. In nanodiagnostics, the integration and interrelationship of several technologies are involved and these are taking a more active part in decision-making in future health-care systems [6]. Medicinal uses of nanobiosensors mainly focus on accurately identifying altered cells or places in the body by measuring changes in volume, concentration, displacement, velocity, gravitational, electrical, and magnetic forces, and pressure or temperature of cells in the body. A microbial nanobiosensor integrates microorganism with a physical transducer used to generate a measurable signal proportional to the concentration of analytes. Nanosensors may be able to distinguish or recognize certain cells such as cancer cells, at the molecular level in order to deliver medicine or monitor development of specific places in the body. Nanomedicine, by conjugating tumor-specific antibodies, has an important role in tumor-targeting anti-cancer therapeutics. Antibody-mediated targeting has been used for single bacterial cell quantitation and cell-surface labeling [46]. Recently, a conjugated DNA-QD system has been used to recognize the genomic targets of transcription factors (TFs). The DNA was labeled with an intercalating dye, and proteins were cross-linked to DNA; the complex was then labeled with QDs that were functionalized with antibodies (Abs) against the TFs. It was shown that these methods were used for detecting individual QD-labeled T7-RNA polymerases on the T7 bacteriophage genome. So it is used to read protein-binding position [47]. Nanoparticles are also used in drug-delivery systems to improve several crucial properties of "free" drugs, such as solubility, in vivo stability, pharmacokinetics, biodistribution, and efficacy [48]. Different types of nanoparticles and their role in drug delivery are given in Table 8.1 [49].

Table 8.1 Different types of nanoparticles and their role in drug delivery

S. No.	Structure	Size	Role in drug delivery
1.	Biologically synthesized silver nanoparticles	10–40 nm	Efficiently kills HeLa cells compared to its chemical counterpart.
2.	Cadmium sulfide quantum dots	100 nm	Reduces *E. coli* growth by delaying the lag phase.
3.	Engineered iron oxide nanoparticles	100–130 nm	Targeted drug delivery, sorting of cancer cells and imaging.
4.	Silver nanoparticles	34 nm	DLA tumor cell death.
5.	Liposomal nanoformulation of 5-fluorouracil and doxorubicin	100–130 nm	Efficient reduction in ascetic tumor size compared to bare drug.
6.	Magnetic nanoparticles entrapped in dendrimers	40–50 nm	Targeted drug delivery, destruction of cancer cells and imaging.
7.	Liposomal nanoformulation of Rhodamine 6G	100–120 nm	Multidrug-resistant Gram-negative bacteria inactivation for sewage treatment.

8.5.1.1 In cancer detection

Nanobiosensors are nano-scaled analytical frameworks that comprise nano-conjugated biological materials as a transducing system for the detection of miniscule quanta of any biological, chemical, or physical analytes. The constructs utilize optical, electrochemical, thermometric, piezoelectric, magnetic, or micromechanical methods to convey the relevant information in the form of signals. The generated signals precisely depend on the principle of selective biorecognition of cancer-cell-associated intracellular or surface biomarkers by the attached antibody or bioligands.

Biorecognition-based DNA/RNA nanobiosensors are diagnostic devices that are based on the measurement of the responses generated by the nucleic acid conversion processes. An immune nanobiosensor consists of a surface-immobilized antigen or antibody that participates in biospecific interaction with the target molecules to generate a measurable electrochemical, optical, or mechanical signal. Quantum dots are used in the treatment of cancer. Quantum dots glow when exposed to UV light. When injected, they seep into cancer, tumor and the surgeon can see the glowing tumor [50].

8.5.2 Environmental Applications

Nanobiosensors have broader environmental applications. The environment undergoes so many rapid changes almost every second by human influences and natural ecological processes. Nanobiosensors are used for the detection of heavy metals, pathogens, and other impurities from waste streams. They are also used for monitoring of weather conditions such as the estimation of humidity and many other vital functions. Nanomaterial sensors are very versatile in terms of their detection and monitoring of these tasks at singleplex or multiplex. Environmental monitoring typically involves several steps such as sampling, sample handling, and sample transportation to a specialized laboratory to determine the chemical composition and to establish the toxic effects. A number of nanostructured materials are candidates for assessing the occurrence of water contamination. These include polyamic acid (PAA) metal nanoparticle composite membranes, polyoxy-dianiline membranes, sequestered gold, silver or palladium nanoparticles within electroconducting polymers and under potential deposition of metal films, membranes, and colloids onto solid electrodes [51].

Titanium oxide (TiO_2) is a competitive and promising compound in gas sensor applications due to its optical and electronic properties.

Lanthanum oxide (Ln_2O_3) has gained potential application in nanosensors due to its high sensing ability for different chemical compounds, including H_2S, NH_3, CO, O_3, and Cl_2. Zinc oxide (ZnO) is a typical n-type semiconductor with a direct wide band gap of 3.37 eV and large excitation binding energy of 60 meV. The use of ZnO-based gas sensor showed improved gas-sensing properties, including higher gas response to a variety of reducing or oxidizing gases, low cost, being environment friendly, and low operating temperature [52]. Metal and metal-oxide nanoparticles are co-immobilized with carbon nanomaterials to realize monohybrid composite materials with enhanced electrocatalytic properties. They show the synergistic effects of the metal–carbon nanomaterial composites. Due to the large surface area of carbon nanomaterials, the fast electron transfer activity of the composite minimizes mass transport limitations and the electrocatalytic synergy between the metal and the carbon nanomaterial. These new nanohybrid materials form electrochemical biosensors for environmental analysis due to the cooperative effects of metal nanoparticles and carbon nanomaterials based on the coupling of the most common metal nanoparticles (gold, silver, and platinum) with CNTs and graphene, as carbon nanomaterials [53].

8.5.3 Application in Food Analysis

A research group has developed a nanoparticle-based bioassay that can rapidly identify *E. coli* in food, which causes one of the most dangerous food-borne diseases and could be fatal, especially in elderly or children. Many commercial nanobiosensors are available for the measurement of alcohols, carbohydrates, and acids. These instruments are mostly used in quality assurance laboratories. Biosensors are also used in industries such as wine, beer, yogurt, and soft drinks producers. For the measurement of amino acids, amines, amides, heterocyclic compounds, carbohydrates, carboxylic acids, gases, cofactors, inorganic ions, alcohols, and phenols, enzyme-based biosensors are used. Immunosensors have significant potential in detecting and ensuring food safety when pathogenic organisms are found in fresh meat, poultry, or fish [2]. The role of nanobiosensors in food processing and packaging is depicted in Fig. 8.9 [54].

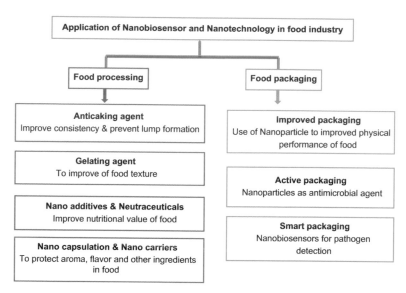

Figure 8.9 Role of nanobiosensors in food processing and packaging.

8.5.3.1 Nanobiosensors in agriculture and agroproducts

As a diagnostic method for determining soil quality and disease: Soil health is predicted by quantitative measurement of differential oxygen consumption in the respiration of different microbes in the soil, which may be good or bad for soil.

As a device to detect contaminants and other molecules: Several nanobiosensors have been designed to detect contaminants, pests, nutrient content, and plant stress due to drought, temperature, or pressure.

As an agent to promote sustainable agriculture: A nanofertilizer refers to a product that delivers nutrients to crops encapsulated within a nanoparticle. There are three ways of encapsulation: nutrient encapsulated inside nanomaterials such as nanotubes or nanoporous materials, coated with a thin protective polymer film, and delivered as particles or emulsions of nanoscale dimensions. Recently, carbon nanotubes have been shown to penetrate tomato seeds and zinc oxide nanoparticles have been shown to enter the root tissue of ryegrass [37].

8.5.4 Miscellaneous Applications

Nanobiosensors can have several other applications in the detection of ions, organic compounds, and pharmaceutical analytes. Nanobiosensors can be used to regulate the feeding of nutrient media and substratum mixtures into the bioreactors for various applications in industrial operations. On an industrial scale, many commercial preparations and separations can be done using these sensors [50].

8.5.4.1 Microorganism detection

Detection of bacteria: The rapid and sensitive detection of pathogenic bacteria is extremely important in medical diagnosis and measures against bioterrorism. Microbial biosensors have been integrated with nanotechnologies and applied to a wide range of detection purposes. For microorganism detection, microbial biosensors are used as analytical devices composed of a microorganism that detects a target substrate and converted into a readable form. A bioconjugated nanoparticle-based bioassay for in situ pathogen quantification can detect a single bacterium within 20 min. The detection of a small number of Salmonella enteric bacteria is achieved due to a change in the surface stress on the silicon nitride cantilever surface in situ upon the binding of bacteria.

Detection of viruses: Rapid, selective, and sensitive detection of viruses is crucial for implementing an effective response to viral infection. Established methods for viral analysis include plaque assays, immunological assays, transmission electron microscopy, and PCR-based testing of viral nucleic acids. These methods can't quickly identify at a single virus level and, therefore, involve a large amount of sample handling, which is inconvenient for infectious agents, but it can be easily done by using nanowires. Nucleic acid sequence-based amplification and Livermore microbial detection array technologies are used for the identification of human viruses. For the detection of rotavirus, nanomaterials such as CNTs, quantum dots, graphene oxide, and silicon nanowire tubes are used. When a virus particle binds to an antibody receptor on a nanowire device, the conductance of that device changes from the baseline value, and when the virus unbinds again, the conductance returns to the baseline value [55].

8.6 Current Trends and Recent Developments in Nanobiosensors

In the last two decades, a wide range of nanobiosensors have been developed. At small concentrations, nanobiosensors have generated a great deal of excitement due to their ability to detect a wide range of materials. Micro-/nanoelectromechanical systems, or MEMS/NEMS, sensors systems exploiting mechanical motion for detecting analytes can generally be divided into two categories: static deflection devices and resonators. Static devices, also known as deflection-based sensors, are typically beams supported at one end by a larger substrate. Static deflection devices employing functionalized cantilevers have been used to detect oligonucleotides. Nanoparticle-based biosensors are particularly attractive because they can be easily synthesized in bulk using standard chemistry techniques and do not require advanced fabrication approaches. They also offer particularly high surface area due to their extremely small size and the fact that they are typically used suspended in solution. Gold nanoparticles suspended in solution exhibit a strong optical absorbance at particular visible optical wavelengths due to plasmon absorption. This property is dependent on the nanoparticle size. Several one-dimensional nanostructures such as nanowires, nanotubes, and nanofibers have been used as nanobiosensors. These devices offer improved sensitivities due to large surface-to-volume ratios, which enable bound analyte molecules to more significantly affect the bulk electrical properties of the structure. In some cases (e.g., CNTs), the inherent electrical properties of the device are particularly extraordinary and lend themselves to improved sensor sensitivity [40].

Biosensors can be categorized based on their biological component or transduction component: (i) optical (colorimetric, luminescent, fluorescent, and interferometric), (ii) electrochemical (amperometeric, voltammetry, conductimetric, and potentiometric), (iii) calorimetric, (iv) mass-based (piezoelectric or magnetic and gravimetric/acoustic wave), and (v) thermometric biosensors [23].

The detection potential of biosensors varies in a broad range of target analytes, from small and tiny molecules (antibody, herbicide, pesticide, and explosive) to medical analytes quantification (such as small and tiny proteins, bacterial and viral pathogens) [56].

In the recent days, nanobiosensors have been developed in the fields of diagnostics, food analysis, environment monitoring, and other industries, including aptamer biosensors, electrochemiluminescence (ECL) nanobiosensors, silica-based electrochemical enzyme biosensors, nano-enabling electrochemical β-A sensors, nucleic-acid-based electrochemical nanobiosensors, miniaturized electrochemical nanobiosensors, and microfluidic devices. Nucleic-acid-based electrochemical nanobiosensors have been established on the principle of synergy between nucleic acid recognition units of high specificity and high sensitivity of electrochemical signal transductions. Analytes are detected at a concentration range of clinical significance through nucleic-acid-based electrochemical biosensors generally on target amplification strategies such as polymerase chain reactions. The main limitation of nucleic-acid-based electrochemical biosensors is complexity and time-consuming character of the amplification methods. But recent advancements in nanotechnology have helped the nucleic-acid-based electrochemical biosensors to act as point-of-care diagnostic tools from the laboratory-based method [57]. In the recent days, aptamer biosensors have captured a remarkable area of interest because they introduce valuable intrinsic properties such as nucleic acids and target-induced conformational changes. The aptamer introduces a straightforward graphene-oxide-based immobilization-free screening method. Analytical performance and commercial application of aptasensors were found to be enhanced with the inclusion of nanomaterials such as carbon materials, metallic nanoparticles, and functional nanospheres [47].

Electrochemiluminescence (ECL) nanobiosensors have been a major part of studies in the recent time for practical purposes of nanobiosensors. Metal nanoclusters (MNCs) are in huge use today for biosensor fabrication, construction, development, and design of direct or label-free ECL nanobiosensors. Examples of nanomaterials, nanostructured materials (NSMs), nanocomposites (NCs), and metal nanoclusters (MNCs) are as follows:

Nanomaterials or NSMs: AuNPs (gold nanoparticles), AgNPs (silver nanoparticles), Fe_3O_4 (Iron II, III oxide), CFQDs (cadmium-free quantum dots), graphene, CNTs, and fullerenes

MNCs: Au (gold), Ag (silver), and Pt (platinum)

NCs: From metallic and nonmetallic origin

In the recent days of diagnostics, ECL nanobiosensors have been developed and used in the specialized and practical field of personalized medical monitoring. They include ECL nano-immunosensors, enzymatic ECL nanobiosensors, and ECL aptasensors with potential in a wide range of applications. In the future, the use of ECL-based nanobiosensors in diagnosing various pathologies and diseases will eliminate lengthy laboratory assays on bodily fluids (urine, blood, saliva, and tears) [58].

The current state-of-the-art diagnostic biosensors are based on several technologies, often including either ELISA or the amplification of a sample by polymerase chain reaction (PCR), which allows detection at very small concentration. As an early detection system, PCR-ELISA can detect β-A. PCR-ELISA is 100-fold more sensitive than the conventional PCR method. PCR-ELISA serves as a powerful detection method from the medical sector to the food and agriculture sectors. However, recently, an antibody-based electrochemical immunosensing approach against β-A has been found to detect β-A at pM levels within 30–40 min compared to 6–8 h of ELISA test. The introduction of nano-enabling electrochemical sensing technology could enable rapid detection of β-A at POC, which may lead to rapid and personalized health-care delivery. The research on nanobioelectronics and biosensors aims at the integration of nanoelectronics, tools, and materials into low-cost, user-friendly, and efficient sensors and biosensors, with interest in several fields such as diagnostics, food analysis, environment monitoring, and other industries [7, 59]. Electrochemical nanobiosensors have low cost, ease of use, portability, and ability to perform both screening and real-time monitoring of the area under investigation as compared to conventional analytical methods, which require high cost and trained personnel [53].

8.7   Future of Nanobiosensors

The development of nanobiosensor nanotechnology has really proved to be a very significant blessing. The use of nanomaterials has significantly improved transduction mechanisms. Nanobiosensors make the overall process fast, easy to execute, and better in terms of performance, making them future tools that are dynamic and versatile [4]. By using nanotechnology, the effectiveness of sensing strategy has been increased, which has afforded an excellent enhancement in diagnosis systems. Nanobiosensing provides a low-cost strategy to identify disorders by offering much better management [60]. Nanobiotics uses nano-sized tools for bacterial diseases and to gain an increased understanding of the complex, underlying patho-physiology of diseases and can be used in the future for cancer treatment. The growth and development of biosensing systems offer noninvasive, transportable, easy-to-use, and economical diagnosis devices with higher specificity, level of responsiveness, and stability for biomarkers [61]. It is expected that CNT-based biosensors have a variety of future developments and applications. CNT-based biosensors and novel fluorescence CNT-based nanosensors are attractive instances of living biological tissues [27]. Nanobiosensors offer advanced diagnostic tools with increased sensitivity, specificity, and reliability [34]. This may even spur a dramatic increase in the number of point-of-care diagnostics, as well as diagnostic tools that can be used by patients on their own. Whether a single nanobiosensor architecture will become dominant or several will transition to commercial devices is yet to be seen. However, it is almost certain that these sensors will allow the detection of pathogens and diseases like never before [40].

References

1. Malhotra, S., Verma, A., Tyagi, N., and Kumar, V. (2017). Biosensors: Principle, types and applications. *Int. J. Adv. Res. Innov. Ideas Educ.*, 3(2), pp. 3639–3644.

2. Dede, S. and Altay, F. (2018). Biosensors from the first generation to nano-biosensors. *Int. Adv. Res.eng. J.*, 2(2), pp. 200–207.

3. Bhalla, P. and Singh, N. (2016). Generalized drude scattering rate from the memory function formalism: An independent verification of the Sharapov–Carbotte result. *Eur. Phys. J. B.*, **89**(2), pp. 1–49.

4. Malik, P., Katyal, V., Malik, V., Asatkar, A., Inwati, G., and Mukherjee, T. K. (2013). Nanobiosensors: Concepts and variations. *ISRN Nanomater.*, pp. 1–9.

5. Kara, S. (2012). *A Roadmap of Biomedical Engineers and Milestones* (BoD–Books on Demand).

6. Prasad, S. (2014). Nanobiosensors: The future for diagnosis of disease. *Nanobiosens. Dis. Diagn.*, **3**, pp. 1–10.

7. Sagadevan, S. and Periasamy, M. (2014). Recent trends in nanobiosensors and their applications: A review. *Rev. Adv. Mater. Sci.*, **36**, pp. 62–69.

8. Andle, J. and Vetelino, J. (1995). Acoustic wave biosensors. In: *IEEE Int. Ultrason. Symp.*, pp. 451–460.

9. Liu, T., Tang, Ja., and Jiang, L. (2004). The enhancement effect of gold nanoparticles as a surface modifier on DNA sensor sensitivity. *Biochem. Bioph. Res. Co.*, **313**(1), pp. 3–7.

10. Holzinger, M., Le Goff, A., and Cosnier, S. (2014). Nanomaterials for biosensing applications: A review. *Front. Chem.*, **2**, pp. 1–63.

11. Chemla, Y. R., Grossman, H., Poon, Y., McDermott, R., Stevens, R., Alper, M., and Clarke, J. (2000). Ultrasensitive magnetic biosensor for homogeneous immunoassay. *Proc. Natl. Acad. Sci.*, **97**(26), pp. 14268–14272.

12. Richardson, J., Hawkins, P., and Luxton, R. (2001). The use of coated paramagnetic particles as a physical label in a magneto-immunoassay. *Biosens. Bioelectron.*, **16**(9–12), pp. 989–993.

13. Lud, S. Q., Nikolaides, M. G., Haase, I., Fischer, M., and Bausch, A. R. (2006). Field effect of screened charges: Electrical detection of peptides and proteins by a thin-film resistor. *ChemPhysChem,* **7**(2), pp. 379–384.

14. Grieshaber, D., MacKenzie, R., Vörös, J., and Reimhult, E. (2008). Electrochemical biosensors: Sensor principles and architectures. *Sensors.*, **8**(3), pp. 1400–1458.

15. Cai, H., Xu, C., He, P., and Fang, Y. (2001). Colloid Au-enhanced DNA immobilization for the electrochemical detection of sequence-specific DNA. *J. Electroanal. Chem.*, **510**(1–2), pp. 78–85.

16. Janoutová, J., Janáčková, P., Šerý, O., Zeman, T., Ambroz, P., Kovalová, M., Vařechová, K., Hosák, L., Jiřík, V., and Janout, V. (2016). Epidemiology

and risk factors of schizophrenia. *Neuroendocrinol. Lett.*, **37**(1), pp. 1–8.

17. Borman, S. (1987). Biosensors: Potentiometric and amperometric. *Anal. Chem.*, **59**(18), pp. 1091–1098.

18. Cosio, M., Scampicchio, M., and Benedetti, S. (2012). Electronic noses and tongues. (Academic Press: Boston, MA, USA).

19. Zuber, A. A., Klantsataya, E., and Bachhuka, A. (2019). Biosensing. In: *Comprehensive Nanoscience and Nanotechnology*, 2nd ed. (Elsevier), pp. 105–126.

20. Dziąbowska, K., Czaczyk, E., and Nidzworski, D. (2017). Application of electrochemical methods in biosensing technologies. In: *Biosensing Technologies for the Detection of Pathogens: A Prospective Way for Rapid Analysis* (BoD – Books on Demand), pp. 216.

21. Stradiotto, N. R., Yamanaka, H., and Zanoni, M. V. B. (2003). Electrochemical sensors: A powerful tool in analytical chemistry. *J. Braz. Chem. Soc.*, **14**(2), pp. 159–173.

22. Dzyadevych, S. and Jaffrezic-Renault, N. (2014). Conductometric biosensors. In: *Biological Identification* (Elsevier), pp. 153–193.

23. Yang, L. and Guiseppi-Elie, A. (2008). Impedimetric biosensors for nano and microfluidics. *Encyclopedia of Microfluidics and Nanofluidics* (Springer: Boston, MA), pp. 811–823.

24. Pohanka, M. and Skládal, P. (2008). Electrochemical biosensors: Principles and applications. *J. Appl. Biomed.*, **6**(2), pp. 57–64.

25. Tîlmaciu, C.-M. and Morris, M. C. (2015). Carbon nanotube biosensors. *Front. Chem.*, **3**, pp. 1–59.

26. Yang, N., Chen, X., Ren, T., Zhang, P., and Yang, D. (2015). Carbon nanotube based biosensors. *Sensor. Actuat. B: Chem.*, **207**, pp. 690–715.

27. Ijeomah, G., Obite, F., and Rahman, O. (2016). Development of carbon nanotube-based biosensors. *Int. J. Nano. Biomater.*, **6**(2), pp. 83–109.

28. Wang, J. (2005). Carbon-nanotube based electrochemical biosensors: A review. *Electroanalysis*, **17**(1), pp. 7–14.

29. De Volder, M. F., Tawfick, S. H., Baughman, R. H., and Hart, A. J. (2013). Carbon nanotubes: Present and future commercial applications. *Science*, **339**(6119), pp. 535–539.

30. Ma, P.-C., Siddiqui, N. A., Marom, G., and Kim, J.-K. (2010). Dispersion and functionalization of carbon nanotubes for polymer-based nanocomposites: A review. *Compos. Part A: Appl. S.*, **41**(10), pp. 1345–1367.

31. Tkachenko, A. V. (2002). Morphological diversity of DNA-colloidal self-assembly. *Phys. Rev. Lett.,* **89**(14), pp. 148303.

32. Ambhorkar, P., Wang, Z., Ko, H., Lee, S., Koo, K.-I., and Kim, K. (2018). Cho D-iD: Nanowire-based biosensors: From growth to applications. *Micromachines,* **9**(12), pp. 679.

33. Yeom, S.-H., Kang, B.-H., Kim, K.-J., and Kang, S.-W. (2011). Nanostructures in biosensor: A review. *Front. Biosci.,* **16**, pp. 997–1023.

34. Razavi, H. and Janfaza, S. (2015). Medical nanobiosensors: A tutorial review. *Nanomed. J.,* **2**(2), pp. 74–87.

35. Baltzer, N. and Copponnex, T. (2014). *Precious Metals for Biomedical Applications* (Elsevier).

36. Pumera, M. (2011). Graphene in biosensing. *Appl. Mater.,* **14**(7–8), pp. 308–315.

37. Rai, V., Acharya, S., and Dey, N. (2012). Implications of nanobiosensors in agriculture. *J. Biomater. Nanobiotechnol.,* **3**, pp. 1–10.

38. Pohanka, M. and Leuchter, J. (2017). Biosensors based on semiconductors: A review. *Int. J. Electrochem. Sci.,* **12**(7), pp. 6611–6621.

39. Saah, A. J. and Hoover, D. R. (1997). "Sensitivity" and "specificity" reconsidered: The meaning of these terms in analytical and diagnostic settings. *Ann. Intern. Med.,* **126**(1) pp. 91–94.

40. Bellan, L. M., Wu, D., and Langer, R. S. (2011). Current trends in nanobiosensor technology. *Wires. Nanomed. Nanobi.,* **3**(3), pp. 229–246.

41. Bai, Y., Mora-Sero, I., De Angelis, F., Bisquert, J., and Wang, P. (2014). Titanium dioxide nanomaterials for photovoltaic applications. *Chem. Rev.,* **114**(19), pp. 10095–10130.

42. Lim, C. T., Han, J., and Guck, J. (2010). Micro and nanotechnology for biological and biomedical applications. *Med. Biol. Eng. Comput.,* **48**(10), pp. 941–943.

43. Abu-Salah, K. M., Zourob, M. M., Mouffouk, F., Alrokayan, S. A., Alaamery, M. A., and Ansari, A. A. (2015). DNA-based nanobiosensors as an emerging platform for detection of disease. *Sensors,* **15**(6), pp. 14539–14568.

44. Demidov, W. (2004). Nanobiosensors and molecular diagnostics: A promising partnership. *Expert. Rev. Mol. Diagn.,* **4**(3), pp. 267–268.

45. Chamorro-Garcia, A. and Merkoçi, A. (2016). Nanobiosensors in diagnostics. *Nanomed-Nanotechnol.*, **3**, pp. 35–74.

46. Yadav, A., Ghune, M., and Jain, D. K. (2011). Nano-medicine based drug delivery system. *J. Adv. Pharm. Educ. Res.*, **1**(4), pp. 201–213.

47. Dolatabadi, J. E. N., Mashinchian, O., Ayoubi, B., Jamali, A. A., Mobed, A., Losic, D., Omidi, Y., and de la Guardia, M. (2011). Optical and electrochemical DNA nanobiosensors. *Trac-Trend Anal. Chem.*, **30**(3), pp. 459–472.

48. Nasimi, P. and Haidari, M. (2013). Medical use of nanoparticles: Drug delivery and diagnosis diseases. *Int. J. Green Nanotechnol.*, **1**, pp. 892–978.

49. Girigoswami, K. and Akhtar, N. (2019). Nanobiosensors and fluorescence based biosensors: An overview. *Int. J. Nano.*, **10**(1), pp. 1–17.

50. Shandilya, R., Bhargava, A., Bunkar, N., Tiwari, R., Goryacheva, I. Y., and Mishra, P. K. (2019). Nanobiosensors: Point-of-care approaches for cancer diagnostics. *Biosens. Bioelectron.*, **130**, pp. 147–165.

51. Sadik, O. A. (2014). Advanced nanosensors for environmental monitoring. In: *Nanotechnology Applications for Clean Water*, 2nd ed. (Elsevier), pp. 21–45.

52. Julkapli, N. M. and Bagheri, S. (2018). Nanosensor in gas monitoring: A review. *Environ. Sci. Nano.*, **1**, pp. 443–472.

53. Mazzei, F., Favero, G., Bollella, P., Tortolini, C., Mannina, L., Conti, M. E., and Antiochia, R. (2015). Recent trends in electrochemical nanobiosensors for environmental analysis. *Int. J. Environ. Health Res.*, **7**(3), pp. 267–291.

54. Singh, T., Shukla, S., Kumar, P., Wahla, V., Bajpai, V., and Rather, I. (2017). Application of nanotechnology in food science: Perception and overview. *Front. Microbiol.*, **8**, pp. 1501.

55. Agrawal, S. and Prajapati, R. (2012). Nanosensors and their pharmaceutical applications: A review. *Int. J. Pharm. Sci. Technol.*, **4**, pp. 1528–1535.

56. Hassanpour, S., Baradaran, B., Hejazi, M., Hasanzadeh, M., Mokhtarzadeh, A., and de la Guardia, M. (2018). Recent trends in rapid detection of influenza infections by bio and nanobiosensor. *Trac-Trend Anal. Chem.*, **98**, pp. 201–215.

57. Abi, A., Mohammadpour, Z., Zuo, X., and Safavi, A. (2018). Nucleic acid-based electrochemical nanobiosensors. *Biosens. Bioelectron.*, **102**, pp. 479–489.

58. Rizwan, M., Mohd-Naim, N. F., and Ahmed, M. U. (2018). Trends and advances in electrochemiluminescence nanobiosensors. *Sensors, 18*(1), pp. 1–166.

59. Kaushik, A., Jayant, R. D., Tiwari, S., Vashist, A., and Nair, M. (2016). Nano-biosensors to detect beta-amyloid for Alzheimer's disease management. *Biosens. Bioelectron., 80*, pp. 273–287.

60. Mishra, R. K. and Rajakumari, R. (2019). Nanobiosensors for biomedical application: Present and future prospects. In: *Characterization and Biology of Nanomaterials for Drug Delivery*, 1st ed. (Elsevier), pp. 1–23.

61. Mukherjee, A., Bhattacharya, J., and Moulick, R. G. (2020). Nanodevices: The future of medical diagnostics. *NanoBioMedicine*, pp. 371–388.

Multiple Choice Questions

1. Who is known as the "father of biosensors"?
   (a) Griffin and Nelson   (b) Leland C. Clark
   (c) M. Cremer   (d) Sumio Iijima

2. In the working process of nanobiosensors, identify the unmarked component.

   (a) Sensitive biological element (b) Amplifier
   (c) Transducer   (d) Microprocessor

3. Which of the following detection techniques is not used in nanobiosensor technology?
   (a) Electrochemical and electrical detection
   (b) Optical detection
   (c) Photovoltaic detection
   (d) Mass-based detection

4. Which of the following is not a subtype of surface-generated acoustic waves?
   (a) Bulk acoustic   (b) Lamb wave
   (c) Liquid-guided acoustic wave (d) Bluestein-Gulyaev

5. Electrochemical biosensor principle is related to
   (a) Ionization of atoms or molecule

(b) Agglutination

(c) Antigen–antibody interaction

(d) Enzymatic catalysis of a reaction that produces or consumes electron

6. Carbon nanotubes often refer to as _____

(a) Buck tubes
(b) Piston tubes

(c) Plunger tubes
(d) Gear balls

7. Which of the following bonding is present in carbon nanotube functionalization?

(a) Hydrogen bonding
(b) Electrostatic bonding

(c) Van der Waals bonding
(d) Covalent bonding

8. The size of gold nanoparticles is about

(a) 1–500 nm
(b) 1–200 nm

(c) 1–300 nm
(d) 1–100 nm

9. Nanowires are used in_____

(a) Transistors
(b) Transducers

(c) Capacitors
(d) Resistors

10. The size range of PEBBLE nanobiosensors is

(a) 20–200 nm
(b) 1–100 nm

(c) 5–50 nm
(d) 10–500 nm

11. The most commonly used semiconductor nanowires are made of

(a) Ytterbium
(b) Zirconium

(c) Silver
(d) Silicon

12. The principle used in indirect biosensing assay is

(a) Antigen directly absorbed on surface where the biosensing is performed

(b) Antigen absorbed on surface where the biosensing is performed by an intermediate

(c) Interaction between capturing antibody and detecting antibody

(d) None of the above

13. Cadmium sulfide quantum dots are used to reduce the growth of which bacteria?

(a) *Vibrio cholera*
(b) *Escherichia coli*

(c) *Clostridium tetani*
(d) *Salmonella typhi*

14. Nanobiosensors are widely used in which of the following areas?
    (a) Environmental application
    (b) Application in food analysis
    (c) Biological, biomedical, and diagnostic applications
    (d) All of the above

15. An example of metal nanoclusters (MNCs) is
    (a) Graphene   (b) Carbon nanotubes
    (c) Ag, Au, or Pt   (d) Fe_3O_4

16. The size of biologically synthesized silver nanoparticles is
    (a) 10–40 nm   (b) 20–80 nm
    (c) 10–200 nm   (d) 50–100 nm

17. The role of nanobiosensors in food processing and packaging is
    (a) Used as a gelating agent   (b) Used as an anticaking agent
    (c) Used in smart packaging   (d) All of the above

18. For environmental sensors with high sensitivity or selectivity to feel the gaseous chemicals, we use
    (a) SO_2 (b) NO_3
    (c) NO_2 (d) NO

19. Graphene structure is related to____
    (a) Honeycomb sheet of carbon atoms
    (b) Orthorhombic structure
    (c) Tetrahedral structure
    (d) Hexagonal structure

20. Graphene is composed of
    (a) Entirely of carbon
    (b) 50% carbon and 50% silicon
    (c) 50% carbon, 20% silicon, and 30% silver
    (d) 80% carbon and 20% germanium

21. The commercial production of graphene
    (a) Has been prohibited by most governments
    (b) Is in various trial stages in several countries
    (c) Is already done by more than 100 companies worldwide
    (d) Has not yet begun

22. The following is not an example of a transducer.

    (a) Analogue voltmeter   (b) Photovoltaic cell

    (c) Photoelectric cell   (d) Potentiometric cell

23. How many types of measurement can be done by impedimetric-based electrochemical biosensors?

    (a) 2   (b) 3

    (c) 4   (d) 5

24. Who discovered the carbon nanotubes?

    (a) Tony Turner   (b) Sumio Iijima

    (c) James Hock   (d) Antony Peter

25. The principle used in amperometric electrochemical biosensors is

    (a) Measuring the indirect current generated by enzymatic reaction

    (b) Measuring the potential difference

    (c) Measuring the direct current generated by enzymatic reaction

    (d) Measuring the conductivity

Answer Key

1. (b)	2. (c)	3. (c)	4. (a)	5. (d)	6. (a)	7. (d)
8. (d)	9. (a)	10. (a)	11. (b)	12. (b)	13. (b)	14. (d)
15. (c)	16. (a)	17. (d)	18. (d)	19. (a)	20. (a)	21. (c)
22. (a)	23. (a)	24. (b)	25. (c)			

Short Answer Questions

1. What is the difference between a biosensor and a nanobiosensor?
2. Write the working principle of a nanobiosensor.
3. What are the main detection techniques used in nanobiosensors?
4. Write the type of acoustic wave sensors that are mostly used.
5. Write the use of carbon nanotubes in nanobiosensor technology.
6. What are the various methods for the functionalization of colloidal quantum dots?
7. Describe the PEBBLE nanobiosensors.
8. Define accuracy and precision. What does it mean for a sensor to be precise but to have low accuracy?

9. Give examples of nanomaterials (NMs) or nanostructured materials (NSMs) and nanocomposites (NCs).

10. Describe the role of nanobiosensors in the diagnosis of diseases.

Long Answer Questions

1. Describe the main components of biosensors and explain their function.

2. Describe the chemical and physical functionalizations of CNTs.

3. Describe various electrochemical biosensors on the basis of working type.

4. Which type of nanowires is mostly used? Describe their function.

5. Describe the future prospective of nanobiosensors.

Chapter 9

Nanofertilizers: Applications and Future Prospects

Manju Bernela,[a]* Ruma Rani,[b]* Parth Malik,[c] and Tapan K. Mukherjee[d]*

[a]Department of Bio and Nanotechnology, Guru Jambheshwar University of Science and Technology, Hisar 125001, Haryana, India
[b]ICAR-National Research Centre on Equines, Hisar 125001, Haryana, India
[c]School of Chemical Sciences, Central University of Gujarat, Gandhinagar 382030, India
[d]Department of Internal Medicine, University of Utah, Salt Lake City, Utah, USA
tapan400@gmail.com

Fertilizers have been used for the past many years in agriculture for the benefit of farmers. Traditional fertilizers are expensive as well as harmful to human beings and the environment. Therefore, there is a need for developing environment-friendly fertilizers having high nutrient value as well as compatibility with soil and environment. Nanotechnology is rising as a promising alternative in the form of nanofertilizers to enhance the qualitative attributes therein. A nanofertilizer comprises nanoformulations of nutrients deliverable

---

*These authors have contributed equally to this chapter.

Nanotechnology: Principles and Applications
Edited by Rakesh K. Sindhu, Mansi Chitkara, and Inderjeet Singh Sandhu
Copyright © 2021 Jenny Stanford Publishing Pte. Ltd.
ISBN 978-981-4877-43-5 (Hardcover), 978-1-003-12026-1 (eBook)
www.jennystanford.com

to plants, enabling sustained and homogeneous absorption. Previous researches have shown that nanofertilizers enable plant productivity to increase the nutrient usage, reduce soil toxicity, mitigate possible adverse effects of excessive use of chemical fertilizers, as well as fertilizer application frequency. Moreover, the use of nanofertilizers drastically reduces waste, thereby saving money and protecting the environment. Furthermore, nanofertilizers in combination with microorganisms, i.e, nanobiofertilizers offer great benefits and open a new paramount approach toward sustainable agriculture. The eco-friendly products have been expected to reduce the usage of conventional fertilizers by 50%. Although nanofertilizers have a lot of advantages, their consequences during and after application should always be carefully examined and kept in mind to make them more advantageous.

9.1 Introduction

The global population is growing at an alarming rate, which has increased the demand for food continuously and is predictable to rise by 70% up to 2050 [1]. This prediction may range from 100% to 170% for developing countries, where about 800 million people have been placed under the food-insecure category and about 2 billion people are under the hidden hunger category, which has contributed significantly to high mortality rates [2, 3]. The wastage of food (1.3 billion tonnes) each year during the supply chain is causing economic losses of $750 billion [4]. It was predicted that 40% of losses occur at the post-harvest and processing levels in developing countries, and also approximately 40% of losses occur at the retail and consumer levels in developed countries [5]. People are engaged in agriculture applications for improving farming methods so as to increase the production of food crops and other agriculture goods like fibers, timber, fuels, fertilizers, flowers, ornamental, nursery plants, and raw materials for industries etc. However, the shortage of agricultural land, water, climatic changes, crop pests, low nutrient use efficiency, and limited crop productivity are the significant hindrances toward the achievement of global food security. Because of the increasing global population, the demand for food is increasing day by day, which has compelled the growers toward large-scale use of fertilizers.

Fertilizers, also called plant food elements, are materials formed to supply elements or nutrients that are vital for plant growth and development. Ideal crop nutrition is an essential prerequisite for crop production, which means perfect fertilization has a protruding role in the field of agriculture. The yield of crop is extremely reliant on the involvement of macronutrients (N, P, K, S, Ca, Mg) and micronutrients (B, Fe, Mn, Cu, Zn, Mo, and Cl) to agricultural lands [6]. They provide the necessary nutrients to plants for optimal growth and yield; still, current agriculture practices cannot accomplish the increasing food demand without dependence on the extensive use of fertilizers. World consumption of the three main fertilizer nutrients—nitrogen (N), phosphorus (P), and potassium (K)— was estimated at 186.67 million tonnes in 2016, which increased by 1.4% from the level consumed in 2015. Moreover, the demand for N, P, and K grew annually on average by 1.5%, 2.2%, and 2.4%, respectively, from 2015 to 2020 and is expected to further increase in the next 5 years [7]. The use of chemical fertilizers increases the cost and reduces the profit margin for growers. Moreover, the rigorous use of conventional fertilizers over prolonged periods of time has instigated severe environmental restraints worldwide such as groundwater pollution, chemical burn, water eutrophication, soil quality degradation, and air pollution [8, 9]. Conventional fertilizers result in low nutrient use efficiencies (NUEs) due to high release rates, which may be overwhelming the actual nutrient absorption rate of plants, and/or transformation of nutrients that are not bioavailable to crops [10]. The physical and chemical properties of soil, gaseous losses, leaching, runoff, and fertilizer characteristics play a significant role in the NUE of crop plants (Fig. 9.1), and it was estimated that in all agro-ecosystems, the NUE of crop plants was lower than 50% [11].

The enhancement of crop production via improving nutrient use efficacy is one of the foremost pillars of sustainable agriculture and environmental health [12–14]. Presently, the horticulture division is facing a pressure for attaining substantial productivity in crops and food by using chemical fertilizers as an alternative. Various strategies such as fertigation, precise fertilization, limited application, and use of nanofertilizers instead of conventional fertilizers have been planned for the enhancement of fertilizer use efficiency [15, 16]. Nanotechnology plays a substantial role in the production of crops to meet the increasing food demands of population with environmental integrity, environmental sustainability, and economic stability [17–19]. The use of nanoencapsulation in the field of agriculture has enhanced the efficacy by encapsulation of fertilizers within the polymer matrix and preventing the volatilization/degradation of fertilizers under adverse environment conditions [20]. Various reports have been published regarding the application of nanotechnology in the field of agriculture, including seed treatment, germination, plant growth and development, fertilizer delivery, etc. [21–24].

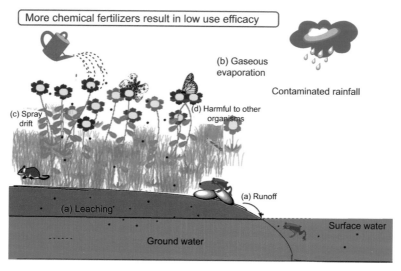

Figure 9.1 Depiction of different ways of contamination of the environment due to overuse of fertilizers in fields. (a) Leaching can cause contamination of groundwater, (b) runoff with rainfall, (c) spray drift into soil, (d) harmful to other organisms, (e) gaseous evaporation can lead to contaminated rainfall.

The chapter, thus, summarizes the concerns regarding the use of conventional fertilizers and the need for nanofertilizers along with discussions on the types, production, and advantages and disadvantages of nanofertilizers. We will also draw a conclusion and discuss the future perspectives of using nanofertilizers.

9.2 Nanofertilizers

Nanofertilizers are made from conventional fertilizers, bulk materials for fertilizers, or extracted from different plant or plant parts by encapsulating/coating them with nanomaterials for controlled and slow release of nutrients for the development of soil fertility, productivity, and quality of agricultural products [25]. Considering the unique physicochemical properties and advantages of nanomaterials, the encapsulated nutrients can be released in nanosized form in a controlled manner to improve the efficiency of crop plants along with minimum impact on the environment. Owing to the high reactivity of nanomaterials, they interact with fertilizers, which results in an improved and effective absorption of nutrients for plants [26]. The use of nanofertilizers in a correct way can feed plants slowly in such a manner that increases the NUE, prevents leaching, minimizes volatilization, and diminishes overall environmental risks [27–29]. Nanofertilizers improve the bioavailability of nutrients owing to high specific surface area, mini size, and more reactivity. The encapsulation of nutrients with nanomaterials can be done in three different ways [30].

- Entrapped/encapsulated within the nanomaterials
- Coated with a layer of nanomaterials
- Delivered in the form of nanoemulsions

The effectiveness of nanofertilizers depends on three factors: intrinsic factors, extrinsic factors, and route of administration. Intrinsic factors include method of preparation of nanoformulation, particle size of nanoformulation, and surface coating. While extrinsic factors include soil depth, soil pH, soil texture, temperature, organic matter, and microbial activity, which may also affect the potential use of nanofertilizers [25, 31–33]. Moreover, the route of administration/mode of application through plant roots or leaves (foliar) also plays a significant role in the absorption, behavior, and bioavailability of nanofertilizers. Figure 9.2 outlines the comparative application of different fertilizers from conventional to nanofertilizers.

Nanofertilizers have been classified into three groups: (1) nanoformulation of micronutrients, (2) nanoformulation of macronutrients, and (3) nutrients-loaded nanomaterials [34]. Out of the three categories, nanomaterials or nanocarriers of nutrients are more popular as compared to nanomaterials made up of nutrients. The benefit of using nutrients-loaded nanomaterials is that they are safe to workers and environment friendly. Moreover, fertilizers encapsulated in the nanocarriers can release fertilizers in a precise manner according to the requirement. Various kinds of nanomaterials have been used for encapsulation and controlled release of fertilizers, such as polymeric nanoparticles, carbon-based nanomaterials, nanoclays, mesoporous silica, and other nanomaterials [20, 35–37]. Controlled-release nanocarriers have also been employed for many other applications, including pesticides, food, and drug delivery [38–41].

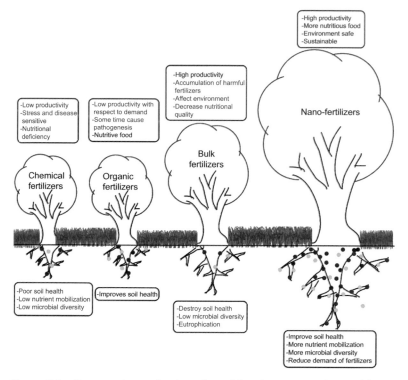

Figure 9.2 Comparative evaluation of possible advantages, gains, and losses of chemical fertilizers, organic fertilizers, bulk fertilizers, and nanofertilizers on plant growth and soil rhizosphere. [Red and blue text respectively denotes adverse and good impacts].

9.3 Manufacturing of Nanofertilizers

Nanomaterials for nanofertilizers can be prepared by different approaches: physical (top-down), chemical (bottom-up), and biological (biosynthetic) approaches (Fig. 9.3). The top-down approach is based on size reduction of relatively large particles into smaller particles of nanoscale by mechanical attrition. Examples of top-down approach are pearl/ball milling, microfluidizer technology, high-pressure homogenization, nanomorph technology, nanocochleate technology, and controlled-flow cavitation technology [42, 43]. The limitation in this approach is the low control on the size of nanoparticles (NPs) and a greater quantity of impurities. In the bottom-up approach, one starts with molecules in the solution and moves via association of these molecules to form NPs using chemical reactions. Examples of bottom-down approach are precipitation method, hydrosol methods, spray freezing into liquid, or supercritical fluid technology and self-assembly. Other methods based on different types of nanomaterials used are ionic gelation, polyelectrolyte complex formation, solvent diffusion, solvent evaporation, complex coacervation, coprecipitation, self-assembly, solid-lipid NPs, and nanostructured lipid carrier suspension [44]. It is a chemically controlled synthetic process; therefore, particle size can be controlled and impurities reduced [45, 46]. In addition to top-down and bottom-up approaches, NPs can also be synthesized biologically. There are several natural sources such as plants, fungi, yeast, and bacteria. The most favorable features of NPs are control of toxicity and particle size [47, 48]. The preparation of nanofertilizers should be done with proper care so that it can prove to be efficient as well as cost effective.

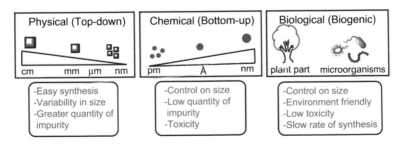

Figure 9.3 Representation of methods for the production of nanofertilizers: physical (top-down), chemical (bottom-up), and biological approaches and advantages and limitations represented in the lower text box with green and red color, respectively.

9.4   Types of Nanofertilizers

Nanoparticulate carriers basically modify the role of fertilizers and help to improve crop yield. Various types of NPs can act as fertilizer or delivery vehicles for fertilizers.

Basically, they modify the role of fertilizers, thereby improving the crop yield. Depending on the type of nutrient, nanofertilizers can be broadly divided into three types: macronutrient based, micronutrient based, and biofertilizers based. Macronutrients can further be divided into primary and secondary. Although primary macronutrients (N, P, and K) are consumed in higher quantities, secondary macronutrients (Ca, Mg, and S) are also very vital for plant growth, which includes calcium, magnesium, and sulfur.

9.4.1   Macronutrient-Based Nanofertilizers

For plant nutrition, sufficient amount of macro- and micronutrients is necessary, including carbon, oxygen, hydrogen, nitrogen, phosphorus, potassium, calcium, sulfur, and magnesium. Out of these, the first three are structural elements and extracted from the environment, while the remaining six are extracted from soil. Though all the macronutrients are important, yet primary macronutrients are consumed in higher quantities in comparison to secondary ones. These primary macronutrients (nitrogen, phosphorus, potassium) are considered fertilizer elements as the familiar "N-P-K" identified on fertilizer labels.

9.4.1.1   Nitrogen nanofertilizers

Nitrogen is the first and foremost nutrient essential for plant growth as it is important for energy metabolism and protein synthesis. Often the extensive use of conventional nitrogenous fertilizers, for example urea, has raised several environmental concerns, because the excess nitrogen is lost through denitrification, volatilization, and leaching. The current NPK ratio is 8.2:3.2:1 in comparison to the optimal ratio of 4:2:1, which has resulted in groundwater pollution and eutrophication in aquatic systems [49]. This has resulted in the demand for development of delivery systems that can release fertilizers at a slow rate to achieve sustained release of N during crop period. Various studies have inferred that nitrogen-based fertilizers hold more potential to increase the production in contrast to conventional mineral urea while minimizing the disadvantages of conventional ones [50]. Nanocarriers such as zeolites, chitosan, or clay can coordinate with plant's demand and release the fertilizer at a sustained rate, thereby resulting in enhanced plant uptake and justified use of N [51, 52]. Owing to the properties such as large surface area and capability to synchronize the release of nitrogen, nanozeolites and their blends have been extensively employed for designing nanofertilizers. Manikandan and Subramanian [49] developed intercalated N nanofertilizer formulations and found a consistent increase in growth, yield, quality, and nutrient uptake in maize crop with respect to conventional urea. Dwivedi et al. [53] have also reported enhancement in solubility and availability of P using zeolites as a carrier. Another study also supported this fact where zeolite-based nitrogen nanofertilizer not only indicated higher accumulation of N in plants but also the post-effect of application in soil exhibited better pH, moisture, and available nitrogen than the conventional fertilizer [54].

9.4.1.2 Phosphorus nanofertilizers

Apart from its role in root growth and flowering stage, P is essential for transporting and storing energy, photosynthesis execution, and organic compound formation. It aids in the plant's ability to survive in unpleasant climatic conditions. Although the amount of P in soils may be much higher than the one desired for plant growth, several factors can limit its availability to plant [55]. For example, its complexes with iron, aluminum hydroxides, and calcium in the soil or its immobilization with clay particles in the soil restricts its availability [56]. Only 10–20% of the supplied P fertilizers are taken up by plants. The boosted use of N fertilizers has further worsened the condition by modifying the microbial biomass of P and its ratios with microbial biomass of N and C [57]. So to overcome these problems, several researchers have formulated and evaluated nanotechnology-based approach for phosphorus fertilizers. Increases of 32.6% and 20.4% were observed in the growth rate and seed yield of soybean plants, respectively, for nanohydroxyapatite-based fertilizer with respect to regular P fertilizers [58]. The use of hydroxyapatite NPs as a carrier for fertilizers has also been studied on *Adansonia digitata*, and it was found that hydroxyapatite NPs led to enhanced plant growth parameters, chemical contents, and anticancer activity of leaves in comparison to different sources of P nanofertilizers [59]. Surface-modified zeolites have also been reported to endorse the phosphorous use efficiency, which barely exceeds 20% in the case of a conventional system [60]. A group of Danish scientists trying to encapsulate P in biodegradable NPs found that P is absorbed directly through leaves in plants, thereby eliminating the need of binding P to the soil. The objective of their study was to reduce the annual consumption of P because excess P gets accumulated in soil only with no benefit to crop [61]. Polysaccharides and lignin have also been explored to coat water-soluble triple superphosphate granular fertilizers. Three polysaccharides—sodium alginate, carrageenan and carboxymethyl cellulose along with lignin—were prepared with phosphate fertilizers in different mass ratios and it was found that coated fertilizers showed sustained release [62].

9.4.1.3   Potassium nanofertilizers

The processes involving the role of potassium include regulation of water, transport of the plant's reserve substances, enhancement of photosynthesis capacity, strengthening of cell tissue, triggering of the absorption of nitrates, stimulation of flowering, and synthesis of carbohydrates and enzymes. Although no clear reason was found for the need for the development of potassium nanofertilizers, several researchers have formulated them and concluded that potassium nanofertilizers give better performance than conventional. Ghahremani et al. observed that when applied at a concentration of 6/1000 in *Ocimum basilicum*, nano-K was most effective in increasing the leaf area, grain yield, biological yield, harvest index, potassium percentage, and chlorophyll content [63]. Gerdini et al. evaluated nano-K fertilizer for foliar application on *Cucurbita pepo* and reported increased number of leaves, improved product quality, disease and pest resistance, and drought tolerance owing to improved nutrient absorption [64]. A nanofertilizer Lithovit has been reported to enhance the plant growth and productivity due to increased natural photosynthesis by means of supplying carbon dioxide (CO_2) at optimum concentration. It was stated that Lithovit particles persist as a thin layer on the surface of leaves and penetrate often after getting wet by dew at night [65]. It has been found that even if administered in equal amounts, higher contents of K are found in nanofertilizers-treated plants [66]. NPs of chitosan and methacrylic acid were used to encapsulate N, P, and K for evaluation on garden pea, and it was noticed that the root-elongation rate reduced in a dose-dependent manner. Although upregulation of some major proteins was observed at lower concentrations, all the concentrations showed genotoxic effects [67]. Potassium nanofertilizer application at the level of 150 + 150 ppm resulted in a significant increase in nutrients content in shoots and seeds of peanut plants in comparison to other treatments [68].

9.4.1.4 Calcium nanofertilizers

Calcium plays a significant role in various processes such as cell wall stabilization, mineral retention in soil and their transportation, neutralizing toxic substances, and seed formation. Foliar application of Ca has shown potential to increase Ca concentration in fruits but still has low efficiency in various instances, which can be ascribed to restrictions in Ca uptake, fruits penetration, epidermal characteristics, composition and presence of cuticle, and low translocation rates of Ca in the phloem [69–72]. A study performed on the yield and quality of pomegranate for lower concentrations of calcium nanofertilizer versus higher concentrations of calcium chloride revealed that the nanofertilizer significantly reduced fruit cracking and increased the yield, while foliar application of calcium did not show any significant effects on crop production, fruits per tree count, and average fruit weight [73]. Ranjbar et al. have [74] assessed the effects of calcium nanofertilizer and calcium chloride spraying on the quantitative and qualitative characteristics of apple fruit at pre-harvest stage. Nanocalcium treatment significantly improved the fruit quality as well as quantity, the highest being at a concentration of 2% [74]. Foliar application of nano-$CaCO_3$ has also been evaluated for its effect on growth and flowering of lisianthus. Spraying of nanofertilizer at a concentration of 500 mg/L resulted in flowering about 15 days prior to control plants along with 56.3% increase in the number of flowers [75].

9.4.1.5 Magnesium nanofertilizers

Magnesium is a vital element for plant growth as it composes the core of the chlorophyll molecule, thus becoming crucial for photosynthesis. It also acts as an enzyme activator. Nonetheless, Mg has been an underestimated mineral nutrient in the last decades and, thus, has been termed "the forgotten element" [76]. The reason may be assigned to the fact that Mg deficiency is seldom identified, and hence there is a research gap in this area. Mg is frequently lost from soil mainly by mobilization, leaching, and biased use of fertilizers. The uptake of magnesium is affected by the presence of other cations such as NH_4, Ca, and K [77]. Delfani et al. [78] evaluated Mg and Fe nanoparticulate solutions for foliar application in black-eyed pea. Enhancement in almost all the investigated attributes was noticed for the combination of these two elements [78]. Magnesium hydroxide NPs have also been explored for their efficacy in seed germination as well as in vitro and in vivo plant growth promotion on *Zea mays*. The particles were found to exhibit 100% seed germination and improved growth at a concentration of 500 ppm. The amounts of Mg content in the leaves and roots of in vitro plants were found to be 131.45 mg/kg and 103.52 mg/kg, respectively, while the amounts were 132.58 mg/kg and 114.58 mg/kg, respectively, for plants grown in vivo [79].

9.4.1.6 Sulfur nanofertilizers

Sulfur contributes to chlorophyll formation and increases nitrogen efficiency as well as plant defenses. Yet, sulfur has been overlooked in the current agronomic practices, thereby affecting both crop yields and quality. The most common sources of S include sulfate (SO_4^{2-}) or elemental sulfur (S_8). Sulfate salts can be quickly uptaken by plants but possess low S concentrations that are not capable of meeting large input requirements of crops. Furthermore, leaching problems of SO_4^{2-} result in significant loss and environmental issues. Elemental sulfur (S_8) has much higher concentrations (>90% S in products) of S, but it can be taken up by plants only after its biological oxidation by soil microorganisms, which is strongly influenced by the particle size of fertilizers [80]. So reduction in particle size may significantly influence the oxidation rate [81]. Thus, the development of sulfur nanofertilizers can be a useful approach. Alipour (2016) investigated the effect of sulfur nanofertilizers on the growth and nutrition of *Ocimum basilicum* in response to salt stress and found no significant effect of sulfur on the examined traits [82]. Green synthesis of sulfur NPs has been achieved using *Ocimum basilicum* leaves extract and were applied in different concentrations (12.5, 25, 50, 100, and 200 µM) to *Helianthus annuus* seeds and irrigated with 100 mM $MnSO_4$ for a pot study. It was observed that sulfur NPs decreased Mn uptake, enhanced S metabolism, elevated the water content of seedlings, and eliminated physiological drought, indicating that sulfur nanofertilizers can limit the deleterious effects of Mn stress [83].

9.4.2 Micronutrient-Based Nanofertilizers

Micronutrients include Fe, Mn, Zn, Cu, Mo, and Ni. In comparison with the macronutrients, only trace levels of micronutrients are required for the healthy growth of crops and other plants.

9.4.2.1 Iron nanofertilizers

Iron acts as an important cofactor for enzymes dealing with numerous biological processes in plants [84]. Soil is rich in iron because approximately 5% of the weight of the earth's crust is iron [85]. But a large amount of iron is not available to plants due to its presence in insoluble forms. A promising approach to make iron available to plants is the use of highly stable and slow-release nanoformulations. Iron chelate nanofertilizer is highly stable and provides slow release of iron in a broad pH range. Another positive aspect of iron-based nanofertilizers is that their structure is free from ethylene-based compounds, which leads to early aging and senescence of plants [86]. Foliar treatment of black-eyed peas with iron NPs (500 mg/L) significantly improved the number of pods per plant by 47%, weight of 1000 seeds by 7%, Fe content in leaves by 34%, and chlorophyll content by 10% as compared to the controls [78]. In another study, the application of iron oxide NPs (Fe_2O_3) in plants treated with varying concentrations for 70 days resulted in a significant increase in growth parameters, photosynthetic pigments, and total protein contents with the maximum amount at a concentration of 30 µM [87]. The evaluation of γ-Fe_2O_3 NPs (20–100 mg/L) in watermelon and *Zea mays* increased the Cl content

[88]. Lower concentrations (0–0.75 g/L) of ferrous oxide NPs were reported to enhance the Cl content and the levels of lipids and proteins, whereas reduction in these parameters was achieved at higher concentrations (0.75–1.0 g/L) in soybean plants [89]. A study conducted on the effects of nanoscale zero-valent iron (nZVI) on a terrestrial crop, *Medicago sativa* (Alfalfa), exhibited a higher content of chlorophyll in 20-day-old seedlings, while the carbohydrate and lignin contents decreased slightly [90]. The application of Cornelian cherry fruit extract synthesized Fe_2O_3 NPs displayed statistically significant root and shoot biomass stimulation [91].

9.4.2.2   Zinc nanofertilizers

Zinc is important for the catalytic activity of various metabolic enzymes, such as isomerases, aldolases, dehydrogenases, transphosphorylases, and RNA and DNA polymerases. It is also intricate in different processes such as cell division, tryptophan synthesis, photosynthesis, protein synthesis, and in the maintenance of membrane structure and potential [92]. The limitation of Zn fertilizers is the soil fixation of most of the added Zn, but zinc-based nanofertilizers have indicated a great potential [93]. Application of zinc nanofertilizers to plants can be accomplished by various methods such as by soil mixing, foliar spray, and/or seed-priming method. Out of these, the seed-priming method is simple, more efficient, and cost effective [94–97]. Low concentrations of ZnO NPs (≤100 mg/Kg) when applied to the soil resulted in an enhanced Zn uptake by cucumber plant, but higher concentrations (1000 mg/kg) caused plant growth inhibition [98]. Foliar application of zinc oxide NPs on lisianthus has resulted in enhanced petal anthocyanin and leaf chlorophyll content along with increased number of leaf, lateral branches, and flowers [75].

9.4.2.3   Copper nanofertilizers

Copper is a crucial micronutrient for several important physiological functions, including mitochondrial respiration, cellular transportation, antioxidative activity, protein trafficking, and hormone signaling of plants [99]. Copper fertilizers are mostly employed in formulations intended for use in crop protection as copper plays a key role in the health and nutrition of plants. The use of copper NPs in the form of fertilizers presents a probable way of exposure to the plants [100]. But actual concentration and methods of application need to be determined so as to achieve beneficial effect and avoid the ecotoxic effect of nanofertilizers [101]. Usages of Cu ions in minor amounts have been suggested for serving the purpose of microelement and stimulate plant growth [102]. The application of biosynthesized (using *Citrus medica* L. fruit extract) copper NPs (CuNPs) at doses up to 20 µg/mL improved the mitotic index in actively dividing cells of *Allium cepa* [103]. Improvement in stress tolerance in wheat was achieved with the employment of Cu NPs as indicated by improved levels of proteins involved in starch degradation and glycolysis, superoxide dismutase activity, sugar content, and Cu content in CuNP-treated seeds [104]. A substantial increase in root length, height, fresh and dry weights of pigeon pea (*Cajanus cajan* L.) seedlings was noticed when treated with biogenic CuNPs having 20 nm size [105]. Encapsulation of CuNPs in CS-PVA hydrogels improved the yield, nutraceutical properties, total antioxidant capacity, and higher lycopene content [106]. Treatment of tomato plants with CuNPs has been reported to produce fruits with more firmness along with enhancement in vitamin C, lycopene contents, antioxidant capacity, and activity of superoxide dismutase activity and catalase [107]. Foliar spray of Cu-chitosan NPs or in combination with seed coating has boosted the yield and growth profile of finger millet plants as well as enhanced defense enzymes resulting in the suppression of blast disease [108].

9.4.2.4 Manganese nanofertilizers

Manganese NPs (MnNPs) have been proven as a better micronutrient source of Mn in comparison to commercially available $MnSO_4$ salt. It has been noticed that MnNPs not only increased the growth but also improved photosynthesis in mung bean (*Vigna radiata*). Maximum growth increment was achieved by using MnNPs at 0.05 mg/L and increased effect was noted in shoot length by 10%, in root length by 2%, in fresh biomass by 8%, in number of rootlets by 28%, and in dry biomass by 100% with respect to $MnSO_4$ salt. When used at a concentration of 1 mg/L, $MnSO_4$ salt exhibited an inhibitory effect on plant growth, whereas the response was still positive for MnNPs [109]. MnNPs (0.1, 0.5, 1 mg/L), when evaluated as a nano-priming agent to improve salinity stress (100 mM NaCl during germination) in *Capsicum annuum* L., has been found to significantly improve the root growth in salt-stressed as well as non-salt seedlings [110].

9.4.2.5 Boron nanofertilizers

Boron as a fertilizer element is required in small quantities by plants and has an important role in the formation of cellular walls and assists the movement and transfer of photosynthesis of leaves to active areas [111]. It is also vital for the formation of bark, transfer of some active hormones that affect the growth of the stem and root levels, germination of pollen, flowering, and increasing the level of carbohydrates transferred to the active areas of growth during the reproductive stage. The flowering stage requires its continuous supply, while it should be present in sufficient amount for effective nodulation and nitrogen fixation in legumes [112]. Boron deficiency can be lessened with the help of fertilizers, but the harmful effect of frequent fertilizer application disturbs the soil fertility and results in environmental pollution. Considering this, nanotechnology has been proposed as an alternative technique, which can be effectively used for B acquisition [113]. Boron metal and its NPs, when sprayed at three concentrations (0, 90, and 180 mg/L), boron NPs at a concentration of 90 mg/L, showed higher effect on increasing the plant height, the number of pods, and total seed yield [114]. It has also been indicated that alfalfa can be produced in large quantities with suitable forage quality when boron nanofertilization is used under calcareous conditions [115].

9.4.2.6 Molybdenum nanofertilizers

Molybdenum is required in very small quantities. The range is between 0.3 and 1.5 ppm for most of plant tissue and between 0.01 and 0.20 ppm for a growing medium. Molybdenum deficiency or toxicity is rare, but its deficiency is commonly found in poinsettias (*Euphorbia pulcherrima*) [116]. Molybdenum is important for two enzymes that convert nitrate to nitrite and then to ammonia prior to its use in the synthesis of amino acids within the plant [117]. In legumes, it is required by symbiotic nitrogen-fixing bacteria for the fixation of atmospheric nitrogen [118]. It is also used by plants for the conversion of inorganic phosphorus into organic forms. Due to the alluring benefits of nanofertilizers, attempts have been made to study the effects of molybdenum NPs (MoNPs) as a fertilizer. Taran et al. used MoNPs solution as a micronutrient source of Mo for chickpea and reported that application of MoNPs intact or in combination with microbial treatment had the potential to improve the yield, performance, and disease resistance of legume as well as other crop species [119]. Biosynthesized MoNPs (2–7 nm) using a fungus *Aspergillus tubingensis* TFR29 have shown significant improvement in root area, root length, number of tips, root diameter, beneficial enzymes, and microbial activities in the rhizosphere, biomass, and grain yield at a standardized dose of 4 ppm [116].

9.4.2.7 Nickel nanofertilizers

Although nickel has been known as a trace element, its uptake is very important for different enzyme activities, for maintenance of cellular redox condition and several other activities responsible for development such as physiological, biochemical, and growth responses [120]. Nickel NPs of 5 nm at low concentrations of 0.01 and 0.1 mg/L exhibited no affect or stimulated growth in 10-day wheat seedlings although a little increase was noticed in the content of Chla and Chlb after the application of 0.01 mg/L [121].

No published report could be found for nanofertilizers of chlorine to the best of our search.

9.4.3 Biofertilizers-Based Nanofertilizers

The layman definition of nanobiofertilizer encompasses an intentional coexistence of a biocompatible nanomaterial and a biological-source-driven fertilizer (substantially organic), encompassing high efficacy of both the ingredients. These traits are aimed toward facilitating slow and gradual nutrient release over a long span of crop growth, together contributing to improved nutrient usage as well as promoting crop yield and productivity [122, 123]. Probably, studies in the last decade have conveyed a gradual shift in interest toward nano- and biofertilizers from their chemical counterparts [124]. A biofertilizer is exclusively composed of biologically useful microorganisms such as rhizobium, blue-green algae, mycorrhizae, bacterium azotobacter, azospirillum, phosphate-dissolving bacteria such as *Pseudomonas* and *Bacillus* species. These microorganisms act equivalent to those of catalysts, not only by modulating the characteristic nitrogen-fixing ability but also by improving the solubility of insoluble complex organic matter, through conversion to simpler form, resulting in cumulative enhanced bioavailability. It is due to these microbes only that the moisture-absorbing ability of host soil enhanced the nutrient availability to plants, by maintaining a homogeneous soil chemical texture (through replenishing soil microbial content, augmenting soil aeration and natural fertilization) [125]. Though seeming revolutionary and renewable, these positive traits do not come without a cost. Vulnerability toward nanoscale texture retainment, poor on-field stability, and varying activities under fluctuating environmental conditions (temperature, pH sensitivity, radiation exposure), shortage of beneficial bacterial strains, susceptibility toward desiccation, and substantially high dose requirement for a large area are some of the limiting factors of this technology [126]. Such issues complicate the desired fertilizer availability to the host plant, failing which there is a consistent possibility of environmental quality deterioration. The nanoscale formulation of a biofertilizer (wherein, nanofertilizers terminology) resolves these issues through conferring structural protection to biofertilizer nutrients and plant-growth-promoting bacteria, via nanoencapsulation-mediated coating of nanoscale polymers [127]. The nanoencapsulation approach could be used as a dynamic mechanism to elongate the structural protection of being delivered biofertilizer, enhance its chemical shelf life and dispersion in fertilizer formulation, allowing a controlled release [128]. Besides improving nutrient release characteristics, the technology also betters the field performance and conclusively reduces the economic expenditure (through cost reduction as well as reduced application extents). Improvement in inorganic nutrient utilization (the N, P, and K hypothesis), activity of associated soil systems, bettered crop product quality along with an improved disease resistance are some key advantages of the nanobiofertilizer technology [123]. Figure 9.4 depicts the typical role of nanofertilizers in enhancing plant growth and nutritional upkeep. The diverse impacts of nanobiofertilizers on soil texture and plant system enzymes manifest significant benefits in enabling improved growth and nutritional quality. Nanomaterials such as chitosan, zeolites, and polymers facilitate considerable enhancements in the absorption of organic nutrients through the nanoencapsulation phenomenon, forming a sustainable rich source of nutrients for the plants [129]. The large surface coatings of NPs over the biofertilizer improve the distribution of constituent nutrients.

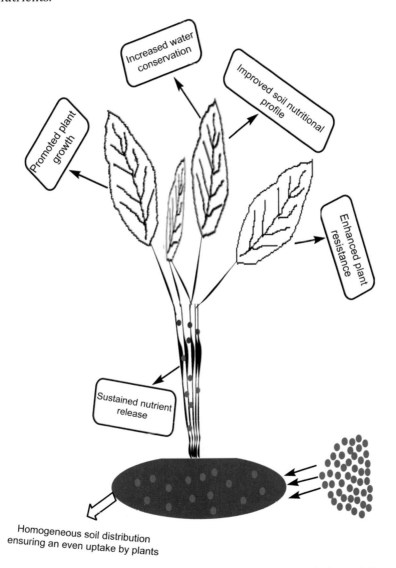

Figure 9.4 Nanofertilizer impacts on the different plant growth determining factors.

Increased SA, nanoscale dimensions, and higher chemical reactivity of NP-coated fertilizers facilitate enhanced interaction and stable chemical texture, allowing for an efficient uptake via improved bioavailability. The steady release from the bound nanocarriers also ensures a long-range availability of administered fertilizers along the different stages of plant growth [48]. The biological content of nanobiofertilizer (plant-growth-promoting rhizobia, PGPR, or some fungal inoculants) is synergistically benefitted via improving the soil nutrient content using multiple mechanisms. Some of these potential mechanisms include atmospheric nitrogen fixation through simultaneous activities of plant roots and rhizobacterium), formation of siderophores (provisions for chelating metals and improving their accessibility to plant root), phosphate solubilization through the activities of P-solubilizing bacterial and fungal strains, besides assuring needful hormonal activities [130, 131]. The wholesome improved response of nanobiofertilizer administration in crop plants has been reported in manifold studies, in terms of improved qualitative as well as quantitative crop growth parameters.

9.4.3.1 Effects on morphological and physiological aspects

Application of nanobiofertilizers has been studied to significantly improve the crop growth through the optimization of photosynthesis, nutrient absorption efficacy, higher photosynthate accumulation and nutrient translocation, enabling enhanced productivity as well as quality. One study has specifically reported the characteristic effect of nanobiofertilizers made via entrapment of biofertilizer (growth-promoting bacteria) within Au and Ag NPs, wherein significantly higher crop growth was witnessed upon the administration of NPs with bacteria compared to those with the NPs alone. Nanostructured fertilizer consisting of neem cake with PGPR provides efficacy toward promoting crop-harvest yields in several leguminous crops through an earlier and greater seed germination as well as effective delivery of doped nutrients [9].

9.4.3.2 Enhanced nutritional security in plant system

Repeated application of conventional/chemical fertilizers gradually subtracts the essential soil nutrients prevailing in the fertile top soil. Excessive soil acidification is the exclusive factor impairing soil fertility [9]. Poor soil fertility and nutrient content are the major challenges encountered by farmers, attributed to below-par crop production and abysmal food-nutrition extents [132]. Nanobiofertilizer offers a sustainable, cost-effective, and competent integrated management of nutrients in resolving such issues, via modulating absorption and assimilation of nutrients by plant alongside minimizing nutrient loss caused by leaching, gasification, or competing with other microbes [133]. All crops do not require the same extent of a particular nutrient in the different stages of their growth, and it is obvious that after initial stages, the requirement of protein-based elements is moderate (as enhancing growth almost ceases). It is highly important that the supporting soil base supports such plant needs, and that is only possible only if the existent pool of soil constituents is not damaged or unduly involved in some undesired chemical activities. The NPs (nanotechnology component) of nanobiofertilizers perform the role of ensuring such needs through safeguarding the non-required nutrient materials at a given time instant and releasing them in a need-based manner through gradient diffusion and hormonal controls. Such an advantage is simply unattainable with chemical fertilizers, which increase the stress load on soil and also damage the underground water table through progressive seepage of non-required nutrient elements.

9.4.3.3 Improved pests and pathogen resistance

Gatahi et al. studied the effect of nanobiofertilizers in bacterial wilt (caused by *Ralstoniasola nacearum*) infected tomato crop and reported its pesticide resistance [134]. Similarly, Gouda et al. studied the effects of PGPR-containing nanobiofertilizers toward fatal fungal and bacterial pathogens within the rhizosphere of leguminous crops. The scientists reported the protective role of PGPR-containing nanobiofertilizers toward various pathogens [135]. The enhanced adhesion of useful bacteria onto the roots of oilseed rape and safety from harmful fungal infection were observed through the application of TiNP-coated nanobiofertilizers [136]. A nanoclay-based biological compound comprising *Trichoderma* and *Pseudomonas* sp was investigated for controlling the fungal nematode disease in rabi crops and was noticed to confer crop resistance against abiotic stress [137].

Mechanism of uptake: Nanofertilizers enter plant tissue either through roots or through upper parts. Although specific uptake and translocation of NPs by plant cells are not known yet, several reports support the statement that uptake of NPs by plants is mainly dependent on size, shape, and interaction behavior of nanoparticles with cell wall. The size exclusion limit of cell wall (5–20 nm) acts as a barrier that restricts the entry of larger particles into plant cells [93]. Enlargement of pores or induction of new pores can be achieved via surface functionalization of NPs. Other possible routes include ion channels, through endocytosis, via complex formation with membrane transporters or root exudates [138]. Nanocarriers protect encapsulated nutrients from soil filtration and keep in soil around the roots. Encapsulated components may enter the soil network via hydrogen bonds, molecular force, surface tension, or viscous force, thus extending their spatial scale [139].

For foliar applications, NPs may enter through stomatal openings or cuticles [138]. The first barrier that restricts NPs of size less than 5 nm is the cuticles of leaves. NPs entering through stomatal pores can pass the vascular system via apoplastic or sympathetic pathway. The particles with a size range of 10–50 nm prefer the sympathetic pathway, while the large one (50–200 nm) prefer the apoplastic pathway. Furthermore, many factors affect the efficacy of NPs such as chemical composition and morphology of the leaves, sunlight, the microsphere of the photosphere, temperature, and humidity [140–142].

Advantages and limitations of nanofertilizers: Though nanofertilizers are a boon to agriculture, yet some researchers have shown their concern regarding the ill effects that may occur with their improper use. Table 9.1 presents various advantages and disadvantages [25, 129] associated with the use of nanofertilizers.

Table 9.1   Advantages and limitation of nanofertilizers

Sr. No.	Properties	Effects
		I. Advantages
1.	Facilitate higher nutrient use efficiency	• Small particle size than pore size of root and leaves leads to more penetration into the plant • Improve uptake and nutrient use efficacy of crop plants • Prevent the loss of nutrients
2.	Nutritional value and health	• Nanofertilizers enhance growth of plant parts and metabolic process such as photosynthesis; improve the yield • More availability of nutrients helps to increase the quality parameters of crops, such as protein, oil content, sugar content, etc. • More availability of nanonutrient to the plant, prevent from disease, nutrient deficiency and other biotic and abiotic stress, which result in better yield and quality food products for human and animal consumption
3.	Controlled release	• Nanofertilizers control the speed and dose of encapsulated nutrient/fertilizers to make more uptake by crop plant. • Increase availability due to slow release of nutrients • Increase actual duration of nutrient supply
4.	Reduce lose and demand of fertilizers	• Nanofertilizers can take up by the plants due to slower rate of release • Nutrients can be taken up by plants without wastage by leaching and/or leaking • Reduce the demand for fertilizers
5.	Improve soil quality	• Improve water-holding capacity and soil quality • Increase microbial activity

(*Continued*)

Table 9.1 (*Continued*)

Sr. No.	Properties	Effects
		II. Disadvantages
1.	Transformation of NPs	• Owing to the property of reactivity, nanomaterials can interact with different components of environment, which leads to transformation and changes in physicochemical properties • Nanomaterials can interact with soil components and may cause toxicity
2.	Accumulation of NPs	• Nanofertilizers can accumulate in plant parts, leading to growth inhibition, generation of reactive oxygen species, and cell death • Can accumulate in food parts and, when consumed, may cause human health problems
3.	Safety concern for farm workers	• Reactivity and variability of nanomaterials have raised safety concerns for workers who may become exposed during their manufacturing and application in the field

9.5 Future Prospects

Despite stout academic rationale and logic, the practical implementation of nanofertilizers still has a long way to go. There are some fundamental reasons and justifications for this unscrupulous underestimated expectant scenario. First and foremost, developing countries such as India and several others have extensive agriculture practices, which are being mitigated in the rural background. Getting the backing of farmers (who are the real stakeholders) in such intriguing circumstances and conservative familial associations are challenge that have perhaps alluded most of the scientific distinctions. So there is a need to make grassroot efforts in awakening the farming community and farmers about the positives of nanocarrier-mediated fertilizer delivery. In many regions, there is a continued setback regarding vulnerable climate pattern to how and when what kind of fertilizers are good for which kind of soil.

Despite significant clarification about greater specific activities of fertilizers, the knowledge of which and how many fertilizers are to be used at which time of year, is still incomplete to the farmers. Therefore, scientists and media personnel must initiate harmonious and committed joint efforts along with reliable governmental support so that the exact scientific rationale for nanofertilizer usage is understood. The prospects regarding how nanocarriers can enable homogeneous distribution of fertilizers and reduce the net quantity of used fertilizers are highly crucial when illustrated from a comparative viewpoint of their normal administration. Policies at the government level should focus less on compensation but more on scientific resolution of incumbent vulnerabilities. This would, in turn, inculcate a big boost in farmers' mind regarding the willingness to accept and embrace the scientific remedies. For instance, how to practice mixed cropping, crop rotation, and how a same soil does not need same fertilizers after harvest of a particular agricultural crop, can be decisive factors in minimizing the risk of natural calamities. Similarly, what are the possible gradual consequences of unaided (without nanocarrier) administered fertilizer to the concerned crop and soil variety can be highly substantial inputs for readiness to accept the technology. Likewise, the benefits of providing higher organic content by nanobiofertilizers can be a significant game-changing strategy. Many other things could be conceptualized, but the success of all depends on committed efforts keeping the nation's development at the top and not at all individual interests.

9.6 Conclusion

The scientific essence of nanofertilizers is to boost agricultural outputs, characterized by correct selection and uniform dispersal of seeds, thorough irrigation and adequate as well as regulated use of fertilizers. Fertilizer distribution efficacy is an important criterion that affects the agricultural growth and economic contribution. Several factors determine this phenomenon, including soil type, chemical combination with other nutrients, leaching effect, and uptake efficiency of plants. Unchecked use of chemical fertilizers has resulted in serious deterioration of soil fertility, enhanced episodes of environmental pollution, pest resistance, and threats to biodiversity and economy as a whole. Thereby, an intensive focus of current agricultural research is to find amicable alternative to chemical fertilizer usage, enabling an environment-friendly approach in agriculture. The diversity of robustly prepared nanomaterials (from microbes to plants) bears enormous potential for developing better, safer, and easily biodegradable fertilizers while improving the distribution of chemical fertilizers. The sole requirement in achieving this task mandates that the chemical makeup of NPs (the core as well as capping agent) should not cross-react with soil environment. It is with optimum combinations of nanomaterial and biofertilizer that the development of low-cost eco-friendly nanobiofertilizers can be boosted. The best scenario toward understanding improved fertilizer distribution and administration through nanomaterials and nanodevices includes the delivery of biologically prepared NPs with fertilizers or even biofertilizers. Here the biologically generated NPs can improve the distribution and prevent overlogging of soil with excessive and undesired chemicals. In a nutshell, nanobiofertilizers hold a great potential to boost the agricultural output at the desired rate when used in optimum concentrations while overcoming the limitations of conventional fertilizers.

References

1. Bindraban, P. S., Dimkpa, C. O., Angle, S., and Rabbinge, R. (2018). Unlocking the multiple public good services from balanced fertilizers. *Food Secur.*, **10**, pp. 273–285.

2. Alexandratos, N. and Bruinsma, J. (2012). *World Agriculture towards 2030/2050: The 2012 Revision. ESA Working Paper No. 12-03.* Rome: FAO.

3. McClafferty, B. and Zuckermann, J. C. (2015). *Cultivating Nutritious Food Systems: A Snapshot Report.* Washington, DC: Global Alliance for Improved Nutrition (GAIN), pp. 47.

4. UN News. UN Report: (2013). One-third of world's food wasted annually, at great economic, environmental cost; https://news.un.org/en/story/2013/09/448652#.VtR44niOeLE (Accessed February 17, 2020).

5. Food and Agriculture Organization (FAO). Save food: Global initiative on food loss and waste reduction; http://www.fao.org/save-food/resources/keyfindings/en/ (Accessed February 17, 2020).

6. Food and Agriculture Organization (FAO). (2015). World fertilizer trends and outlook to 2020, pp. 1-38; http://www.fao.org/3/a-i6895e.pdf (Accessed February 17, 2020).

7. Marschner, H. (2011). *Marschner's Mineral Nutrition of Higher Plants*. Academic Press.

8. Savci, S. (2012). An agricultural pollutant: Chemical fertilizer. *Int. J. Environ. Sci. Dev.*, **3**, pp. 73.

9. Rahman, K. M. and Zhang, D. (2018). Effects of fertilizer broadcasting on the excessive use of inorganic fertilizers and environmental sustainability. *Sustainability*, **10**, pp. 759.

10. Chhipa, H. (2017). Nanofertilizers and nanopesticides for agriculture. *Environ. Chem. Lett.*, **15,** pp. 15–22.

11. Baligar, V. C. and Fageria, N. K. (2015). Nutrient use efficiency in plants: An overview. In *Nutrient Use Efficiency: From Basics to Advances* (Rakshit, A., Singh, H. B., and Sen, A., Eds.), pp. 1–14. Springer.

12. Campos, E. V. R., de Oliveira, J. L., Fraceto, L. F., and Singh, B. (2015). Polysaccharides as safer release systems for agrochemicals. *Agro. Sustain. Dev.,* **35**, pp. 47–66.

13. Manjunatha, S. B., Biradar, D. P., and Aladakatti, Y. R. (2016). Nanotechnology and its applications in agriculture: A review. *J. Farm Sci.*, **29**, pp. 1–13.

14. Morales-Díaz, A. B., Ortega-Ortíz, H., Juárez-Maldonado, A., Cadenas-Pliego, G., González-Morales, S., and Benavides-Mendoza, A. (2017). Application of nanoelements in plant nutrition and its impact in ecosystems. *Adv. Nat. Sci. Nanosci.*, **8**, pp. 013001.

15. Lü, S., Feng, C., Gao, C., Wang, X., Xu, X., Bai, X., Gao. N., and Liu, M. (2016). Multifunctional environmental smart fertilizer based on L-aspartic acid for sustained nutrient release. *J. Agr. Food Chem.*, **64**, pp. 4965–4974.

16. Van Eerd, L. L., Turnbull, J. J. D., Bakker, C. J., Vyn, R. J., McKeown, A. W., and Westerveld, S. M. (2017). Comparing soluble to controlled-release nitrogen fertilizers: Storage cabbage yield, profit margins, and N use efficiency. *Can. J. Plant Sci.*, **98**, pp. 815–829.

17. Liu, R. and Lal, R. (2015). Potentials of engineered nanoparticles as fertilizers for increasing agronomic productions. *Sci. Total Environ.,* **514**, pp. 131–139.

18. Raliya, R., Saharan, V., Dimkpa, C., and Biswas, P. (2017). Nanofertilizer for precision and sustainable agriculture: Current state and future perspectives. *J. Agr. Food Chem.*, **66**, pp. 6487–6503.

19. Feregrino-Perez, A. A., Magaña-López, E., Guzmán, C., and Esquivel, K. (2018). A general overview of the benefits and possible negative effects of the nanotechnology in horticulture. *Sci. Hortic.*, **238**, pp. 126–137.

20. Kumar, R., Ashfaq, M., and Verma, N. (2018). Synthesis of novel PVA-starch formulation-supported Cu–Zn nanoparticle carrying carbon nanofibers as a nanofertilizer: Controlled release of micronutrients. *J. Mater. Sci.*, **53**, pp. 7150–7164.

21. Jyothi, T. V. and Hebsur, N. S. (2017). Effect of nanofertilizers on growth and yield of selected cereals: A review. *Agric. Rev.*, **38**, pp. 112–120.

22. Al-Juthery, H. W., Hassan, A. K. H., Musa, R. F., and Sahan, A. H. (2018). Maximize growth and yield of wheat by foliar application of complete nano-fertilizer and some bio-stimulators. *Res. Crops*, **19**, pp. 387–393.

23. Shebl, A., Hassan, A. A., Salama, D. M., Abd El-Aziz, M. E., and Abd Elwahed, M. S. (2019). Green synthesis of nanofertilizers and their application as a foliar for *Cucurbita pepo* L. *J. Nanomater.*, pp. 1–11.

24. El-Aziz, M. A., Morsi, S. M. M., Salama, D. M., Abdel-Aziz, M. S., Elwahed, M. S. A., Shaaban, E. A., and Youssef, A. M. (2019). Preparation and characterization of chitosan/polyacrylic acid/copper nanocomposites and their impact on onion production. *Int. J. Biol. Macromol.*, **123**, pp. 856–865.

25. Zulfiqar, F., Navarro, M., Ashraf, M., Akram, N. A., and Munné-Bosch, S. (2019). Nanofertilizer use for sustainable agriculture: Advantages and limitations. *Plant Sci.*, 110270.

26. Prasad, R., Bhattacharyya, A., and Nguyen, Q. D. (2017). Nanotechnology in sustainable agriculture: Recent developments, challenges, and perspectives. *Front. Microbiol.*, **8**, 1014.

27. Linquist, B. A., Liu, L., van Kessel, C., and van Groenigen, K. J. (2013). Enhanced efficiency nitrogen fertilizers for rice systems: Meta-analysis of yield and nitrogen uptake. *Field Crops Res.*, **154**, pp. 246–254.

28. Solanki, P., Bhargava, A., Chhipa, H., Jain, N., and Panwar, J. (2015). Nano-fertilizers and their smart delivery system. In *Nanotechnologies in Food and Agriculture*, pp. 81–101. Springer, Cham.

29. Chen, J. and Wei, X. (2018). Controlled-released fertilizers as a means to reduce nitrogen leaching and runoff in container-grown plant production. In *Nitrogen in Agriculture: Updates* (Khan, A. and Fahad, S., Eds.), pp. 33–52. IntechOpen.

30. Iqbal, M. A. (2019). Nano-fertilizers for sustainable crop production under changing climate: A global perspective. In *Sustainable Crop*

Production (Hasanuzzaman, M., Fujita, M., Filho, M. C. M. T., and Nogueira, T. A. R., Eds.). IntechOpen.

31. Solanki, P., Bhargava, A., Chhipa, H., Jain, N., and Panwar, J. (2015). Nano-fertilizers and their smart delivery system. In *Nanotechnologies in Food and Agriculture*, pp. 81–101. Springer, Cham.

32. Ma, C., White, J. C., Zhao, J., Zhao, Q., and Xing, B. (2018). Uptake of engineered nanoparticles by food crops: Characterization, mechanisms, and implications. *Annu. Rev. Food Sci. Technol.*, **9**, pp. 129–153.

33. El-Ramady, H., Abdalla, N., Alshaal, T., El-Henawy, A., Elmahrouk, M., Bayoumi, Y., Shalaby, T., Amer, M., Shehata, S., Fári, M., and Domokos-Szabolcsy, E. (2018). Plant nano-nutrition: Perspectives and challenges. In *Nanotechnology, Food Security and Water Treatment*, pp. 129–161. Springer, Cham.

34. Kah, M., Kookana, R. S., Gogos, A., and Bucheli, T. D. (2018). A critical evaluation of nanopesticides and nanofertilizers against their conventional analogues. *Nat. Nanotechnol.*, **13**, pp. 677–684.

35. Roshanravan, B., Soltani, S. M., Rashid, S. A., Mahdavi, F., and Yusop, M. K. (2015). Enhancement of nitrogen release properties of urea-kaolinite fertilizer with chitosan binder. *Chem. Spec. Bioavailab*, **27**, pp. 44–51.

36. Guo, H., White, J. C., Wang, Z., and Xing, B. (2018). Nano-enabled fertilizers to control the release and use efficiency of nutrients. *Curr. Opin. Environ. Sci. Health*, **6**, pp.77–83.

37. Rastogi, A., Tripathi, D. K., Yadav, S., Chauhan, D. K., Živčák, M., Ghorbanpour, M., El-Sheery, N. I., and Brestic, M. (2019). Application of silicon nanoparticles in agriculture. *3Biotech*, **9**, pp. 90.

38. Chopra, M., Kaur, P., Bernela, M., and Thakur, R. (2014). Surfactant assisted nisin loaded chitosan-carageenan nanocapsule synthesis for controlling food pathogens. *Food Control*, **37**, pp. 158–164.

39. Kumar, S., Kaur, P., Bernela, M., Rani, R., and Thakur, R. (2016). Ketoconazole encapsulated in chitosan-gellan gum nanocomplexes exhibits prolonged antifungal activity. *Int. J. Biol. Macromol.*, **93**, pp. 988–994.

40. Sotelo-Boyas, M., Correa-Pacheco, Z., Bautista-Batnos, S., and Gomez, Y. G. Y. (2017). Release study and inhibitory activity of thyme essential oil-loaded chitosan nanoparticles and nanocapsules against foodborne bacteria. *Int. J. Biol. Macromol.*, **103**, pp. 409–414.

41. Bernela, M., Kaur, P., Ahuja, M., and Thakur, R. (2018). Nano-based delivery system for nutraceuticals: The potential future. In *Advances in Animal Biotechnology and Its Applications*, pp. 103–117. Springer, Singapore.

42. Yadav, T. P., Yadav, R. M., and Singh, D. P. (2012). Mechanical milling: A top down approach for the synthesis of nanomaterials and nanocomposites. *Nanosci. Nanotechnol.*, **2**, pp. 22–48.

43. Kumar, S., Dilbaghi, N., Rani, R., Bhanjana, G., and Umar, A. (2013). Novel approaches for enhancement of drug bioavailability. *Rev. Adv. Sci. Eng.*, **2**, pp. 133–154.

44. Kumar, S., Dilbaghi, N., Saharan, R., and Bhanjana, G. (2012). Nanotechnology as emerging tool for enhancing solubility of poorly water-soluble drugs. *BioNanoScience*, **2**, pp. 227–250.

45. Singh, G. and Rattanpal, H. (2014). Use of nanotechnology in horticulture: A review. *Int. J. Agric. Sci. Vet. Med*, **2**, pp. 34–42.

46. Pradhan, S. and Mailapalli, D. R. (2017). Interaction of engineered nanoparticles with the agri-environment. *J. Agric. Food Chem.*, **65**, pp. 8279–8294.

47. Ingale, A. G. and Chaudhari, A. N. (2013). Biogenic synthesis of nanoparticles and potential applications: An eco-friendly approach. *J. Nanomed. Nanotechol.*, **4**, pp. 1–7.

48. El-Ghamry, A., Mosa, A. A., Alshaal, T., and El-Ramady, H. (2018). Nanofertilizers vs. biofertilizers: New insights. *Environ. Biodiver. Soil Security*, **2**, pp. 51–72.

49. Manikandan, A. and Subramanian, K. S. (2016). Evaluation of zeolite-based nitrogen nano-fertilizers on maize growth, yield and quality on inceptisols and alfisols. *Int. J. Plant Soil Sci.*, **9**, pp. 1–9.

50. Milani, N., McLaughlin, M. J., Stacey, S. P., Kirby, J. K., Hettiarachchi, G. M., Beak, D. G., and Cornelis, G. (2012). Dissolution kinetics of macronutrient fertilizers coated with manufactured zinc oxide nanoparticles. *J. Agric. Food Chem.*, **60**, pp. 3991–3998.

51. Aziz, H. M. A., Hasaneen, M. N., and Omer, A. M. (2016). Nano chitosan-NPK fertilizer enhances the growth and productivity of wheat plants grown in sandy soil. *Span. J. Agric. Res.*, **14**, pp. 17.

52. Panpatte, D. G., Jhala, Y. K., Shelat, H. N., and Vyas, R. V. (2016). Nanoparticles: The next generation technology for sustainable agriculture. In *Microbial Inoculants in Sustainable Agricultural Productivity*, pp. 289–300. Springer, New Delhi.

53. Dwivedi, S., Saquib, Q., Al-Khedhairy, A. A., and Musarrat, J. (2016). Understanding the role of nanomaterials in agriculture. In *Microbial Inoculants in Sustainable Agricultural Productivity*, pp. 271–288. Springer, New Delhi.

54. Rajonee, A. A., Nigar, F., Ahmed, S., and Huq, S. I. (2016). Synthesis of nitrogen nano fertilizer and its efficacy. *Can. J. Pure Appl. Sci.*, **10**, pp. 3913–3919.

55. Sohrt, J., Lang, F., and Weiler, M. (2017). Quantifying components of the phosphorus cycle in temperate forests. *Wiley Interdisciplinary Rev. Water*, **4**, e1243.

56. Bindraban, P. S., Dimkpa, C. O., and Pandey, R. (2020). Exploring phosphorus fertilizers and fertilization strategies for improved human and environmental health. *Biol. Fert. Soils*, pp. 1–19.

57. Fan, Y., Lin, F., Yang, L., Zhong, X., Wang, M., Zhou, J., Chen, Y., and Yang, Y. (2018). Decreased soil organic P fraction associated with ectomycorrhizal fungal activity to meet increased P demand under N application in a subtropical forest ecosystem. *Biol. Fert. Soils,* **54**, pp. 149–161.

58. Liu, R. and Lal, R. (2014). Synthetic apatite nanoparticles as a phosphorus fertilizer for soybean (*Glycine max*). *Sci. Rep.*, **4**, pp. 5686.

59. Soliman, A. S., Hassan, M., Abou-Elella, F., Ahmed, A. H., and El-Feky, S. A. (2016). Effect of nano and molecular phosphorus fertilizers on growth and chemical composition of baobab (*Adansonia digitata* L.). *J. Plant Sci.,* **11**, pp. 52–60.

60. Preetha, P. S. and Balakrishnan, N. (2017). A review of nano fertilizers and their use and functions in soil. *Int. J. Curr. Microbiol. App. Sci.*, **6**, pp. 3117–3133.

61. Husted, S. (2018). Innovative approach taken to phosphorus nanofertilizer research. AG ChemiGroup. https://www.agchemigroup.eu/.

62. Fertahi, S., Bertrand, I., Ilsouk, M., Oukarroum, A., Zeroual, Y., and Barakat, A. (2020). New generation of controlled release phosphorus fertilizers based on biological macromolecules: Effect of formulation properties on phosphorus release. *Int. J. Biol. Macromol.*, **143**, pp. 153–162.

63. Ghahremani, A., Akbari, K., Yousefpour, M., and Ardalani, H. (2014). Effects of nano-potassium and nano calcium chelated fertilizers on qualitative and quantitative characteristics of *Ocimum basilicum*. *Int. J. Pharm. Res. Schol.*, **3**, 00167.

64. Gerdini, F. S. (2016). Effect of nano potassium fertilizer on some parchment pumpkin (*Cucurbita pepo*) morphological and physiological characteristics under drought conditions. *Intl. J. Farm Alli. Sci.*, **5**, pp. 367–371.

65. Attia, A. N. E., El-Hendi, M. H., Hamoda, S. A. F., and El-Sayed, O. S. (2016). Effect of nano-fertilizer (Lithovit) and potassium on leaves chemical composition of Egyptian cotton under different planting dates. *J. Plant Production*, **7**, pp. 935–942.

66. Rajonee, A. A., Zaman, S., and Huq, S. M. I. (2017). Preparation, characterization and evaluation of efficacy of phosphorus and potassium incorporated nano fertilizer. *Adv. Nanoparticles*, **6**, pp. 62.

67. Khalifa, N. S. and Hasaneen, M. N. (2018). The effect of chitosan–PMAA–NPK nanofertilizer on *Pisum sativum* plants. *3 Biotech*, **8**, pp. 193.

68. Afify, R. R., El-Nwehy, S. S., Bakry, A. B., and ME, A.E.A. (2019). Response of peanut (*Arachis hypogaea* L.) crop grown on newly reclaimed sandy soil to foliar application of potassium nano-fertilizer. *Sciences*, **9**, pp. 78–85.

69. Wojcik, P. (2001). Effect of calcium chloride sprays at different water volumes on "Szampion" apple calcium concentration. *J. Plant Nutr.*, **24**, pp. 639–650.

70. Conway, W. S., Sams, C. E., and Hickey, K. D. (2001). Pre- and postharvest calcium treatment of apple fruit and its effect on quality. *Int. Symp. Foliar Nutr. Perennial Fruit Plants,* **594**, pp. 413–419.

71. Mengel, K. (2001). Alternative or complementary role of foliar supply in mineral nutrition. *Int. Symp. Foliar Nutr. Perennial Fruit Plants,* **594**, pp. 33–47.

72. Danner, M. A., Scariotto, S., Citadin, I., Penso, G. A., and Cassol, L. C. (2015). Calcium sources applied to soil can replace leaf application in 'Fuji' apple tree. *Pesquisa Agropecuária Tropical*, **45**, pp. 266–273.

73. Davarpanah, S., Tehranifar, A., Abadía, J., Val, J., Davarynejad, G., Aran, M., and Khorassani, R. (2018). Foliar calcium fertilization reduces fruit cracking in pomegranate (*Punica granatum* cv. Ardestani). *Scientia Horticulturae*, **230**, pp. 86–91.

74. Ranjbar, S., Ramezanian, A., and Rahemi, M. (2019). Nano-calcium and its potential to improve 'Red Delicious' apple fruit characteristics. *Hortic. Environ. Biotech*, pp. 1–8.

75. Seydmohammadi, Z., Roein, Z., and Rezvanipour, S. (2020). Accelerating the growth and flowering of *Eustoma grandiflorum* by foliar application of nano-ZnO and nano-$CaCO_3$. *Plant Physiol. Rep.*, **25**, pp. 140–148.

76. Cakmak, I. and Yazici, A. M. (2010). Magnesium: A forgotten element in crop production. *Better Crops*, **94**, pp. 23–25.

77. Gransee, A. and Führs, H. (2013). Magnesium mobility in soils as a challenge for soil and plant analysis, magnesium fertilization and root uptake under adverse growth conditions. *Plant Soil*, **368**, pp. 5–21.

78. Delfani, M., Baradarn Firouzabadi, M., Farrokhi, N., and Makarian, H. (2014). Some physiological responses of black-eyed pea to iron and magnesium nanofertilizers. *Commun. Soil Sci. Plant Anal.*, **45**, pp. 530–540.

79. Shinde, S., Paralikar, P., Ingle, A. P., and Rai, M. (2018). Promotion of seed germination and seedling growth of *Zea mays* by magnesium hydroxide nanoparticles synthesized by the filtrate from *Aspergillus niger*. *Arab. J. Chem.*, **13**, pp. 3172–3182.

80. Valle, S. F., Giroto, A. S., Klaic, R., Guimarães, G. G., and Ribeiro, C. (2019). Sulfur fertilizer based on inverse vulcanization process with soybean oil. *Polym. Degrad. Stabil.*, **162**, pp. 102–105.

81. Germida, J. J. and Janzen, H. H. (1993). Factors affecting the oxidation of elemental sulfur in soils. *Fertilizer Res.*, **35**, pp. 101–114.

82. Alipour, Z. T. (2016). The effect of phosphorus and sulfur nanofertilizers on the growth and nutrition of *Ocimum basilicum* in response to salt stress. *J. Chem. Health Risks*, **6**. DOI: 10.22034/JCHR.2016.544137.

83. Ragab, G. A. and Saad-Allah, K. M. (2020). Green synthesis of sulfur nanoparticles using *Ocimum basilicum* leaves and its prospective effect on manganese-stressed *Helianthus annuus* (L.) seedlings. *Ecotox. Environ. Safe.*, **191**, pp. 110242.

84. Tarafdar, J. C. and Raliya, R. (2013). Rapid, low-cost, and ecofriendly approach for iron nanoparticle synthesis using *Aspergillus oryzae* TFR9. *J. Nanoparticles*, **2013**, 141274.

85. Talaei, A. S. (1998). *Physiology of Temperate Zone Fruit Trees*, pp. 423. Tehran University Press, Tehran.

86. Armin, M., Akbari, S., and Mashhadi, S. (2014). Effect of time and concentration of nano-Fe foliar application on yield and yield components of wheat. *Int. J. Biosci.*, **4**, pp. 69–75.

87. Askary, M., Amirjani, M. R., and Saberi, T. (2017). Comparison of the effects of nano-iron fertilizer with iron-chelate on growth parameters and some biochemical properties of *Catharanthus roseus*. *J. Plant Nutr.*, **40**, pp. 974–982.

88. Hu, J., Wu, C., Ren, H., Wang, Y., Li, J., and Huang, J. (2018). Comparative analysis of physiological impact of γ-Fe_2O_3 nanoparticles on

dicotyledon and monocotyledon. *J. Nanosci. Nanotechnol.*, **18**, pp. 743–752.

89. Sheykhbaglou, R., Sedghi, M., and Fathi-Achachlouie, B. (2018). The effect of ferrous nano-oxide particles on physiological traits and nutritional compounds of soybean (*Glycine max* L.) seed. *Anais da Academia Brasileira de Ciências*, **90**, pp. 485–494.

90. Kim, J. H., Kim, D., Seo, S. M., and Kim, D. (2019). Physiological effects of zero-valent iron nanoparticles in rhizosphere on edible crop, *Medicago sativa* (Alfalfa), grown in soil. *Ecotoxicology*, **28**, pp. 869–877.

91. Rostamizadeh, E., Iranbakhsh, A., Majd, A., Arbabian, S., and Mehregan, I. (2020). Green synthesis of Fe_2O_3 nanoparticles using fruit extract of *Cornus mas* L. and its growth-promoting roles in barley. *J. Nanostruct. Chem.*, **10**, pp. 125–130.

92. Chaudhuri, S. K. and Malodia, L. (2017). Biosynthesis of zinc oxide nanoparticles using leaf extract of *Calotropis gigantea*: Characterization and its evaluation on tree seedling growth in nursery stage. *Appl. Nanosci.*, **7**, pp. 501–512.

93. Wang, P., Lombi, E., Zhao, F. J., and Kopittke, P. M. (2016). Nanotechnology: A new opportunity in plant sciences. *Trends Plant Sci.*, **21**, pp. 699–712.

94. Narendhran, S., Rajiv, P., and Sivaraj, R. (2016). Influence of zinc oxide nanoparticles on growth of *Sesamum indicum* L. in zinc deficient soil. *Int. J. Pharm. Pharm. Sci.*, **8**, pp. 365–371.

95. Sharifi, R., Mohammadi, K., and Rokhzadi, A. (2016). Effect of seed priming and foliar application with micronutrients on quality of forage corn (*Zea mays*). *Environ. Exp. Biol.*, **14**, pp. 151–156.

96. Khanm, H., Vaishnavi, B. A., and Shankar, A. G. (2018). Rise of nano-fertilizer era: Effect of nano scale zinc oxide particles on the germination, growth and yield of tomato (*Solanum lycopersicum*). *Int. J. Curr. Microbiol. Appl. Sci.*, **7**, pp. 1861–1871.

97. Munir, T., Rizwan, M., Kashif, M., Shahzad, A., Ali, S., Zahid, R., Alam, M. F. E., and Imran, M. (2018). Effect of zinc oxide nanoparticles on the growth and Zn uptake in wheat (*Triticum aestivum* L.) by seed priming method. *Dig. J. Nanomater. Bios.*, **13**, pp. 315–323.

98. Moghaddasi, S., Fotovat, A., Khoshgoftarmanesh, A. H., Karimzadeh, F., Khazaei, H. R., and Khorassani, R. (2017). Bioavailability of coated and uncoated ZnO nanoparticles to cucumber in soil with or without organic matter. *Ecotox. Environ. Safe*, **144**, pp. 543–551.

99. Priyanka, N., Geetha, N., Ghorbanpour, M., and Venkatachalam, P. (2019). Role of engineered zinc and copper oxide nanoparticles in promoting plant growth and yield: Present status and future prospects. In *Advances in Phytonanotechnology*, pp. 183–201. Academic Press.

100. Rawat, S., Pullagurala, V. L., Hernandez-Molina, M., Sun, Y., Niu, G., Hernandez-Viezcas, J. A., Peralta-Videa, J. R., and Gardea-Torresdey, J. L. (2018). Impacts of copper oxide nanoparticles on bell pepper (*Capsicum annum* L.) plants: A full life cycle study. *Environ. Sci. Nano.*, **5**, pp. 83–95.

101. Shukla, Y. M. (2019). Nanofertilizers: A recent approach in crop production. In *Nanotechnology for Agriculture: Crop Production & Protection*, pp. 25–58. Springer, Singapore.

102. Rajput, V. D., Minkina, T., Suskova, S., Mandzhieva, S., Tsitsuashvili, V., Chapligin, V., and Fedorenko, A. (2018). Effects of copper nanoparticles (CuO NPs) on crop plants: A mini review. *BioNanoScience*, **8**, pp. 36–42.

103. Nagaonkar, D., Shende, S., and Rai, M. (2015). Biosynthesis of copper nanoparticles and its effect on actively dividing cells of mitosis in *Allium cepa. Biotechnol. Prog.*, **31**, pp. 557–565.

104. Yasmeen, F., Raja, N. I., Razzaq, A., and Komatsu, S. (2017). Proteomic and physiological analyses of wheat seeds exposed to copper and iron nanoparticles. *Biochimicaet Biophysica Acta-Proteins Proteomics*, **1865**, pp. 28–42.

105. Shende, S., Rathod, D., Gade, A., and Rai, M. (2017). Biogenic copper nanoparticles promote the growth of pigeon pea (*Cajanus cajan* L.). *IET Nanobiotechnol.*, **11**, pp. 773–781.

106. Hernandez, H. H., Benavides-Mendoza, A., Ortega-Ortiz, H., Hernández-Fuentes, A. D., and Juárez-Maldonado, A. (2017). Cu nanoparticles in chitosan-PVA hydrogels as promoters of growth, productivity and fruit quality in tomato. *Emir. J. Food Agric*, **29**, pp. 573–580.

107. López-Vargas, E. R., Ortega-Ortíz, H., Cadenas-Pliego, G., de Alba Romenus, K., Cabrera de la Fuente, M., Benavides-Mendoza, A., and Juárez-Maldonado, A. (2018). Foliar application of copper nanoparticles increases the fruit quality and the content of bioactive compounds in tomatoes. *Appl. Sci.*, **8**, pp. 1020.

108. Sathiyabama, M. and Manikandan, A. (2018). Application of copper-chitosan nanoparticles stimulate growth and induce resistance in finger millet (*Eleusine coracana* Gaertn) plants against blast disease. *J. Agric. Food Chem.*, **66**, pp. 1784–1790.

109. Pradhan, S., Patra, P., Das, S., Chandra, S., Mitra, S., Dey, K. K., Akbar, S., Palit, P., and Goswami, A. (2013). Photochemical modulation of biosafe manganese nanoparticles on *Vigna radiata*: A detailed molecular, biochemical, and biophysical study. *Environ. Sci. Technol.*, **47**, pp. 13122–13131.

110. Ye, Y., Cota-Ruiz, K., Hernandez-Viezcas, J. A., Valdes, C., Medina-Velo, I. A., Turley, R. S., Peralta-Videa, J. R., and Gardea-Torresdey, J. L. (2020). Manganese nanoparticles control salinity-modulated molecular responses in *Capsicum annuum* L. through priming: A sustainable approach for agriculture. *ACS Sustain. Chem. Eng.*, **8**, pp. 1427–1436.

111. Ahmad, W., Niaz, A., Kanwal, S., and Khalid, M. (2009). Role of boron in plant growth: A review. *J. Agric. Res.*, **47**, pp. 329–338.

112. Shil, N. C., Noor, S., and Hossain, M. A. (2007). Effects of boron and molybdenum on the yield of chickpea. *J. Agr. Rural Dev.*, pp. 17–24.

113. Shireen, F., Nawaz, M. A., Chen, C., Zhang, Q., Zheng, Z., Sohail, H., Sun, J., Cao, H., Huang, Y., and Bie, Z. (2018). Boron: Functions and approaches to enhance its availability in plants for sustainable agriculture. *Int. J. Mol. Sci.*, **19**, pp. 1856.

114. Ibrahim, N. K. and Al Farttoosi, H. A. K. (2019). Response of mung bean to boron nanoparticles and spraying stages (*Vigna Radiata* L.). *Plant Archives*, **19**, pp. 712–715.

115. Taherian, M., Bostani, A., and Omidi, H. (2019). Boron and pigment content in alfalfa affected by nano fertilization under calcareous conditions. *J. Trace Elem. Med. Bio.*, **53**, pp. 136–143.

116. Thomas, E., Rathore, I., and Tarafdar, J. C. (2017). Bioinspired production of molybdenum nanoparticles and its effect on chickpea (*Cicer arietinum* L). *J. Bionanosci.*, **11**, pp. 153–159.

117. Mendel, R. R. and Hänsch, R. (2002). Molybdoenzymes and molybdenum cofactor in plants. *J. Exp. Bot.*, **53**, pp. 1689–1698.

118. Self, W. T., Grunden, A. M., Hasona, A., and Shanmugam, K. T. (2001). Molybdate transport. *Res. Microbiol.*, **152**, pp. 311–321.

119. Taran, N. Y., Gonchar, O. M., Lopatko, K. G., Batsmanova, L. M., Patyka, M. V., and Volkogon, M. V. (2014). The effect of colloidal solution of molybdenum nanoparticles on the microbial composition in rhizosphere of *Cicer arietinum* L. *Nanoscale Res. Lett.*, **9**, pp. 289.

120. Yusuf, M., Fariduddin, Q., Hayat, S., and Ahmad, A. (2011). Nickel: An overview of uptake, essentiality and toxicity in plants. *Bull. Environ. Contam. Toxicol.*, **86**, pp. 1–17.

121. Zotikova, A. P., Astafurova, T. P., Burenina, A. A., Suchkova, S. A., and Morgalev, Y. N. (2018). Morphophysiological features of wheat (*Triticum aestivum* L.) seedlings upon exposure to nickel nanoparticles. *Sel'skokhozyaistvennaya Biologiya*, **53**, pp. 578–586.

122. Duhan, J. S., Kumar, R., Kumar, N., Kaur, P., Nehra, K., and Duhan, S. (2017). Nanotechnology: The new perspective in precision agriculture. *Biotechnol. Rep.*, **15**, pp. 11–23.

123. Thirugnanasambandan, T. (2019). Advances and trends in nanobiofertilizers, *SSRN*, p. 59. https://doi,org/10.2139/ssrn.3306998.

124. Dhir, B. (2017). Biofertilizers and biopesticides: Ecofriendly biological agents. In *Advances in Environmental Biotechnology* (Kumar, R., Sharma, A., and Ahluwalia, S., Eds.), pp. 167–188. Springer, Singapore.

125. Itelima, J. U., Bang, W. J., Onyimba, I. A., and Oj, E. (2018). A review: Biofertilizer - a key player in enhancing soil fertility and crop productivity. *J. Microbiol. Biotechnol. Rep.*, **2**, pp. 22–28.

126. Mishra, C., Keswani, C., Abhilash, P. C., Fraceto, L. F., and Singh, H. B. (2017). Integrated approach of agri-nanotechnology: Challenges and future trends. *Front. Plant Sci.*, **8**, pp. 471.

127. Golbashy, M., Sabahi, H., Allahdadi, I., Nazokdast, H., and Hossein, M. (2017). Synthesis of highly intercalated urea-clay nanocomposite via domestic monmorillonite as eco-friendly slow-release fertilizer. *Arch. Agron. Soil Sci.*, **63**, pp. 1.

128. Vejan, P., Abdullah, R., Khadiran, T., Ismail, S., and Boyce, A. N. (2016). Role of plant growth promoting rhizobacteria in agricultural sustainability: A review. *Molecules*, **21**, pp. 1–17.

129. Qureshi, A., Singh, D. K., and Dwivedi, S. (2018). Nano-fertilizers: A novel way for enhancing nutrient use efficiency and crop productivity. *Int. J. Curr. Microbiol. App. Sci.*, **7**, pp. 3325–3335.

130. Ahemad, M. and Kibret, M. (2014). Mechanisms and applications of plant growth promoting rhizobacteria: Current perspective. *J. King Saud Univ. Sci.*, **26**, pp. 1–20.

131. Mala, R., Celsia, A. V., Bharathi, S. V., Blessina, S. R., and Maheswari, U. (2017). Evaluation of nano structured slow release fertilizer on the soil fertility, yield and nutritional profile of *Vigna radiata. Recent Pat. Nanotechnol.*, **11**, pp. 50–62.

132. Monreal, C. M., De Rosa, M., Mallubhotla, S. C., Bindraban, P. S., and Dimpka, C. (2016). Nanotechnologies for increasing the crop use

efficiency of fertilizer-micronutrients. *Biol. Fertil. Soils*, **52**, pp. 423–437.

133. Janmohammadi, M., Navid, A., Segherloo, A. E., and Sabaghnia, N. (2016). Impact of nano-chelated micronutrients and biological fertilizers on growth performance and grain yield of maize under deficit irrigation conditions. *Biologija*, **62**, pp. 134–147.

134. Gatahi, D. M., Wanyika, H., Kihurani, A. W., Ateka, E., and Kavoo, A. (2015). Use of bio-nanocomposites in enhancing bacterial wilt plant resistance and water conservation in greenhouse farming. In *2015 JKUAT Scientific Conference. Agricultural Sciences, Technologies and Global Networking*, **41**, pp. 52.

135. Gouda, S., Kerry, R. G., Das, G., Paramithiotis, S., Shin, H.-S., and Patra, J. K. (2018). Revitalization of plant growth promoting rhizobacteria for sustainable development in agriculture. *Microbiol. Res.*, **206**, pp. 131–140.

136. Mishra, V. K. and Kumar, A. (2009). Impact of metal nanoparticles on the plant growth promoting rhizobacteria. *Digest J. Nanomater. Biostruct.*, **4**, pp. 587–592.

137. Mukhopadhyay, R. and De, N. (2014). Nano clay polymer composite: Synthesis, characterization, properties and application in rainfed agriculture. *Global J. Bio Biotechnol.*, **3**, pp. 133–138.

138. Nair, R., Varghese, S. H., Nair, B. G., Maekawa, T., Yoshida, Y., and Kumar, D. S. (2010). Nanoparticulate material delivery to plants. *Plant Sci.*, **179**, pp. 154–163.

139. Cai, D., Wu, Z., Jiang, J., Wu, Y., Feng, H., Brown, I. G., Chu, P. K., and Yu, Z. (2014). Controlling nitrogen migration through micro-nano networks. *Sci. Rep.*, **4**, pp. 1–8.

140. Varshney, R. K., Pandey, M. K., and Chitikineni, A. (2018). Plant genetics and molecular biology: An introduction. In *Plant Genetics and Molecular Biology*, pp. 1–9. Springer, Cham.

141. Yang, J., Jiang, F., Ma, C., Rui, Y., Rui, M., Adeel, M., Cao, W., and Xing, B. (2018). Alteration of crop yield and quality of wheat upon exposure to silver nanoparticles in a life cycle study. *J. Agric. Food Chem.*, **66**, pp. 2589–2597.

142. Talan, A., Mishra, A., Eremin, S. A., Narang, J., Kumar, A., and Gandhi, S. (2018). Ultrasensitive electrochemical immuno-sensing platform based on gold nanoparticles triggering chlorpyrifos detection in fruits and vegetables. *Biosens. Bioelectron.*, **105**, pp. 14–21.

Multiple Choice Questions

1. Nanofertilizers are better than fertilizers in
   (a) Reducing the amount of fertilizer used
   (b) Reducing the time taken for a desired improved crop yield
   (c) Both (a) and (b)
   (d) Reducing the chemical pressure on the soil

2. Most suitable nanomaterial for making nanofertilizers is
   (a) Chemically prepared metal NPs
   (b) Biologically prepared metal NPs
   (c) Carbon nanotubes and fullerenes
   (d) None of these

3. With the implementation of nanofertilizers, the dependence of agriculture on rain in countries like India
   (a) Reduces   (b) Increases
   (c) Remains unaffected   (d) None of these

4. Which of the following is the most appropriate mechanism by which nanotechnology could improve fertilizer administration?
   (a) Through delivering pesticides using nanocarriers
   (b) Through including nanomaterials in conventionally delivered fertilizers
   (c) Through including microorganisms and metallic precursors in conventionally administered fertilizers
   (d) By tilling and irrigating the soil through nanomaterials

5. Nanobiofertilizers are better than biofertilizers in
   (a) Improved nutrient absorption from soil and concurrent distribution
   (b) In reducing the extent of biofertilizers used
   (c) Both (a) and (b)
   (d) None of these

6. Nanomaterial inclusion in the fertilizer ensures
   (a) Their slow and gradual release, lasting till greater time durations
   (b) Better absorption and utilization of nutrients (in general) and water

(c) That soil fertility is not irreversibly damaged even after harvest of a grown crop

(d) All of these

7. Most crucial concerns for amalgamating nanotechnology with agriculture is regarding

(a) Terrestrial toxicity of nanomaterials

(b) End product of used nanomaterials

(c) Hindrance in proper irrigation extent

(d) None of these

8. The shape of the nanomaterials that is best suited to enhance the fertilizer activity is

(a) Spherical   (b) Rod like

(c) Triangular   (d) None of these

9. Some studies propose the use of fertilizers dispersed in nanoemulsions for use as fertilizers. Can this be considered a form of nanobiofertilizer?

(a) Yes

(b) No

(c) Depends on the ingredients of the administered nanoemulsions

(d) None of these

10. Recent US initiatives have aimed to administer nanofertilizers to maize crops. In this context, the acronyms CRF and EQIP denote

(a) Central Reserve Force and Economic Quality Initiative Perusal

(b) Controlled Release Fertilizers and Environmental Quantity Inspection Protocol

(c) Controlled Release Fertilizers and Environmental Quality Incentives Program

(d) None of these

11. Initial trials on administration of urea with hydroxyapatite NPs to rice (in Sri Lanka) revealed how much benefit in reducing the expenditure on urea?

(a) 15%   (b) 30%

(c) 45%   (d) 50%

12. The major reason for Brazil's initiative program (EMBRAPA), utilizing polycaprolactone (PCL), a nanopolymer for co-delivery with urea to wheat crop, was

(a) Biocompatibility of PCL

(b) Slow degradation by bacteria and fungi

(c) Both (a) and (b)

(d) Slow degradation by bacteria and fungi along with cheaper manufacturing

13. Studies on nano-clay based fertilizer formulations (comprising zeolites and 30–40 nm montmorillonite) have shown the nitrogen-releasing capability of this system as

(a) Same as that of conventional fertilizer

(b) Thrice as that of conventional fertilizer

(c) Twice as that of conventional fertilizer

(d) Ten times that of conventional fertilizer

14. Microorganisms *Mycorrhizae* and *Azospirillum* are common in the sense that

(a) Both are responsible for nitrogen fixation

(b) Both are involved in optimized nutrient and water uptake

(c) Both possess antagonistic activities against plant pathogens

(d) Both restore the P deficiency in plants

15. The nanomaterial coating of fertilizer contents ensures

(a) No aggregation and instantaneous chemical mismatch of ingredients

(b) Homogeneous and sustained delivery of similar constituent texture

(c) No chemical clogging of host soil

(d) Both (a) and (b)

16. The mechanism of preparing nanobiofertilizer relies on

(a) Adsorption-driven interactions of nanomaterial with the fertilizer component

(b) Physicochemical interactions of nanomaterial and fertilizers

(c) Chemical interactions of nanomaterial and fertilizers

(d) None of these

17. Modern agriculture relies on

(a) Traditional practices of crop rotation, mixed cropping, and manure supplementation

(b) Genetic engineering of the seeds through recombinant DNA technology

(c) Scientific mechanisms, initially optimized on laboratory scale

(d) All of these, as per the concerned farmer's awareness and accessibility

18. The chief advantage of biofertilizers over their chemical counterparts is

(a) Less chemical effect on the host soil

(b) Easier preparation, processing and enhanced organic content (humus) of the soil concerned

(c) No ill effects on soil salinity

(d) Biofertilizers are not more beneficial than chemical fertilizers

19. Not much industries are known for the production of nanobiofertilizers, inferring a risk of their commercial feasibility. This is primarily due to

(a) Non-homogeneous results of nanomaterials on the laboratory scale

(b) Insufficient data about the environmental fate of NPs with varying chemical compositions

(c) Societal conservations and non-readiness of farmers to accept this science

(d) All of these

20. What is the most feasible strategy to improve the nitrogen-fixing attributes of rhizobium species be bettered?

(a) Through genetic engineering (enrichment culture growth) of concerned microbial species

(b) Using microbial cultures supplemented with biologically prepared NPs along with biological grade fertilizers

(c) Adjusting the pH, salt content, and chemical texture of the soil concerned

(d) None of these

21. Biofertilizers could be prepared using plenty of natural resources, which collectively bestow

(a) Ubiquitous natural resources as dispersion promoting agents

(b) Rich in organic content, vis-à-vis amino acid and oligosaccharide contribution

(c) Simplicity and ease of preparation

(d) All of these

22. Most important element used as fertilizer is
    (a) Sodium
    (b) Potassium
    (c) Nitrogen
    (d) Phosphorous

23. The major problems associated with inorganic nutrient supplementation to the soil are
    (a) Post-harvest groundwater pollution
    (b) No improvement in soil structure alongside decreased organic matter mediated nutrient imbalance and soil acidification
    (c) Not easily available to farmers throughout the year and is costly
    (d) All of these

24. The amino acid essential for improved seed germination rate is
    (a) Histidine
    (b) Leucine
    (c) Tryptophan
    (d) Phenylalanine

25. Which environment is more suitable for the best availability of tryptophan and organic moieties as nutrient?
    (a) Strongly acidic
    (b) Moderately acidic
    (c) Alkaline
    (d) Neutral pH

Answer Key

1. (c)   2. (b)   3. (a)   4. (c)   5. (c)   6. (d)   7. (a)
8. (b)   9. (c)   10. (c)   11. (d)   12. (d)   13. (c)   14. (b)
15. (d)   16. (b)   17. (c)   18. (b)   19. (d)   20. (b)   21. (d)
22. (b)   23. (d)   24. (c)   25. (c)

Short Answer Questions

1. What are macronutrient fertilizers and describe their importance?
2. Are micronutrient fertilizers equally important as macronutrient fertilizers? Please elaborate with examples.
3. What are the differences between conventional fertilizers and nanofertilizers?
4. Discuss in brief the advantages and limitations of nanofertilizers.
5. What is the difference between nanofertilizers and nanobiofertilizers?
6. Why NPK fertilizers are used more frequently and what is the role of these three in plant growth?

7. Which fertilizers are helpful to develop salt stress tolerance in plants?

8. What are macro- and micronutrients? Please discuss in brief their importance in plant health.

Long Answer Questions

1. Define fertilizers and discuss their types.

2. Discuss in detail the mechanism of uptake of nanofertilizers.

3. What are the different methods that can be used for the preparation of nanofertilizers?

4. Discuss the current trends in the use of fertilizers.

5. Describe the risks associated with the excessive use of fertilizers and possible alternatives.

Chapter 10

Nanotechnology in Food Production

Kanishka Rawat

Applied Nuclear Physics Division, Saha Institute of Nuclear Physics,
Sector 1, AF Block, Bidhan Nagar, Kolkata 700064, West Bengal, India
kanishka.rawat.phy@gmail.com

Nanotechnology has become increasingly important in the agriculture, medicine, and food sectors. Nanotechnology provides a lot of opportunities in the food industry to improve the quality, healthiness, and shelf life of food. It can be used in the food industry for production, processing, and packaging of food. Nanotechnology improves food processes, thereby increasing the health benefits of food. The novel features of nanomaterials can improve the quality of foods by providing a better flavor, color, and appearance. Nanoencapsulation helps to control the degradation of flavors during food processing and storage. The nanosensors can detect harmful components in food, which can be recognized with the help of a good packaging system. Nanotechnology used for packaging would be able to repair small holes and tears and respond to environmental conditions such as moisture and temperature changes so that customers can be alerted about the food. This can be achieved by

Nanotechnology: Principles and Applications
Edited by Rakesh K. Sindhu, Mansi Chitkara, and Inderjeet Singh Sandhu
Copyright © 2021 Jenny Stanford Publishing Pte. Ltd.
ISBN 978-981-4877-43-5 (Hardcover), 978-1-003-12026-1 (eBook)
www.jennystanford.com

modifying the barrier properties and improving the heat-resistance properties. Despite many advantages, there are safety concerns with food nanotechnology. Many countries have adopted regulations and certification systems for food nanotechnology. In this chapter, we discuss the nanomaterials used in food, use of nanotechnology in food processing, packaging, safety issues, regulations in food nanotechnology, and prospects.

10.1 Introduction

Nanotechnology is a field of research that deals with atoms or molecules of approximately 1–100 nm in size to create materials with novel properties and utilize them. The novel properties of nanomaterials are strength, color, solubility, diffusivity, toxicity, magnetic and thermodynamic properties, and also high surface-to-volume ratio [1, 2]. Nanotechnology has brought a revolution in medicine, agriculture, and food sectors across the globe. Therefore, it offers a wide range of opportunities for the development of structures in various areas such as agriculture, food, medicine, and many more.

Due to the rise in the consumer demand for enhancing food quality and health benefits, researchers are finding solutions without disturbing the nutritional value of food. Since nanoparticle-based materials are found to be nontoxic [3], their demand has increased in the food sector as many of them contain essential elements that can be consumed easily. They are also found to be stable at high temperature and pressure [4]. So nanotechnology offers a complete solution in the food sector in manufacturing, processing, and packaging. Nanomaterials enhance not only the food quality but also the health benefits associated with food. Many researchers, laboratories, organizations, and industries are coming up with useful methods, techniques, and products that have a direct application of nanotechnology in the food sector [5].

The applications of nanotechnology in the food sector can be categorized into two main groups: those that utilize food nanostructured ingredients and those that are based on food nanosensing. The ingredients of food nanotechnology include food processing and food packaging. In food processing, the nanostructured materials can be used as food additives, a medium for smart delivery of nutrients, fillers to improve mechanical strength, antimicrobial agents, and to enhance the durability of packaging material. Food nanosensing can be used to get better food quality and for safety evaluation [6]. In the next section, we discuss the nanomaterials in food with their examples.

10.2 Nanomaterials in Food

Atoms and molecules combine to form various structures that are the building blocks of every organism. All the organisms are a combination of various nanoscale-sized structures; humans, hormones, cell membranes, and DNA are examples of such nanostructures. The existence of every organism is because of the interaction of various nanostructures [7]. Even food molecules such as carbohydrates, proteins, and fats are formed due to the merger between nanosized sugars, amino acids, and fatty acids [8].

Biological molecules such as sugars or proteins are the target-recognition groups for nanostructures that can be used, e.g., as biosensors on foods [9]. These biosensors detect food pathogens and other contaminants. Nanotechnology can be used in encapsulation systems to protect against environmental conditions. It can also be used to design food ingredients such as antioxidants and flavors [10] so that the function of such ingredients can be improved.

Many of the applications of nanotechnology in the food sector are either too expensive or too impractical to be implemented on a commercial scale. So the techniques that are most cost effective are food formulations, development of new functional materials, food processing at microscale and nanoscale, and development of products and their storage [11]. Nanoparticles are more biologically active because of their greater surface area per mass unit, which makes them more useful for food applications. Because of the reduced particle size, nanotechnology can help in improving the properties of bioactive compounds, such as solubility, delivery, and efficient absorption through cells [12]. Omega-3 and omega-6 fatty acids, prebiotics, probiotics, vitamins, and minerals act as bioactive compounds that have their applications in food nanotechnology [13]. Bioactive compounds can also be found naturally in certain foods that have physiological benefits and may help in reducing the risk of certain deadly diseases such as cancer. Nanomaterials such as biopolymeric nanoparticles, liposomes, nanoemulsions, and cubosomes have many applications, including food safety. We now discuss nanoemulsions and their applications in food science.

10.2.1   Nanoemulsions

These emulsions are droplets of the order of 100 nm in size and contain oil, water, and an emulsifier. An emulsifier is included to create small-sized droplets so that the interfacial tension and, hence, surface energy per unit area between the oil and water phases of the emulsion can be decreased. The emulsifier also stabilizes nanoemulsions through steric hindrance and repulsive electrostatic interactions [14]. Nanoemulsions are kinetically stable, i.e., given sufficient time, nanoemulsion phase separates [15]. Proteins and lipids have been effective in the preparation of nanoemulsions [16]. Nanoemulsions can be prepared through mainly two methods: high-energy and low-energy methods [17]. For the formation of small droplets, high-energy methods such as high-pressure homogenization (HPH) and ultrasonication [18] consume much energy ($\sim$108–1010 W/kg), whereas low-energy methods such as phase inversion temperature (PIT) and emulsion inversion point (EIP) [19, 20] consume less energy ($\sim$103 W/kg) and exploit specific system properties to make small droplets. Nanoemulsions are also formed due to bubble bursting at the oil/water interface and evaporative ripening [21, 22]. They can be used for drug delivery to deliver hydrophobic drugs. They are also utilized in the food industry as well as in the cosmetic industry to test skin hydration [23]. We discuss nanoparticles in the next sub-section.

10.2.2   Nanoparticles

Nanoparticles improve food's flow property, stability, and color. Their effectiveness depends on their bioavailability in a system [24]. Nanoparticles in the form of plastic films such as zinc oxide, silicate nanoparticles, and titanium oxide are utilized to minimize the flow of oxygen in packaging containers [25] and also to reduce moisture leakage, thereby maintaining the freshness of food for a longer time [25]. Nanoparticles also help in the removal of chemicals or pathogens from food [26]. The most commonly used nanoparticle silicon dioxide is used as an anticaking and a drying agent that helps to absorb water molecules in food [27]. Another mostly used nanoparticle, titanium dioxide, acts as a food colorant [28] and is known as a photocatalytic disinfecting agent. It is also used as a food whitener for dairy products [27]. It is effective for UV protection in food packaging. Silver nanoparticles have larger surface area, which can be easily dispersed in food, readily ionized, and chemically active, thereby acting as a good antibacterial agent. Hence, silver nanoparticles protect food from microbial infestation [26]. Silver nanoparticles are effective antimicrobials since they have a broader spectrum of activity [26]. Silver is a stable element and does not cause any major threat to the biological system if used within the limits as assigned by the Food and Drug Administration (FDA) standards [26]. Silver is the best antimicrobial than the rest of the antimicrobials available in the market because it can penetrate through biofilms [26] and can be incorporated easily into packaging materials. As per reports, silver infiltrates the microbial system and spoils the ribosomal activity causing hindrance in the production of several crucial enzymes [26]. Silver nanoparticles extend the shelf life of fruits and vegetables due to the absorption and decomposition of ethylene [26]. Carbon nanotubes are also used in food packaging. But their usage is limited as the toxicity levels are higher. Polymeric nanoparticles are effectively used as delivery systems and are made of polymers and surfactants, polylactic coglycolic acid, alginic acid, and chitosan [27]. Titanium dioxide is also antimicrobial, but its use is limited since it can be easily photocatalyzed [28] and is active only in UV light. Thus, it acts as an active bactericide against many pathogens in the presence of UV light. It causes the peroxidation of phospholipids in the cell membrane of the bacterial cell wall. Titanium dioxide photosensitizes the reduction of methylene blue with the irradiation from UV light. The particles bleach in the presence of oxygen and UV, thereby changing their color to blue. Some other nanoparticles that have been reported are antimicrobial, e.g., copper and copper oxide, magnesium oxide, zinc oxide, selenium, telluride, cadmium, chitosans, and single-walled carbon nanotubes. Nanoparticles in cooking oil increase the shelf life of food and keeps the food crispier. Inorganic nanoceramic is used in cooking oil to deep fry food [29]. We discuss nanocomposites in the next sub-section.

10.2.3 Nanocomposites

Nanocomposites are made up of polymers in combination with nanoparticles and enhance the property of polymers [30]. Nanocomposites are used for the development of high barrier properties because they provide a highly versatile chemical functionality [31]. Nanocomposites help to keep food products fresh and protect from microbial infestation for quite a long time. They minimize the leakage of CO_2 from carbonated beverage bottles since they act as gas barriers [31], hence increasing the shelf life of a product. If manufacturers use a layer of nanocomposites in the bottles, leakages can be prevented, and then there is no need to make expensive cans and heavy glass bottles. For example, nanoclay is a nanocomposite used to create gas barriers and is formed from a polymer in combination with nanoparticles. Nanoclays, also known as phyllosilicates, are naturally occurring aluminum silicates and are stable, inexpensive, and ecofriendly in nature [32]. Nanoclay nanocomposites are found in two categories: intercalated nanocomposites and exfoliated nanocomposites [33]. Intercalated nanocomposites with ordered multilayer polymeric structure have alternate polymeric layers formed because of the penetration of polymer chains in the interlayer region of the clay [34]. Exfoliated nanocomposites are randomly scattered clay layers in the polymer matrix and are formed because of the extensive polymer penetration. Nanoreinforcements of cellulose result in inexpensive, light-weight nanocomposites [35]. The cellulose reinforcements are stabilized by hydrogen bonds and are grown in plants in the form of microfibrils. Such nanocomposites are flexible and provide low permeability of the polymer matrix. Silicon dioxide copolymerizes with single-walled nanotubes to give an excellent gas barrier [36]. Some examples of nanoclays are Aegis, Durethan, and Imperm [36] and are biodegradable in nature, transparent, and have good flow, low density, and better surface properties. Aegis improves the barrier property of clay by acting as oxygen scavenger, thereby retaining CO_2

in carbonated drinks [36]. Durethan, made of polyamide, provides stiffness to containers made from paperboard to contain fruit juices [36]. Another commercialized nanoclay called Imperm is a polymer made up of nylon and nanoclay and acts as oxygen scavenger [36]. Nanocor is a nanoclay-based polymer and acts as a gas barrier, which is used to manufacture plastic beer bottles to prevent the escape of CO_2 from the beverage [37]. Nanocoatings such as nanolaminates are the example of nanoencapsulation, which are used to coat cheese, meats, baked products, fruits, and vegetables. Polymers reinforced with metals act as antimicrobials in the form of nanomagnesium oxide and nanozinc oxide.

Nanocomposites with silver coating also act as an antimicrobial agent. Silver gets attached to the surface of cell, which degrades the lipopolysaccharide resulting in increased permeability, thereby causing irreversible damage to the bacterial DNA. Garlic oil nanocomposite coatings help to control pests at stores that spoil packaged food materials [38]. As reported, bionanocomposites made from cellulose and starch derivatives, polycaprolactone, polyhydroxybutyrate (PHB), poly lactic acid, and polybutylene succinate are efficient as layering materials for packaging [39]. Another commercialized nanocomposite known as Guard IN Fresh helps in the ripening of fruits and vegetables by scavenging ethylene gas [40]. Nanocomposites are useful for food packaging because they are biodegradable and ecofriendly. For example, Top Screen DS13 is a water-based nanocomposite, which is easily recyclable [41]. Another example of an ecofriendly nanocomposite coating is NanoCream PAC, which helps in the rapid absorption of unpleasant components causing foul odor and repulsive taste [42]. Enzyme-based nanocomposites are not very good since they are sensitive to a number of degrading factors such as high temperatures, unfavorable pH, or the presence of proteases. Cholesterol or lactase reductase in packaging is helpful to those consumers who are deficient in these enzymes in their body [43]. The enzymes are better than conventional systems of coatings as they provide faster transfer rates and a larger surface area. The enzymes incorporated with nanoclays are also used for packaging of food [44]. In the next sub-section, we discuss nanostructured materials.

10.2.4 Nanostructured Materials

Nanostructured materials (NSMs) range from 1 to 100 nm and consist of interfacial layers, having ions, organic and inorganic molecules that alter the properties of matter. NSMs are of four types: zero-dimensional, one-dimensional, two-dimensional, and three-dimensional NSMs (0D NSM, 1D NSM, 2D NSM, 3D NSM) [45]. Zero-dimensional NSMs have size within the nanoscale range and are amorphous, crystalline, or polycrystalline in nature. These are found in either individual or agglomerated form. One-dimensional NSMs include nanotubes, nanorods, and nanowires. These have significant impact in nanocomposites, nanodevices, and nanoelectronic materials. Two-dimensional NSMs have plate-like structures. They also exhibit a single-layer or multilayer structure deposited on the surface of substrate. Three-dimensional NSMs are bulk materials and are not restrained to the nanoscale. They exhibit nanocrystalline structure and can be formed by multiple arrangements of nanosized crystals. These NSMs contain bundles of nanowires or nanotubes and nanoparticle dispersion.

These NSMs are used in nanofoods, which means they are used at any stage of food development process, including cultivation, production, processing, and packaging. NSMs also help to achieve a longer shelf life of a product, introduce new flavors, and promote health benefits. Sometimes a nanocarrier system is added to protect the ingredients and additives during the processing of food products. Some of the prominent examples of nanofoods are fortified fruit juices, NanoSlim beverage, benzoic and ascorbic acid, nanoceutical slim shakes, isoflavanones, omega-3 fatty acids, some supplements such as vitamins A and E, oat nutritional drinks, and many more [46]. Now we discuss nanotechnology in food processing in the next section.

10.3 Nanotechnology in Food Processing

Biomolecules that have size in nanometers, such as proteins and carbohydrates, are either formed because of the transformation of preliminary products or naturally present in food. Nanoparticles can be used to improve the flow properties, nutritional quality of food, color, flavor, shelf life, and protecting ingredients such as antimicrobials or vitamins. For example, nanoparticles called chitosan enhance food safety since these have antimicrobial properties [47]. In food processing, the processes like preservation, prevention from pathogens, toxins removal, and enhancement of food consistency occur. Food processing has a significant impact on the physicochemical properties of food such as flavor, texture, stability, and release profile [48]. Properties of nanomaterials change with the change in material. For example, the properties of peptide and polysaccharide nanomaterials vary from the properties of metal and metal-oxide nanoparticles. Hence several nanostructured materials have been developed that act as anticaking and gelating agents and enhance taste, texture, and consistency [49]. They increases the shelf life of food products and reduce the wastage of food due to microbial infestation [50]. Nanocarriers, without disturbing the elementary morphology, act as delivery systems to carry food additives in food. Nanocapsules also act as a delivery system and play an important role in the food-processing sector. Nanostructured lipids act as a liquid carrier, which are insoluble in water. Fats such as nanodrops prevent the transportation of cholesterol from the digestive system to bloodstream [51]. Nanoencapsulation improves the release and retention of flavors [52]. Nanocarriers are fabricated for the protection of bioactive molecules and delivery. For example, rutin is one such example but has limited applications due to its poor solubility. Rutin trapped with ferritin nanocages enhances solubility and has better thermal and UV radiation [53]. The most important advantage of nanomaterials in food is that they are the carriers of fragrances and flavors and can release encapsulated compounds over longer time periods [54]. Because of their subcellular size, they provide a favorable improvement in the bioavailability of nutraceuticals [55]. We discuss the applications of nanotechnology in improving the texture, flavor, and appearance of food in the following sub-section.

10.3.1 Improvement in Texture, Flavor, and Appearance of Food

Flavors play a major role in evaluating the quality, consumption of food and are very important ingredients in any food. It is extremely difficult to maintain the flavors during the manufacturing and storage process [56]. The stability of flavors in different food products depends on the acceptability and quality of the food [50]. In order to prevent the degradation of flavors during processing and storage, it is better to encapsulate the flavors before using in the food, thereby providing a controlled release and improving the chemical stability. Encapsulation with a protective carrier guard prevents the interactions among flavors, oxidation, and reactions induced by light [50]. The most common carriers are biopolymers such as carbohydrates, gums, proteins, and chitosan [57]. For designing an encapsulation system, the important factors are carrier (viscosity) that should not react with flavors and physiochemical properties of the flavor (solubility) [50].

Nanoencapsulation fills the substance into nanocarriers and produces a final product that has a function of controlled release of the core materials [58]. With proper controlled release functions such as burst and sustained release, flavors can be released at the desired rate and time [59]. Such a system releases flavors very slowly in solvated conditions but bursts their release due to changes in pH or temperature or ionic strength when a food comes in contact with saliva. Nanocarrier encapsulation maintains the flavor quality during storage since it provides a sustained release of flavors. By encapsulating a compound in a proper nanocarrier that maintains physical stability in a certain performing condition and duration, a sustained release can be attained. The factors that influence the release mechanisms depend on the type of carrier for encapsulation, the preparation method, and the environment in which the flavors are released. During the release mechanism, processes of degradation, diffusion, osmosis, and melting may also be important [60]. The interactions between core and carrier materials influence the controlled release of flavor compounds. The next sub-section is about the importance of nanotechnology in enhancing the nutritional value of food.

10.3.2 Nutritional Value of Food

Most of the bioactive compounds such as carbohydrates, lipids, proteins, and vitamins are sensitive to enzymes activity and the high acidic environment of the stomach and duodenum. Encapsulation enables such compounds to become resistant to adverse conditions and assimilate easily in food products, which is very hard in non-capsulated form because of the low water solubility of these compounds. The tiny nanoparticles-based edible capsules improve the delivery of medicines, fragile micronutrients, or vitamins in daily food products so that they can provide significant health benefits. Nano-emulsification, nanocomposite, and nanostructuration are the different techniques that encapsulate substances in miniature forms so that they can deliver nutrients such as antioxidants and proteins effectively and precisely for targeted health and nutritional benefits. Polymeric nanoparticles protect and transport bioactive compounds to target functions and are suitable for the encapsulation of bioactive compounds such as flavonoids and vitamins. We discuss the importance of nanotechnology in improving the shelf life of food products.

10.3.3 Improvement in Shelf Life

Due to the hostile environment, bioactive components get degraded and inactivated eventually in functional foods. Nanoencapsulation of such components either slows down the degradation processes or prevents the degradation until the product is delivered to the target, thereby extending the shelf life of food. The edible nanocoatings on food products provide a barrier to moisture and gas exchange and deliver flavors, colors, enzymes, antioxidants, and anti-browning agents. The nanocoatings can also increase the shelf life of manufactured food products even after the packaging is opened. The chemical degradation process is slowed down by the encapsulation of functional components, by modifying the properties of the interfacial layers around them. For example, curcumin is the least stable and most active bioactive component of turmeric and shows reduced antioxidant activity on encapsulation. It is stable also to pasteurization at different ionic strengths. We discuss nanoencapsulation in the following sub-section.

10.3.4 Nanoencapsulation

Nanoencapsulation provides several benefits such as change in flavors, ease of handling, protection against oxidation, enhanced stability, retention of volatile ingredients, moisture-triggered controlled release, pH-triggered controlled release, taste making, consecutive delivery of multiple active ingredients, and enhanced bioavailability [61, 62]. It can be defined as a nanovesicular system that possesses a typical core–shell structure to keep the drug in a confined reservoir or in a cavity surrounded by a polymer membrane [63]. The cavity may contain the active substance in solid or liquid form or as a molecular dispersion. Nanoencapsulation is carried out with the help of nanocapsules that deliver the desired component and entrap unwanted components and odors in the food products, preserving the food. Nanocapsules have the tendency to carry food supplements via the gastrointestinal tract, thereby increasing the bioavailability of the substance. Nanocapsules can be prepared by six basic ways: emulsion-diffusion, double emulsification, nanoprecipitation, layer-by-layer emulsion, coacervation, and polymer coating [64]. Nanoemulsion does not change the appearance of food when added, which makes it different from the conventional emulsion. Nanocapsules can also be used for the delivery of fertilizers, pesticides, and vaccines to plants. They also deliver health supplements such as vitamins and minerals in food, fatty acids, and growth hormones, thereby increasing the nutrients in food [65]. Basically, encapsulation protects the hidden component and delivers it precisely to the target site even in unfavorable conditions. An example of a nanobased carrier used for nanoencapsulation is liposome, which helps in the controlled delivery of many components such as nutrients, enzymes, nutraceuticals, vitamins, and antimicrobials within the system [66]. A new encapsulation technique in which zein fibers are loaded with gallic acid using electrospinning protects lipids from degrading within the system before they reach the target. It can be utilized by the food-packaging industry. Lipid-based encapsulation is much more efficient due to the better solubility and specificity of the components encapsulated in it. This kind of encapsulation prevents the component from interacting with the food product, so the original feature of the food remains intact, keeping the component to be delivered unaltered within the biological system. Capsules such as colloidosome have hollow shell-like structure having a very small size, i.e., less than a quarter of a human cell [67]. Many components can be placed inside the shell that acts as a good carrier of drugs and food supplements in the biological system. Similarly, nanocochleates improve the quality of processed food. These are nanocoils wrapped around micronutrients, thereby stabilizing them. Nanoencapsulation of probiotics helps to regulate the immune response and is also an emerging field in nanotechnology. With the help of nanoencapsulation, probiotics are well preserved and efficiently delivered to the gastrointestinal tract. According to several reports, starch-like nanoparticles help to preserve lipid bodies and are delivered to the target site efficiently [68]. Archaeosomes help in the delivery of antioxidants and are an example of nanoencapsulation. They are prepared from archaeobacterial membrane lipids, which are thermostable and resistant to stress [69]. As per some reports, with the nanoencapsulation of α-tocopherol in fat droplets, milk can be protected from degradation [62]. Some commercialized food products such as canola active oil, by a company called Shemen, in Haifa, Israel, are used for the nanoencapsulation of fortified phytosterols [70]. Some food products like fortified fruit juice is used for the nanoencapsulation of fortified vitamin, theanine, lycopene, and sun active iron. These products are manufactured by a company called High Vive in USA. Other products such as NanoResveratrol, which is a plant-based lipid, are manufactured by a company called Life Enhancement. Health Plus International produces a spray for Life Vitamin Supplements that are used for the nanoencapsulation of fortified vitamin beverage. Daily Boost, which is used for the nanoencapsulation of fortified vitamin, is manufactured by a company called Jamba Juice Hawaii. RBC Life Sciences Inc. manufactures Nanoceuticals Slim Shake Chocolate, which is used for the nanoencapsulation of nanoclusters to enhance the flavor of the shake without adding sugar to the drink [71, 72]. The manufacturer of Nanotea is Qinhuangdao Taiji Ring Nano-Products Co. Ltd. (China) [73]. We now discuss nanotechnology in food packaging.

10.4 Nanotechnology in Food Packaging

Nanotechnology-based food packaging has many improved properties than conventional food packaging, such as enhanced durability, temperature and flame resistance, recycling, processability due to lower viscosity resulting in a better delivery of active materials into the biological systems, and minimum costs with less environmental issue. Such properties make it ideal for the development of nanomaterials and to be widely used in food packaging for cheese, cereals, confectionery, processed meat, and boil-in-the-bag foods. It also helps in extrusion-coating applications for dairy products and fruit juices or co-extrusion processes for the manufacture of beer bottles and carbonated drinks [74].

Honeywell International has developed a mixture of nylon6 nanoclay composite for packaging of flavored alcoholic beverages [75]. A company named Nanocor has made polymer nanocomposites for food packaging. It has improved properties such as moisture and gas barrier, strength, toughness, abrasion and chemical resistance, which is required for good packaging [76]. Mitsubishi Gas Chemical Company Inc. USA collaborated with Nanocor and developed nylon nanocomposites named Imperm, which applies a barrier layer for multilayer polyethylene terephthalate (PET) bottle and is used for liquors and small carbonated soft-drink beverages. Another nylon6-based nanocomposite, NanoTuff, exhibits improved barrier properties to O_2, H_2O, and CO_2 [77]. We now discuss the different types of packaging in the next sub-sections.

10.4.1 Biobased Packaging

Biopolymers have attracted much attention for the replacement of oil-based plastic packaging materials. Bioplastic is a better choice than conventional plastics since it is biobased and biodegradable [78]. Biobased polymers can be derived from plant materials such as cellulose, starch, proteins, and other polysaccharides or animal products such as proteins and polysaccharides or microbial products such as PHB or polymers synthesized from polylactide and bio-polyethylene.

Biopolymers have typically low mechanical and barrier properties than conventional plastics, due to which they have limited usage in the food-packaging industry. The most challenging part is the development of moisture barrier in biopolymers.

Some of the biobased nanomaterials such as nano- or microfibrillated cellulose as well as cellulose nanocrystals are extracted from natural fibers that can improve the gas barrier and strength but are very sensitive to moisture. Their application in fillers, stand-alone films, and coatings in nanocomposite systems has been explored, though not much amount of research has been performed on these materials [79].

Research is still going on to remove all the flaws of biopolymer-based packaging materials. Nanomaterial-biopolymer materials have the tendency to meet the requirements for packaging applications that could increase the use of biopolymer.

10.4.2   Improved Packaging

Nanocomposites, a combination of traditional food-packaging materials with nanoparticles, have gained enough attention in the food-packaging sector due to their remarkable antimicrobial spectrum, tough resistant characteristics, and good mechanical performance [80]. Nanocomposites are formed from the amalgamation of a matrix in the continuous phase and a nano-dimensional material in the discontinuous phase. The nano-dimensional phase depends on the nanomaterial and is characterized into nanospheres or nanoparticles, nanotubes, nanowhiskers or nanorods and nanosheets, or nanoplatelets [81]. The nanosized phases increase the mechanical properties of polymer, and the elastic strain is shifted to the nanoreinforced material. Due to this, nanocomposites are good for improvising the barrier and mechanical characteristics of polymers and add active or smart properties to the packaging system [82, 83]. Nanotechnology applications in polymer science can make routes in improving the cost–price competence and characteristic features of packaging materials [84].

Polymer nanocomposites (PNCs) are made up of mixtures of polymers with organic or inorganic fillers having particular geometries, such as flakes, fibers, and spheres that have been recently introduced as packaging materials [85]. The ratio of the largest to the smallest dimension of filler (aspect ratio) plays a significant role. Fillers with the higher aspect ratios have more specific surface area, with high reinforcing properties [86]. Various nanomaterials such as silica, clay, graphene, carbon nanotubes, and ZnO_2 are being used as fillers [87–91].

10.4.3 Active Packaging

After many developments on blends, monolayer bottles with oxygen-scavenging properties have been produced, which consist of blends of PET and oxygen-absorbing polymers such as polyamides or co-polyesters. The co-polyesters are separated by processing and compatibilization of the two polymers into nanoscalar inclusions in the matrix, and the reaction is catalyzed with cobalt dissolved in the polymer matrix. To retain good transparency and improve the properties of the blend, a huge reduction in the size of the polymer domains is required [92]. To attain good transparency, the domain size must be below the wavelength of visible light, i.e., in the nanometer range.

Antimicrobial packaging has got huge attention in research in nanotechnology applications. For oxygen-scavenging applications, blending can be used to provide antimicrobial properties to a material, through controlled release of synthetic or natural antimicrobial agents [93].

Some metal and metal-oxide-based nanomaterials have strong antimicrobial properties, and their incorporation into polymers forms nanomaterial-polymers, which help to preserve packaged foodstuffs for a long time by stopping the growth of microorganisms at the contact of food surface. For example, as discussed earlier, nanosilver has antimicrobial and anti-odorant properties than plastic food containers and bags [94]. The antimicrobial properties of nanomagnesium oxide and nanozinc oxide can be used for antimicrobial food-packaging materials [95]. A plastic wrap with nanozinc oxide is available in some countries, which keeps the packaging surfaces hygienic under indoor-lighting conditions [62].

Another way to improve packaging is to include active functionality through nanotechnology, by grafting active components such as antimicrobial, antioxidant, and scavengers on clay nanoplates. This approach improves the other properties in the presence of clay and efficiently disperses the active compound in the matrix.

10.4.4 Smart Packaging

Smart or, in other words, intelligent packaging boosts the communication aspect of a package. It helps in recognizing the features of food and then apply different mechanisms to convey information on the existing quality of food in terms of its safety and digestibility. Such a system utilizes very innovative communication methods such as time–temperature indicators, freshness indicators, nanosensors, oxygen sensors, etc. [96]. The nanosensors used in food-packaging systems detect pathogens, spoilage-associated changes, and chemical contaminants, thereby providing the exact information on the freshness of food [97]. Nanosensors are nanotechnology-enabled sensors that append an intelligent function to food packaging and can be applied as coatings or labels to ensure the integrity of the package through the detection of leaks, time–temperature variation indications, or microbial safety [98]. Gas sensors are used for revealing the gaseous analyte information in the package. Optical O_2 sensors work on the principle of absorbance or luminescence quenching changes caused by the contact with analyte. Optochemical sensors check the quality of products by sensing gas constituents such as CO_2, hydrogen sulfide, and volatile amines [99]. Because of the unique electro-optical and chemical properties of nanoscale particles to detect environmental changes, degradation or microbial contamination, such a technology would benefit industry stakeholders, food regulators, and consumers [100].

10.4.5 Detection of Pathogens with Nanosensors

Nanosensors can detect food spoilages; an array of thousands of nanoparticles has been designed to fluoresce in different colors after coming in contact with food pathogens. The main aim of nanosensors is to reduce the detection time of pathogens from days to hours to minutes [85]. Such nanosensors can be placed directly into the packaging material to detect chemicals released during food spoilage [101]. Other types of nanosensors are based on microfluidic devices [102], which also detect pathogens and have high sensitivity. A major advantage of using microfluidic sensors is that they require microliters of sample volumes and their miniature format, which has led to their widespread applications in medical and chemical analysis [103].

The devices produced with the so-called nanoelectromechanical system (NEMS) technology contain moving parts ranging from nanometer to millimeter scale, serving as developing tools in food preservation that can control the storage environment. NEMS consists of advanced transducers that detect chemical and biochemical signals. For example, a digital transform spectrometer (DTS) made by Polychromix (USA) uses NEMS technology to detect trans-fat in food products [104]. The utilization of micro- and nanotechnologies (MNTs) has many advantages such as smart communication through various frequency levels, portable instrumentation with quick response, and low costs. MNTs are particularly suitable for food safety and quality because they can detect and monitor any adulteration in packaging and storage [105]. Another class of biosensors called nanocantilevers has the ability to detect biological binding interactions, such as between antibody and antigen, receptor and ligand, and enzyme and substrate, through electromechanical or physical signaling [106]. It consists of tiny pieces of silicon-based materials that can recognize proteins, viruses, and pathogenic bacteria [107]. Nanocantilevers have already succeeded in the studies of molecular interactions and detection of toxins, contaminant chemicals, and antibiotic residues in food products [108]. The pathogens are detected because of their ability to vibrate at various frequencies and also depending on the biomass of the pathogens. For example, the silicon surface of nanocantilevers can be engineered to attach pathogens that can be detected due to the change in the resonant frequency and mass. Bio-Finger, a European Union-funded project, developed a nanocantilever to diagnose cancer. It can also detect pathogens in food and water based on sensing the ligand–receptor interactions [109]. We discuss the safety issues related to nanotechnology in food in the next section.

10.5 Safety Issues

Normally, the food and beverages industry is cautious about the utilization of nanotechnology, despite a lot of potential applications of nanotechnology in food processing and packaging. Because of the high surface area, nanomaterials may have toxic effects [110]. Also there might be unexpected risks of using them in the food industry. With the full application of nanotechnology in the food and beverages industry, there will be a wide range of food products in the market, but there are some serious questions regarding the safety of their use. The generally recognized as safe (GRAS) list of additives will have to be reassessed with the application of nanotechnology [111]. It has been observed that rats breathing nanoparticles from single-walled carbon nanotubes caused acute pulmonary inflammation and stress. Also when rats were exposed to multiwalled carbon nanotubes for 3 months, they showed granulomatous-type inflammation [112]. The main concern is regarding toxicity since nanoparticles are more mobile, reactive, and likely to be more toxic. Nanoparticles have strong tendency to cause oxidative stress, which can produce free radicals, resulting in DNA mutation and diseases such as cancer [113]. Due to their unique properties, nanomaterials could participate in most biological reactions that can have harmful effects on human health or environment. First, their size should be considered in assessing the safety of nanomaterials to be used in food products. Second, the chemical composition should be considered because during the production of nanoparticles, many reagents are used, which may be toxic. Third, surface structure affects cytotoxicity and is one of most important factors that affect the interaction of nanoparticles and biological systems. Fourth, solubility is also crucial in adding the toxicity of nanoparticles. For example, hydrophilic titanium oxide nanoparticles are more toxic than insoluble titanium oxide nanoparticles. Some hydrophilic nickel compounds are recognized as carcinogenic agents. Thus, in order to understand the toxicity and biological activity of nanoparticles, we need to understand these and many other factors before applying nanotechnology in the food industry. Or it is must to investigate the factors regarding the toxicity and environmental activity of nanoparticles before their application [114]. The rise in the demand for nanomaterials as flavor or color additives has attracted much attention of the public and government sectors. With nanoencapsulation, a direct contact of nanomaterials with humans can be made through oral intake. A study on TiO_2 in sugar-coated chewing gum found that over 93% of TiO_2 is nanosized. SiO_2 nanomaterials act as carriers of fragrances or flavors in food products. Now, there are two major concerns, allergy and heavy metal release, which have been reported here [115]. As the report suggests, SiO_2 nanoparticles can induce allergen-specific Th2-type allergic immune responses in vivo, as evident from a study on female BALB/c mice exposed to nanoparticles, further resulting in lipid peroxidation and DNA damage. We now discuss the regulations for food nanotechnology in the next section.

10.6 Regulations for Food Nanotechnology

As per the FDA in the United States, it is mandatory for manufacturers to show that food ingredients and products are not injurious to health. But this regulation does not explicitly cover nanoparticles, which can be harmful in nanotechnology applications. No regulations exist for the use of nanotechnology in the food industry [116]. The regulations suggested by the European Union are yet to be accepted and imposed. FDA regulations emphasize that many products already under regulation contain particles in the nanosize and expect that many nanoproducts will come under the jurisdiction of many of its centers; thereby the Office of Combination Products will take any relevant responsibility [117]. In 2010, the European Food and Safety Authority (EFSA) created a network between Europe Union (EU) member states to assess the risk related to nanotechnology in the food industry. The European Commission implemented Regulation No. 1333/2008, which mandates a new authorization process and safety evaluation of any change in the production method or starting material for a food additive, including particle size. The Europe Union also implemented essential changes to their existing food-labeling law in 2014 that requires all engineered nano materials (ENMs) to be included on the list of ingredients, followed by the term [nano] [118]. A UK-based institute, The Institute of Food Science and Technology (IFST), says that size matters and suggests to regard nanomaterials to be potentially harmful until proven safe by tests [119]. It suggested to use conventional E-numbering system for labeling along with subscript "n" for using nanoparticles in food [120]. The UK Government agreed with the Royal Society and the Royal Academy of Engineering to mark the ingredient labels to aware consumers. Hence, an updated version of ingredient labeling will be an important requirement. A Swiss center for technology assessed the circumstances related to the use of nanotechnology in food in Switzerland [121]. Certain guidelines and rules for nanomaterials have been established by regulatory authorities around the world. The regulatory bodies should be informed, but due to lack of essential safety data, there is an uncertainty over the regulation of nanoproducts [122]. Efforts are being made to facilitate international collaboration to confirm the approval and usage of nanotechnology [123]. So the international organizations are collecting information to decide how to proceed [124]. Several Asian countries have come forward with their own definitions, regulations, and committees for the application of nanotechnology in the food industry. Among them, China is leading the path, followed by Japan and South Korea. Assessment responsibilities have been taken by the State Food and Drug Administration (SFDA), China, and the regulatory body Chinese National: Nanotechnology Standardization Technical Committee (NSTC) has taken the responsibility of regulation of nanotechnology and its application in the food industry. Even then the safety levels are lower than in the United States or any EU country [125]. The Japanese Government supports the development of nanotechnology industry. Among southern countries, Nacional de Vigilância Sanitaria (ANVISA), an entity formed by the Brazil Government, has taken the lead to regulate the use of nanotechnology. ANVISA has established procedures for the registration of new food products in Brazil but included no specific requirements for the use of nanotechnology [126].

10.7   Conclusion and Future Prospects

The introduction of nanotechnology in the food industry has been a revolution, starting from additives to pesticides to nanosensors. The use of nanomaterials and their availability in the market will increase inevitably. There has been 40% rise in publications and 90% patents in the field of nanotechnology over the past two decades. Many companies have R&D focus on nanotechnology-based products [127]. In the past few years, nanotechnology has produced numerous foods and food-packaging products. There are more than 300 nanofood products available in the market across the globe, which has motivated to invest in the R&D of nanotechnology for the food industry. Nanotechnology has an important role in the food industry ranging from manufacturing food to food processing and from nutrition to intelligent packaging [128]. Nanotechnology can benefit from processed foods in numerous ways. The ultimate dream of consumers will be programmable foods, which will have designer food features built into them. Consumers can make a product of their own desired flavor, color, and nutrition using specially programmed microwave ovens. The trick is to formulate the food at the manufacturer's end with millions of nanoparticles of different flavors, colors, and nutrients and programmed in the oven set by the consumer, which is based on his or her preferences. Only selective particles are activated, while others are kept inactive, giving the desired product [129]. Nanotechnology in food packaging will help to upgrade food safety, amplify nourishment life, aware customers that food is sullied or destroyed, repair tears in packaging, and uniformly release added substances to grow the life of food in the package [130]. Nanotechnology is one of the promising fields to maintain leadership in the food industry in the future. The first step will be to improve the safety and quality of food. Nanotechnology allows changes in food systems to ensure safety and enhance the nutritional quality of food.

The future of nanotechnology-based applications in the food industry can be seen in the following areas: using food matrices as delivery tools for bioactive compounds, thereby promoting health; developing ingredients with enhanced functionality to get food with unique flavors and textures; developing unique nano-packaging materials with enhanced barrier properties; developing materials that have self-sanitizing properties; novel processing technologies; and development of nanosensors for food safety [131]. The development of all these areas should be cost effective in the near future.

Despite several benefits, the use of nanotechnology in the food industry seems to be a conflicting topic in the modern world. On one hand, nanotechnology increases the shelf life of food, improves food packaging using nanosensors, prevents antimicrobial contamination, and uses nanocarriers for the delivery of nutrition in food, which are some of the advantages of using nanotechnology in food. On the other hand, the major concern in using nanotechnology is the potential toxicity, thereby raising questions on the side effects of nanofoods on human health and environment. Only proper research, information, and regulation of nanofood can solve this problem. Several bodies have been formed to govern every aspect of nanofood, from their synthesis to their usage in the food industry. Laws and regulations implemented by the US-FDA, ESFA, etc. need to be followed so that nanofood can be safe for human consumption.

Though the integration of nanotechnology with the food industry is still in the developing stage, there is lot of enthusiasm surrounding this technology. The food industry has to recognize the opportunities and discoveries in nanotechnology to enhance production, processing, and packaging of food products. It is very difficult to see the long-term impact of nanotechnology on the food industry. On one hand, the potential dangers due to its misuse cannot be ignored; on the other hand, nanotechnology has the potential to contribute significantly toward the improvement in life through changes in the food and agricultural system. Both the agriculture and food sectors could gain a competitive position with the help of nanotechnology, whereas consumers may benefit from nanotechnology applications that contribute to the safety and nutritional value of food.

Acknowledgments

The contributor thanks Jenny Stanford Publishing and Dr. Rakesh Sindhu for giving an opportunity to get this chapter published.

References

1. Gade, A., Rai, M., and Yadav, A. (2009). Silver nanoparticles as a new generation of antimicrobials, *Biotechnol. Adv.*, **27**, pp. 76–83.

2. Eral, H. B., Doyle, P. S., Gupta, A., et al. (2016). Nanoemulsions: Formation, properties and applications, *Soft Matter*, **12**, pp. 2826–2841.

3. Britti, M. S., Finamore, A., Garaguso, I., et al. (2003). Zinc oxide protects cultured enterocytes from the damage induced by *Escherichia coli*, *J. Nutri.*, **133**, pp. 4077–4082.

4. Sawai, J. (2003). Quantitative evaluation of antibacterial activities of metallic oxide powders (ZnO, MgO and CaO) by conductimetric assay, *J. Microbiol. Method*, **54**, pp. 177–182.

5. Dasgupta, N., Kumar, A., Mundekkad, D., et al. (2015). Nanotechnology in agrofood: From field to plate, *Food Res. Int.*, **69**, pp. 381–400.

6. Anandharamakrishnan, C., Chhanwal, N., Ezhilarasi, P. N., et al. (2013). Nanoencapsulation techniques for food bioactive components: A review, *Food Bioprocess Technol.*, **6**, pp. 628–647.

7. Colin, M. and Powell, M. (2008). Nanotechnology and food safety: Potential benefits, possible risks? *CAB Rev. Perspect. Agric. Vet. Sci. Nutr. Nat. Res.*, **3**, pp. 123–142.

8. Campbell-Platt, G., ed. (2009). *Food Science and Technology* (Chichester: University of Reading, UK; Wiley-Blackwell).

9. Charych, D., Cheng, Q., Kuziemko, G., et al. (1996). A 'litmus test' for molecular recognition using artificial membranes, *Chem. Biol.*, **3**, pp. 113–120.

10. Imafidon, G. I. and Spanier, A. M. (1994). Unraveling the secret of meat flavor, *Trends Food Sci. Technol.*, **5**, pp. 315–321.

11. Li, N., Mädler, L., Nel, A., et al. (2009). Toxic potential of materials at the nanolevel, *Science*, **311**(5761), pp. 622–627.

12. Chen, L. (2006). Food protein-based materials as nutraceuticals delivery systems, *Trends Food Sci. Technol.*, **17**, pp. 272–283.

13. Watanabe, J. (2005). Entrapment of some compounds into biocompatible nano sized particles and their releasing properties, *Colloids Surf. B Biointerfaces*, **42**, pp. 141–146.

14. Chang, C., Graves, S., Mason, T. G., et al. (2006). Nanoemulsions: formation, structure, and physical properties, *J. Phys.: Condens. Matter*, **18**, R635.

15. McClements, D. J. (2012). Nanoemulsions versus microemulsions: terminology, differences, and similarities, *Soft Matter*, **8**, pp. 1719–1729.

16. Bibette, J., Cates, M. E., Couffin, A. C., et al. (2011). How to prepare and stabilize very small nanoemulsions, *Langmuir*, **27**, pp. 1683–1692.

17. Azemar, N., Garcia-Celma, M., Izquierdo, P., et al. (2005). Nano-emulsions, *Curr. Opin. Colloid Interface Sci.*, **10**, pp. 102–110.

18. Fryd, M. M. and Mason, T. G. (2010). Time-dependent nanoemulsion droplet size reduction by evaporative ripening, *J. Phys. Chem. Lett.*, **1**, pp. 3349–3353.

19. Esquena, J., Forgiarini, A., Gonzalez, C., et al. (2000). Studies of the relation between phase behavior and emulsification methods with nanoemulsion formation, *Trends Colloid Interface Sci.*, **XIV**, pp. 36–39.

20. Esquena, J., Forgiarini, A., González, C., et al. (2001). Formation and stability of nano-emulsions in mixed nonionic surfactant systems, *Trends Colloid Interface Sci.*, **XV**, pp. 184–189.

21. Fryd, M. M. and Mason, T. G. (2012). Advanced nanoemulsions, *Annu. Rev. Phys. Chem.*, **63**, pp. 493–518.

22. Arnaudov, L. N., Feng, J., Gurkov, T. D., et al. (2014). Nanoemulsions obtained via bubble bursting at a compound interface, *Nat. Phys.*, **10**, pp. 606–612.

23. Nuchuchua, O., Puttipipatkhachorn, S., Ruktanonchai, U., et al. (2009). Characterization and mosquito repellent activity of citronella oil nanoemulsion, *Int. J. Pharm.*, **372**, pp. 105–111.

24. Coma, V. (2008). Bioactive packaging technologies for extended shelf life of meat-based products, *Meat Sci.*, **78**(2), pp. 90–103.

25. Horner, S. R., Mace, C. R., Miller, B. L., et al. (2006). A proteomic biosensor for enteropathogenic *E. coli*, *Biosens. Bioelectron.*, **21**(8), pp. 1659–1663.

26. Mirkin, C. A., Nam, J. M., and Thaxton, C. S. (2003). Nanoparticle-based bio-bar codes for the ultrasensitive detection of proteins, *Science*, **301**(5641), pp. 1884–1886.

27. Halley, P. J., Torley, P., and Zhao, R. (2008). Emerging biodegradable materials: Starch- and protein-based bio-nanocomposites, *J. Mater. Sci.*, **43**(9), pp. 3058–3071.

28. Acosta, E. (2009). Bioavailability of nanoparticles in nutrient and nutraceutical delivery, *Curr. Opin. Colloid Interface Sci.*, **14**(1), pp. 3–15.

29. Arora, A. and Padua, G. W. (2010). Review: Nanocomposites in food packaging, *J. Food Sci.*, **75**(1), pp. R43–R49.

30. Bouwmeester, H., Dekkers, S., Noordam, M. Y., et al. (2009). Review of health safety aspects of nanotechnologies in food production, *Regul. Toxicol. Pharmacol.*, **53**(1), pp. 52–62.

31. Marinkova, D., Yaneva, S., and Yotova, L. (2013). Biomimetic nanosensors for determination of toxic compounds in food and agricultural products (review), *J. Chem. Technol. Metall.*, **48**(3), pp. 215–227.

32. Gururani, S. K., Pandey, S., and Zaidih, M. G. H. (2013). Recent developments in clay-polymer nano composites, *Sci. J. Rev.*, **2**(11), pp. 296–328.

33. Chen, S. H., Davis, D., Guo, X., et al. (2013). Gold nanoparticle-modified carbon electrode biosensor for the detection of listeria monocytogenes, *Ind. Biotechnol.*, **9**(1), pp. 31–36.

34. Gowri, A. S., Ramachandran, S., and Thirumurugan, A. (2013). Combined effect of bacteriocin with gold nanoparticles against food spoiling bacteria: An approach for food packaging material preparation, *Int. Food Res. J.*, **20**(4), pp. 1909–1912.

35. Flanagan, J. and Singh, H. (2006). Microemulsions: A potential delivery system for bioactives in food, *Crit. Rev. Food Sci. Nutr.*, **46**(3), pp. 221–237.

36. Gao, J. Q., He, C. X., and He, Z. G. (2010). Microemulsions as drug delivery systems to improve the solubility and the bioavailability of poorly water-soluble drugs, *Exp. Opin. Drug Deliv.*, **7**(4), pp. 445–460.

37. Huang, Q., Kokini, J. L., Liu, S., et al. (2003). Nanotechnology: A new frontier in food science, *Food Technol.*, **57**(12), pp. 24–29.

38. Baeumner, A. (2004). Nanosensors identify pathogens in food, *Food Technol.*, **58**(8), pp. 51–55.

39. Gupta, S. and Moulik, S. P. (2008). Biocompatible microemulsions and their prospective uses in drug delivery, *J. Pharm. Sci.*, **97**(1), pp. 22–45.

40. Chang, C. B., Graves, S. M., Mason, T. G., et al. (2006). Nanoemulsions: Formation, structure, and physical properties, *J. Phys.: Cond. Matter*, **18**(41), pp. R635–R666.

41. Avérous, L., Bordes, P., and Pollet, E. (2009). Nano-biocomposites: Biodegradable polyester/nanoclay systems, *Prog. Polym. Sci.*, **34**(2), pp. 125–155.

42. Brody, A. L. (2003). Nano food packaging technology, *Food Technol.*, **57**(12), pp. 52–54.

43. Burdo, O. G. (2005). Nanoscale effects in food-production technologies, *J. Eng. Phys. Thermophys.*, **78**(1), pp. 90–96.

44. Hett, A. (2004). *Nanotechnology: Small Matter, Many Unknowns*. Swiss Reinsurance Company: Zürich, Switzerland.

45. Sekhon, B. S. (2010). Food nanotechnology: An overview, *Nanotechnol. Sci. Appl.*, **3**.

46. Hochella, M. F. (2008). Nanogeoscience: From origins to cutting-edge applications, *Elements*, **4**(6), pp. 373–379.

47. Sahu, S. C. and Wallace, H. A. (2017). Nanotechnology in the food industry: A short review. *Food Safety Magazine*, February/March 2017.

48. Cientifica Report Nanotechnologies in the Food Industry, P.A.A.a.h.w.c.c.w.d.p.i.a.O. (2006). Assessment of the potential use of nanomaterials as food additives or food ingredients in relation to consumer safety and implication for regulatory controls.

49. Barla, A., Ojha, N., Pradhan, N., et al. (2015). Facets of nanotechnology as seen in food processing, packaging, and preservation industry, *Biomed Res Int.*, 365672.

50. Estevinho, B. N. and Rocha, F. (2017). A key for the future of the flavors in food industry. In *Nanotechnology Applications in Food* (Grumezescu, A. M. and Oprea, A. E., Eds.) (Academic Press, Cambridge, MA, USA), pp. 1–19.

51. Abbas, K. A., Mohd Azhan, N., Saeed, M. E., et al. (2009). The recent advances in the nanotechnology and its applications in food processing: A review, *J. Food Agri. Environ.*, **7**(3&4), pp. 14–17.

52. Dingman, J. (2008). Nanotechnology: Its impact on food safety, *J. Environ. Health,* **70**(6), pp. 47–50.

53. Nakagawa, K. (2014). Nano- and microencapsulation for foods. In *Nano- and Micro-Encapsulation of Flavor in Food Systems* (Kwak, H. S., Ed.) (Oxford: John Wiley & Sons), pp. 249–272.

54. Gao, Y., Sun, G., Xu, J., et al. (2015). Synthesis of homogeneous protein-stabilized rutin nanodispersions by reversible assembly of soybean (Glycine max) seed ferritin, *RSC Adv.,* **5**(40), pp. 31533–31540.

55. Bokkers, B. G., Dekkers, S., Krystek, P., et al. (2011). Presence and risks of nanosilica in food products, *Nanotoxicology*, **5**(3), pp. 393–405.

56. Bajpai, V. K., Kumar, P., Shukla, S., et al. (2017). Application of nanotechnology in food science: Perception and overview, *Front Microbiol.,* **8**, pp. 1501.

57. Desobry, S., Jacquot, M., Madene, A., et al. (2006). Flavour encapsulation and controlled release: A review, *Int. J. Food Sci. Technol.*, **41**(1), pp. 1–21.

58. Chambin, O., Gharsallaoui, A., Roudaut, G., et al. (2007). Applications of spray-drying in microencapsulation of food ingredients: An overview, *Food Res. Int.*, **40**(9), pp. 1107–1121.

59. Camacho-Díaz, B. H., Meraz-Torres, L. S., Quintanilla-Carvajal, M. X., et al. (2010). Nanoencapsulation: A new trend in food engineering processing, *Food Eng. Rev.*, **2**(1), pp. 39–50.

60. Alves, A., Estevinho, B. N., Rocha, F., et al. (2013). Microencapsulation with chitosan by spray drying for industry applications: A review, *Trends Food Sci. Technol.*, **31**(2), pp. 138–155.

61. Bugusu, B. and Marsh, K. (2007). Food packaging: Roles, materials, and environmental issues: Scientific status summary, *J. Food Sci.*, **72**(3), pp. R39–R55.

62. Blackburn, J., Chaudhry, Q., Scotter, M. et al. (2008). Applications and implications of nanotechnologies for the food sector, *Food Additives Contaminants*, **25**(3), pp. 241–258.

63. Sekhon, B. S. (2010). Food nanotechnology: An overview, *Nanotechnol. Sci. Appl.*, **3**(1), pp. 1–15.

64. Aitken, R. J., Butz, T., Maynard, A. D. et al. (2006). Safe handling of nanotechnology, *Nature*, **444**(7117), pp. 267–269.

65. Dreher, K. L. (2004). Health and environmental impact of nanotechnology: Toxicological assessment of manufactured nanoparticles, *Toxicol. Sci.*, **77**(1), pp. 3–5.

66. Bradley, K. A., Chopra, K., Godwin, H. A. et al. (2009). The University of California Center for the environmental implications of nanotechnology, *Environ. Sci. Technol.*, **43**(17), pp. 6453–6457.

67. Alvarez, P. J. J., Batley, G. E., Klaine, S. J. et al. (2008). Nanomaterials in the environment: Behavior, fate, bioavailability, and effects, *Environ. Toxicol. Chem.*, **27**(9), pp. 1825–1851.

68. Azeredo, H. M. C. (2009). Nanocomposites for food packaging applications, *Food Res. Int.*, **42**(9), pp. 1240–1253.

69. Alfadul, S. M. and Elneshwy, A. A. (2010). Use of nanotechnology in food processing, packaging and safety: Review, *Af. J. Food Agr., Nutr. Dev.*, **10**(6).

70. Kruif, C. G. and Graveland-Bikker, J. F. (2006). Unique milk protein based nanotubes: Food and nanotechnology meet, *Trends Food Sci. Technol.*, **17**(5), pp. 196–203.

71. Bodmeier, R., Chen, H. G., and Paeratakul, O. (1989). A novel approach to the oral delivery of micro- or nanoparticles, *Pharm. Res.*, **6**(5), pp. 413–417.

72. Cobb, M. D. and Macoubrie, J. (2004). Public perceptions about nanotechnology: Risks, benefits and trust, *J. Nanoparticle Res.*, **6**(4), pp. 395–405.

73. Donaldson, K. and Seaton, A. (2007). The janus faces of nanoparticles, *J. Nanosci. Nanotechnol.*, **7**(12), pp. 4607–4611.

74. Bumbudsanpharoke, N. and Ko, S. (2015). Nano-food packaging: An overview of market, migration research, and safety regulations, *J. Food Sci.*, **80**, pp. R910–R923.

75. Cooper, T. A. (2013). Developments in plastic materials and recycling systems for packaging food, beverages and other fast-moving consumer goods. In *Trends in Packaging of Food, Beverages and Other Fast-Moving Consumer Goods (FMCG): Markets, Materials and Technologies* (Farmer, N., Ed.) (Philadelphia, PA: Woodhead Publishing Limited).

76. Cruz-Romero, M., Cushen, M., Cummins, E., et al. (2012). Nanotechnologies in the food industry: Recent developments, risks and regulation, *Trends Food Sci. Technol.*, **24**, pp. 30–46.

77. Nanocor. (2008). Nanocor Product Lines. Available online at: http://www.nanocor.com/products.asp.

78. Bras, J., Dufresne, A., and Siqueira, G. (2010). Cellulosic bionanocomposites: A review of preparation, properties and applications, *Polymers,* **2**(4), pp. 728–765.

79. Li, F., Mascheroni, E., and Piergiovanni, L. (2015). The potential of nanocellulose in the packaging field: A review, *Packaging Technol. Sci.,* **28**(6), pp. 475–508.

80. Harifi, T. and Montazer, M. (2017). New approaches and future aspects of antibacterial food packaging: From nanoparticles coating to nanofibers and nanocomposites, with foresight to address the regulatory uncertainty. In *Food Package* (Grumezescu, A. M., Ed.) (Academic Press), pp. 533–559.

81. Bratovcic, A., Ćatić, S., Odobašić, A., et al. (2015). Application of polymer nanocomposite materials in food packaging, *Croat. J. Food Sci. Technol.,* **7**, pp. 86–94.

82. Othman, S. H. (2014). Bio-nanocomposite materials for food packaging applications: Types of biopolymer and nano-sized filler, *Agr. Sci. Procedia,* **2**, pp. 296–303.

83. Duncan, T. V. (2011). Applications of nanotechnology in food packaging and food safety: Barrier materials, antimicrobials and sensors, *J. Colloid Interface Sci.,* **363**, pp. 1–24.

84. De Azeredo, H. M. C., McHugh, T. H., and Mattoso, L. H. C. (2011). Nanocomposites in food packaging: A review. In *Advances in Diverse Industrial Applications of Nanocomposites* (Reddy, B. S. R., Ed.) (InTech), pp. 57–78.

85. Gupta, R. K. and Thakur, P. V. K. (2016). Recent progress on ferroelectric polymer-based nanocomposites for high energy density capacitors: Synthesis, dielectric properties, and future aspects, *Chem. Rev.,* **116**, pp. 4260–4317.

86. Cavaille, J. Y., Chazeau, L., Dalmas, F., et al. (2007). Visco elastic behavior and electrical properties of flexible nanofiber filled polymer nanocomposites. Influence of processing conditions, *Composites Sci. Technol.,* **67**, pp. 829–839.

87. Bracho, D., Dougnac, V. N., Palza, H., et al. (2012). Fictionalization of silica nanoparticles for polypropylene nanocomposite applications, *J. Nanomater.,* 2012:263915.

88. Groschel, A. H., Kalo, H., Linkebein, T., et al. (2011). Shear stiff, surface modified, mica-like nanoplatelets: A novel filler for polymer nanocomposites, *J. Mater. Chem.,* **21**, pp. 12110–12116.

89. Han, H., Khan, S. B., Kim, D., et al. (2013). Preparation and characterization of poly (propylene carbonate)/exfoliated graphite nanocomposite films with improved thermal stability, mechanical properties and barrier properties, *Polym. Int.,* **62**, pp. 1386–1394.

90. Pradhan, A. K., Sahu, H. S., and Swain, S. K. (2013). Synthesis of gas barrier starch by dispersion of functionalized multiwalled carbon nanotubes, *Carbohydr. Polym.,* **94**, pp. 663–668.

91. Esthappan, S. K., Joseph, R., Katiyar, P., et al. (2013). Polypropylene/ zinc oxide nanocomposite fibers: Morphology and thermal analysis, *J. Polym. Mater,* **30**, pp. 79–89.

92. Lange, J. and Wyser, Y. (2003). Recent innovations in barrier technologies for plastic packaging: A review, *Packag. Technol. Sci.,* **16**(4), pp. 149–158.

93. Lacoste, A., Schaich, K. M., Yam, K. L., et al. (2005). Advancing controlled release packaging through smart blending, *Packag. Technol. Sci.,* **18**(2), pp. 77–87.

94. Duncan, T. V. (2011). Applications of nanotechnology in food packaging and food safety: Barrier materials, antimicrobials and sensors, *J. Colloid Interface Sci.,* **363**(1), pp. 1–24.

95. Ding, Y., Jiang, Y., Povey, M., et al. (2007). Investigation into the antibacterial behaviour of suspensions of ZnO nanoparticles (ZnO nanofluids), *J. Nanoparticle Res.,* **9**(3), pp. 479–489.

96. Hogan, S. A., Kerry, J. P., and Ogrady, M. N. (2006). Past, current and potential utilisation of active and intelligent packaging systems for meat and muscle-based products: A review, *Meat Sci.,* **74**, pp. 113–130.

97. Chen, C., Liao, F., and Subramanian, V. (2005). Organic TFTs as gas sensors for electronic nose applications, *Sens. Actuators B Chem.,* **107**, pp. 849–855.

98. Mahalik, N. P. and Nambiar, A. N. (2010). Trends in food packaging and manufacturing systems and technology, *Trends Food Sci. Technol.,* **21**, pp. 117–128.

99. Biji, K. B., Gopal, T. K. S., Mohan, C. O., et al. (2015). Smart packaging systems for food applications: A review, *J. Food Sci. Technol.,* **52**, pp. 6125–6135.

100. Bhattacharya, S. (2007). Biomems and nanotechnology based approaches for rapid detection of biological entities, *J. Rapid Methods Auto. Microb.,* **15**, pp. 1–32.

101. Lange, D. (2002). Complementary metal oxide semiconductor cantilever arrays on a single chip: Mass-sensitive detection of volatile organic compounds, *Anal. Chem.*, **74**, pp. 3084–3095.

102. Garcia, M. (2006). Electronic nose for wine discrimination, *Sensors Actuat. B*, **113**, pp. 911–916.

103. Baeummer, A. (2004). Nanosensors identify pathogens in food, *Food Technol.*, **58**, pp. 51–55.

104. Vo-Dinh, T. (2001). Nanosensors and biochips: Frontiers in biomolecular diagnostics, *Sensors Actuat. B*, **74**, pp. 2–11.

105. Mabeck, J. T. and Malliaras, G. G. (2006). Chemical and biological sensors based on organic thin-film transistors, *Anal. Bioanal. Chem.*, **384**, pp. 343–353.

106. Ritter, S. K. (2005). An eye on food, *Chem. Eng. News*, **83**, pp. 28–34.

107. Canel, C. (2006). Micro and nanotechnologies for food safety and quality applications. *Proceedings of MNE1906 Micro- and Nano-Engineering, 5C-3INVMicrosystems and Their Fabrication*, Barcelona, Spain.

108. Hall, R. H. (2002). Biosensor technologies for detecting microbiological food borne hazards, *Microb. Infect.*, **4**, pp. 425–432.

109. Kumar, C. S. S. R. (2006). *Nanomaterials for Biosensors* (Weinheim: Wiley-VCH).

110. Ramirez Frometa, N. (2006). Cantilever biosensors, *Biotecnol. Appl.*, **23**, pp. 320–323.

111. Jain, K. K. (2008). *The Handbook of Nanomedicine* (Totowa: Humana/Springer).

112. Gfeller, K. Y. (2005). Micromechanical oscillators as rapid biosensor for the detection of active growth of *Escherichia coli*, *Biosens. Bioelectron.*, **21**, pp. 528–533.

113. Brody, A. L. (2003). Nano, nano food packaging technology, *Food Technol.*, **57**, pp. 52–54.

114. Dhawan, A., Kumar, A., Shanker, R., et al. (2018). Nanotoxicology: Challenges for Biologists. In *Nanotoxicology: Experimental and Computational Perspectives* (The Royal Society of Chemistry), pp. 1–16.

115. Falck, G., Lindberg, H., Norppa, H., et al. (2010). Nanotechnologies, engineered nanomaterials and occupational health and safety: A review, *Safety Sci.*, **48**(8), pp. 957–963.

116. Sayes, C. M. and Warheit, D. B. (2009). Characterization of nanomaterials for toxicity assessment, *Wiley Interdisciplinary Rev. Nanomed. Nanobiotechnol.*, **1**(6), pp. 660–670.

117. Amini, S. M., Gilaki, M., and Karchani, M. (2014). Safety of nanotechnology in food industries, *Electron. Phys.*, **6**(4), pp. 962.

118. He, X. and Hwang, H. M. (2016). Nanotechnology in food science: Functionality, applicability, and safety assessment, *J. Food Drug Anal.*, **24**(4), pp. 671–681.

119. Chau, C. F., Wu, S. H., and Yen, G. C. (2007). The development of regulations for food nanotechnology, *Trends Food Sci. Technol.*, **18**, pp. 69–280.

120. Tarver, T., *Food Nanotechnology*. Scientific Status Summary synopsis.

121. Kuempel, E. D. and Maynard, A. D. (2005). Airborne nanostructured particles and occupational health, *J. Nanoparticle Res.*, **7**, pp. 587–614.

122. Ireland, F. S. A. O. (2009). *The Relevance for Food Safety of Applications of Nanotechnology in the Food and Feed Industries* (Food Standards Agency UK).

123. Sandoval, B. M. (2009). Perspectives on FDA's Regulation of Nanotechnology: Emerging Challenges and Potential Solutions.

124. Magnuson, B. A. (2009). Nanoscale materials in foods: Existing and potential sources. In *Intentional and Unintentional Contaminants in Food and Feed*, pp. 47–55.

125. Braman, D., Cohen, G., Gastil, J., et al. (2009). Cultural cognition of the risks and benefits of nanotechnology, *Nat. Nanotechnol.*, **4**, pp. 87–90.

126. Alenius, H., Norppa, H., Pylkkänen, L., et al. (2010). Risk assessment of engineered nanomaterials and nanotechnologies: A review, *Toxicology*, **269**(2–3), pp. 92–104.

127. Snir, R. (2014). Trends in global nanotechnology regulation: The public-private interplay, *Vand. J. Ent. Tech. L.*, **17**, pp. 107.

128. Angulo, I., Cruz, J. M., Paseiro Losada, P., et al. (2007). Development of new polyolefin films with nanoclays for application in food packaging, *Eur. Polym. J.*, **43**(6), pp. 2229–2243.

129. Kaynak, C. and Tasan, C. C. (2006). Effects of production parameters on the structure of resol type phenolic resin/layered silicate nanocomposites, *Eur. Polym. J.*, **42**(8), pp. 1908–1921.

130. Dasgupta, N., Mundekkad, D., Ramalingam, C., et al. (2015). Nanotechnology in agro-food: From field to plate, *Food Res. Int.*, **69**, pp. 381–400.

131. Bryant, C., Bugusu, B., Cartwright, T. T. et al. (2006). Report on the First IFT International Food Nanotechnology Conference. June 28–29,

2006, Orlando, FL. Available online at: http://members.ift.org/IFT/ Research/ConferencePapers/firstfoodnano.htm (accessed March 2008).

Multiple Choice Questions

1. Which one is not an example of nanotechnology in food packaging?
   - (a) Active packaging
   - (b) Improved packaging
   - (c) Plastic packaging
   - (d) Smart packaging

2. What is the size of the droplet of a nanoemulsion?
   - (a) 10 nm
   - (b) 100 nm
   - (c) 1000 nm
   - (d) 10,000 nm

3. Which among the following is not a nanoparticle?
   - (a) Silicon dioxide
   - (b) Carbon dioxide
   - (c) Titanium dioxide
   - (d) Silver

4. Which among the following is not an example of nanoclay?
   - (a) Aegis
   - (b) Durethan
   - (c) Magnesium oxide
   - (d) Imperm

5. Which among the following is not an example of nanofood?
   - (a) Benzoic acid
   - (b) Ascorbic acid
   - (c) Fortified fruit juices
   - (d) Nanocurd

6. Which among the following is a food regulatory body in the United States?
   - (a) The Institute of Food Science and Technology
   - (b) Nacional de Vigilância Sanitaria
   - (c) Food and Drug Administration
   - (d) Nanotechnology Standardization Technical Committee

7. Which among the following nanomaterials cannot be used as a filler?
   - (a) Graphene
   - (b) Silica
   - (c) ZnO_2
   - (d) Silver

8. The nanocomposite called nanoTuff does not exhibit an improved barrier property to
   - (a) N_2
   - (b) H_2O
   - (c) O_2
   - (d) CO_2

9. Which among the following bioactive compounds are not sensitive to enzymes activity?

   (a) Saliva

   (b) Carbohydrates

   (c) Proteins

   (d) Vitamins

10. Which among the following is not done by nanoencapsulation?

    (a) Protection against oxidation

    (b) Taste making

    (c) Change the appearance of the food

    (d) pH-triggered controlled release

11. Which company manufactures Nanotea?

    (a) High Vive in the United States

    (b) Life Enhancement

    (c) Qinhuangdao Taiji Ring Nano-Products Co. Ltd

    (d) RBC Life Sciences Inc.

12. Which company manufactures Nanoceuticals Slim Shake Chocolate?

    (a) RBC Life Sciences Inc.

    (b) Health Plus International

    (c) Jamba Juice Hawaii

    (d) Life Enhancement

13. Daily Boost is used for the nanoencapsulation of fortified vitamin. Which company manufactures it?

    (a) Life Enhancement

    (b) Health Plus International

    (c) Jamba Juice Hawaii

    (d) RBC Life Sciences Inc.

14. For what purpose bioplastics are being used.

    (a) Nanoencapsulation

    (b) Packaging

    (c) Nanocomposite

    (d) Nanoclay

15. Which among the following is a food regulatory body for nanotechnology applications in food sector in China?

    (a) Nanotechnology Standardization Technical Committee

    (b) Nacional de Vigilância Sanitaria

    (c) The Institute of Food Science and Technology

    (d) Food and Drug Administration

16. Detection of pathogens can be done with the help of

    (a) Nanocomposites

    (b) Nanosensors

    (c) Nanoclay

    (d) Nanoencapsulation

17. Which company manufactures NanoResveratrol?

    (a) Jamba Juice Hawaii   (b) Health Plus International

    (c) RBC Life Sciences Inc.   (d) Life Enhancement

18. Which among the following is a food regulatory body for nanotechnology applications in the food sector in Brazil?

    (a) The Institute of Food Science and Technology

    (b) Nacional de Vigilância Sanitaria

    (c) Food and Drug Administration

    (d) Nanotechnology Standardization Technical Committee

19. In which year the European Food and Safety Authority (EFSA) created a network between Europe Union (EU) member states to assess the risk related to nanotechnology in food industry?

    (a) 2010   (b) 2008

    (c) 2014   (d) 2012

20. Nanostructured materials do not contain

    (a) Nanowires   (b) Nanotubes

    (c) Nanocubes   (d) Nanoparticle dispersion

21. Top Screen DS13 is a

    (a) Nanocomposite   (b) Nanowire

    (c) Nanoencapsule   (d) Nanotube

22. Nanoencapsulation is done with the help of

    (a) Nanotubes   (b) Nanowires

    (c) Nanocrystals   (d) Nanocapsules

23. Which property of nanomaterials can cause toxicity in food?

    (a) Solubility   (b) Bioactive

    (c) High surface area   (d) Stability

24. What are nanocantilevers?

    (a) Nanocapsules   (b) Nanorods

    (c) Nanocrystals   (d) Nanosensors

25. An example of nanobased carrier used for nanoencapsulation is

    (a) Starch   (b) Cellulose

    (c) Protein   (d) Liposome

Answer Key

1. (c)	2. (b)	3. (b)	4. (c)	5. (d)	6. (c)	7. (d)
8. (a)	9. (a)	10. (c)	11. (c)	12. (a)	13. (c)	14. (b)
15. (a)	16. (b)	17. (d)	18. (b)	19. (a)	20. (c)	21. (a)
22. (d)	23. (c)	24. (d)	25. (d)			

Short Answer Questions

1. What are nanoemulsions? Discuss their preparation methods and uses in the food sector.

2. What are the various nanoparticles and how can they improve the properties of food products?

3. What are nanocomposites? Discuss their applications in the food industry.

4. Discuss briefly how nanotechnology can improve texture, appearance, flavor, and shelf life of food?

5. What is nanoencapsulation? How can it be utilized in the food sector?

6. What are the advantages of using biobased packaging?

7. What is the difference between active and smart packaging?

8. What are the safety issues with nanotechnology in the food sector and discuss the rules and regulations to be followed for it?

Long Answer Questions

1. Discuss the nanomaterials used in food.

2. How nanotechnology has been used in food processing?

3. What are the different nanotechnology-based food packaging?

4. What are the safety issues associated with nanotechnology in food?

5. What are the different regulations formed and required for nanotechnology in food?

Chapter 11

Nanophotocatalysts: Applications and Future Scope

Balwinder Kaur

Department of Chemistry, Punjabi University Patiala,
Punjab, India
balwindertaheem1987@gmail.com

11.1 Introduction

The amalgamation of physics, chemistry, and material science has given birth to a multidisciplinary field called photocatalysis, which harvests photonic energy and facilitates chemical reactions. The exponential research on photocatalysis has opened a gateway for the synthesis of multitudinous products and desired chemical fuels. On comparison with conventional catalytic routes, photocatalysis steps are less time consuming and more economic. It is a green technology and solves enormous environmental crisis. Photocatalysis is one of the most promising research areas in the field of photochemistry. Wastewater treatment and conversion of toxic chemical into

Nanotechnology: Principles and Applications
Edited by Rakesh K. Sindhu, Mansi Chitkara, and Inderjeet Singh Sandhu
Copyright © 2021 Jenny Stanford Publishing Pte. Ltd.
ISBN 978-981-4877-43-5 (Hardcover), 978-1-003-12026-1 (eBook)
www.jennystanford.com

useful products are the two vital achievements that are achieved only by photochemistry. The word photocatalysis itself reveals its mechanism of operation. The photons activate the catalyst by two means: (a) electron transfer and (b) energy transfer [1]. On the absorption of radiations, the photosensitized nanoparticles activate the reactants. In other words, nanoparticles are excited into higher energy levels, which subsequently transfer their energy to low-lying reactants (in ground state). To accompany energy transfer mechanism, the photocatalyst as well as reactants should be in the same multiplicity. On irradiating photocatalyst in the suitable light source (in electron transfer mode) an e^- from valence band (VB) excites into conduction band (CB) leaving behind h^+ i.e., exciton is formed. The generated e^- and h^+ pairs participate in redox reaction. The above-stated mode is highly operated in catalytic applications. The optical properties (electronic transitions) and topographical (surface sites/defects) characteristics collectively attribute toward nano-photocatalysis activity potential.

11.2 Mechanism of Photocatalysis

As stated earlier, electron transfer mechanism is followed by photocatalysis. The electron transfer mechanism is initiated by the exposure of radiation by a suitable light source such as solar light, UV light, visible light, and mercury lamps depending on the energy band gap of nanoparticles. The energy band gap is also called "forbidden energy zone," which is the energy gap between the VB and the CB in semiconductor nanomaterials such as CdS, ZnS, ZnO, and TiO_2. Upon irradiation, the generated e^- and h^+ pairs react with dissolved oxygen or water to give highly reactive and short-lived free radicals OH^- and $OH°$ [2]. These reactive free radicals undergo redox reaction with contaminants such as dyes, drugs, waste, and heavy metals. The following steps are involved in photocatalysis:

- Formation of exciton (separated e^- and h^+ pairs)
- Transfer of exciton to the photocatalyst's surface and initiation of redox reaction [3]

The overall quantum efficiency for interfacial charge transfer is determined by the competition between the charge carrier recombination (exciton) and charge carrier (exciton) trapping in defects on the surface followed by the competition between recombination of charge carriers (photoluminescence) and interfacial charge transfer on the surface. Various charge carrier recombination and interfacial charge transfer processes are shown in Fig. 11.1. The separated e^- and h^+ pairs follow various relaxation pathways; for instance, the separated e^- and h^+ can move on the surface (pathway A) through radiative or nonradiative relaxation processes or recombine in the volume of the semiconductor (pathway B); the recombination of e^- and h^+ pair needs to be retarded for an efficient charge transfer process to occur on the photocatalyst surface; the semiconductor on its surface can donate e^- to reduce accepter (pathway C) and h^+ can migrate to the surface, whereas an e^- from the donor species can combine with the h^+ and oxidize the donor species (pathway D). Charge carrier trappings suppress recombination and increase the lifetime of the separated e^- and h^+ to provide a sufficient time delay for the photoexcited e^- and h^+ to undergo charge transfer to adsorbed species on the semiconductor surface. The generated h^+ oxidizes pollutants either directly, or it can form $OH°$ radical by decomposition of water. The e^- in CB produces superoxide or peroxide by the adsorbed oxygen followed by conversion into $OH°$ radical. The oxidation potential of $OH°$ radical lies in the range of 1.9–2.85 V comparable to NHE.

$$\text{Semiconductor nanomaterial} + h\nu \longrightarrow h^+ + e^- \quad (11.1)$$

The oxidation of water by holes [$H_2O \leftrightarrow H^+ + OH^-$]

$$h^+ + H_2O \longrightarrow OH° \quad (11.2)$$

$$O_2 + e^- \longrightarrow O_2^{-°} \quad (11.3)$$

$$O_2^{-°} + H_2O \longrightarrow OH° + OH^- \quad (11.4)$$

$$H°O_2 + H°O_2 \longrightarrow H_2O_2 + O_2 \quad (11.5)$$

$$H_2O_2 + O_2^{-°} \longrightarrow OH° + OH^- + O_2 \quad (11.6)$$

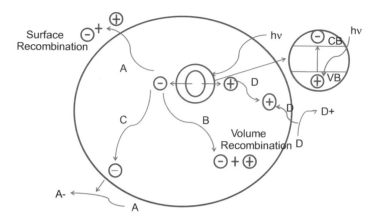

Figure 11.1 Schematic photoexcitation of a semiconductor followed by de-excitation events [4].

11.3   Types of Photocatalysts

Photocatalysts are divided into two categories:

1. **Homogenous photocatalysts:** When the reactant and photocatalyst are in the same physical state, it is termed homogenous photocatalyst.

2. **Heterogeneous photocatalyst:** When the reactant and photocatalyst both are in different physical states, it is termed heterogeneous photocatalyst. The stepwise heterogeneous photocatalysis is accompanied by the following steps:
   - Reactants transfer from the fluid phase to the surface on nanoparticles.
   - Adsorption of the reactants on the nanomaterial surface by stirring in most of the reaction schemes [1].
   - The reaction takes place in the adsorbed phase only.
   - Desorption of the product(s) after completion of photocatalysis.

Heterogeneous photocatalysis is a widely opted catalysis route because after reaction completion, the photocatalyst can be easily recovered and reused. Broadly, heterogeneous photocatalysis is applicable in environment detoxification, polluted water treatment, oxidation, dehydrogenation, exchange of deuterium-alkane isotope, heavy metal remediation, etc. [5].

11.4 Overview of Nanomaterials

Nanomaterials have structural features at the nanoscale, i.e., 1–100 nm. Several physical properties of nanomaterials such as optical, magnetic, mechanical, and chemical properties depend not only on the size parameter but also on the topography, morphology, and special arrangement of atoms. The attraction toward nanotechnology has arisen because various vital biological entities lie at the nanoscale. Nanoscience and nanotechnology act as gateway from the inner world to the outer world. After the biological front, the semiconductor industry has attained miniaturization in its devices by applying nanotechnology principles. Nanomaterials pose the following attractive properties:

- Large surface-to-volume ratio
- High surface energy
- Spatial confinement
- Less surface imperfections
- Melting point depression occurs when the size of nanoparticles decreases due to large surface-to-volume ratio
- High mechanical strength
- Optical properties are different due to plasmon resonance
- Brownian motion is exhibited by colloidal nanoparticles as particles are employing continuously random motions

11.4.1 Inorganic Semiconductor Materials

Inorganic materials can be classified according to their electrical conductivity: conductors, semiconductors, and insulators. In conductors, electrons are free to move from the VB to the CB by supplying thermal energy. Semiconductors have intermediate conductance between conductors and insulators. In insulators, there is restriction to the flow of electrons (Fig. 11.2). In binary semiconductors, the elements can be either form group III and group IV called III–V semiconductors or form group II and group VI called II–VI chalcogenides/ceramic semiconductors. Semiconductor nanomaterials have attracted the interest of the scientific community due to their tunable physical, chemical, and optical properties. Quantum well (2D), quantum wire (1D), and quantum dots (0D) are composed of semiconductors. The word "quantum" comes from the quantum confinement effect. The quantum confinement effect comes into play when the size of semiconductor nanomaterials becomes comparable to the Bohr exciton radius.

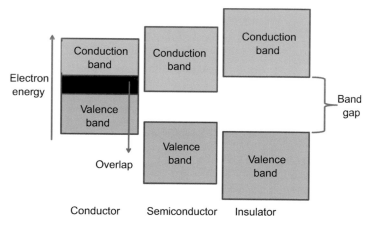

Figure 11.2 Electronic band structure of metals, semiconductors, and insulators.

Each electron in a semiconductor material is bound by covalent bonds and fulfills the octet rule. So electrons are localized within the region of surrounding, i.e., they are not considered to be free. The states situation is exhibited only at the absolute zero Kelvin temperature. At higher temperature, the electrons gain sufficient energy to vacate and become free to move to the crystal lattice and initiate conductions. In semiconductor nanocrystals, there is a well-defined energy gap between the conduction and valence bands. On thermal excitation, an electron from the VB jumps into the CB and creates a hole in the VB. At the absolute zero Kelvin temperature, semiconductor materials behave as insulator.

11.4.2 Quantum Confinement Effect

The quantum confinement effect originates from a very popular theory of physics: quantum mechanics. More clearly, when the size of a particle is reduced from the bulk scale to the nanoscale (1–100 nm), i.e., when the size of a material approaches the Bohr exciton radius (a_B), then the quantum confinement effect is seen. To elaborate, photoexcited carriers (electrons and holes) are provided very little space to move, i.e., they have confined movement. The Bohr exciton radius (a_B) can be calculated by the following equation [7]:

$$a_B = \frac{4\pi\varepsilon_0\varepsilon_\infty\hbar^2}{m_0 e^2}\left[\frac{1}{m_e^*} + \frac{1}{m_h^*}\right] \tag{11.7}$$

where ε_∞ is the high-frequency relative dielectric constant of the medium; m_e^* and m_h^* are the effective masses of the electron and hole, respectively (both in units of electron mass); and m_0 is the mass of the electron at rest. Table 11.1 shows the exciton Bohr radii values for some of the semiconductors.

Table 11.1 Bohr exciton radii for various semiconductors

Semiconductor	Bohr exciton radius (nm)	m_e^*	m_h^*	ε_∞	References
CdS	4	0.14	0.51	8.4	[10, 11]
ZnS	2	0.28	0.50	7.4	[10, 11]
CdSe	6	0.11	0.44	9.7	[10, 11]
CdTe	8	0.09	0.4	10.6	[10, 11]
ZnSe	3	0.16	0.6	7.6	[10, 11]
InP	8	0.07	0.4	9.6	[6]
InAs	34	0.024	0.41	14.5	[11, 12]
InSb	57	0.015	0.39	15.6	[6]
PbS	20	—	—	—	[13]
PbSe	46	—	—	—	[13]
CuCl	1	—	—	—	[13]

Note: Values used to calculate the radii and their references are included. m_e^* is the effective mass of the electron in units of electron mass; m_h^* is the effective mass of the hole in the units of electron mass; and ε_∞ is the relative dielectric constant.

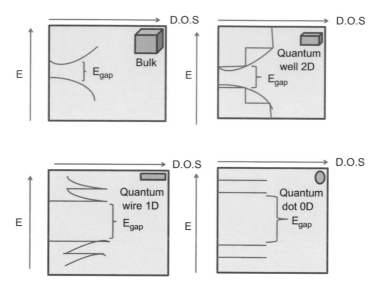

Figure 11.3 Illustration of variation in density of states with increasing confinement of electron–hole pair [6, 14].

The first size-dependent effects of quantum confinement were witnessed in the 1970s, when molecular beam epitaxy made it possible to form very thin layers of semiconductor materials known as quantum wells, i.e., 2D nanostructures [8, 9], because the exciton pair was confined in a 2D plane region. Furthermore, confining two dimensions and keeping one dimension unconfined lead to the synthesis of 1D nanostructures (quantum wires). Confining three dimensions gives way to the synthesis of 0D nanostructures, which are famous by the name of quantum dots. The energy levels and density of states are also modified with size tunability. At the bulk scale, there is continuous energy levels. As one moves from the macroscale/bulk scale to the nanoscale, the electron energy levels become discrete, i.e., the sizes of nanomaterials are of the order of the size of the Bohr exciton radius. Photoexcitation or electrical excitation of nanomaterials creates an e^- and h^+ pair, which is delocalized over the entire volume of nanomaterials. The confinement of exciton in all three dimensions causes the continuous density of state of the bulk solid to collapse into discrete electronic states, as shown in Fig. 11.3 and Fig. 11.4. Through modeling the electronic structure of nanomaterials using the simple effective mass theory, the size effect caused by quantum confinement can be understood. The following approaches quantitatively explain size effects:

- Effective mass approximation
- Tight binding model

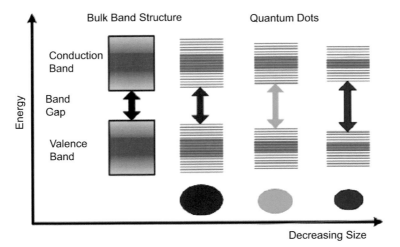

Figure 11.4 Illustration of the effect of confinement on electronic states with decreasing size from bulk to nanoscale [15].

11.4.3 Doped Semiconductor Nanomaterials

Multifunctionality in semiconductor nanomaterials can be achieved by introducing impurity within the lattice. The process of incorporation of impurity in the host matrix/lattice is known as doping. Doping in the host can tune morphology, topography, energy band gap, and various physical as well as chemical properties. Various metals, metalloids, nonmetals, and other elements can be added as impurity in the host matrix depending on the size of cation and anion of the host matrix ions. On the doping of semiconductor nanomaterials with an appropriate dopant, noticeable changes in the optical properties such as band gap and luminescence centers occur, which open up the stage for broad applications. The preliminary optical property is the energy band gap. On doping with an impurity, the energy band gap can be tuned. On the incorporation of an impurity in semiconductors nanomaterials, a Fermi level is introduced between the conduction and valence bands. The position of Fermi level depends on the type of dopant inserted in the host matrix. If p-type impurity is added, the Fermi level will lie closer to or above the VB because p-type is an accepter type impurity. If an n-type impurity is added, the Fermi level will be introduced below the CB. The positions of the Fermi level in intrinsic and extrinsic semiconductors are shown in Fig. 11.5. Pure and doped semiconductors are versatile materials due to the multifunctional properties originating from high surface-to-volume ratio. So these materials have become famous due to their surface active property, i.e., adsorption phenomena. Semiconductor nanoparticles are very efficient adsorbents as they act as catalysts and photocatalysts, whose primary step is adsorption. The surface of semiconductor nanoparticles can be functionalized by physical as well as chemical forces to make their surface highly sensitive and selective for the adsorption of analytes: dyes, drugs, organic/inorganic pollutants, gases, etc.

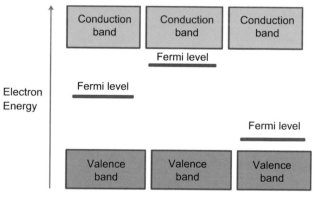

Figure 11.5 Schematic presentation of the Fermi level position in host material, n-type and p-type semiconductors.

11.5 Synthesis of Nanomaterials

The properties of materials originate from the arrangement of their constituent atoms/ions. The properties thus influence the performance of materials. The interfaces between reaction conditions and precursors are controlled by thermodynamics and kinetics of the synthesis. Semiconductor nanomaterials have been synthesized from numerous reaction routes. Intrinsic and extrinsic bulk (microcrystalline) semiconductors are generally synthesized using high-temperature thermal diffusion, molecular deposition techniques such as chemical vapor deposition, atomic layer epitaxy, gas phase deposition, vacuum evaporation, etc. Nanostructured materials can be made by the attrition of parent coarse-grained materials using the top-down approach from the macroscale to the nanoscale, or conversely by the assembly of atoms or molecules using the bottom-up approach. The control over the arrangement of atoms from the macroscale to the nanoscale is definitely the potency of material chemistry. The exotic beauty of chemistry spreads out by the way of accepting its chemical reactions in solid, liquid, and gaseous states. In solid-state synthesis approach, the solid precursors (such as metal oxides or carbonates) are fetched into close contact by grinding and mixing ways and thermally treated at required temperatures to assist dispersion of atoms or ions in the host matrix by chemical reaction.

11.5.1   Synthesis of Nanoparticles

The idea of synthesis of nanoparticles is shared by two incredible approaches (Fig. 11.6): top-down approach and bottom-up approach. These approaches enable the scientific community to synthesize smart nanomaterials.

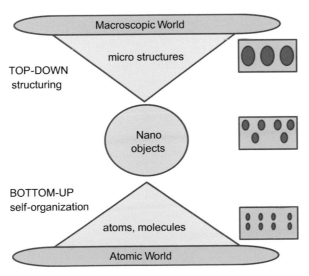

Figure 11.6   Schematic representation of top-down and bottom-up synthesis approaches.

11.5.1.1 Top-down approach

An assertive idea drawn from the top-down approach is to synthesize nanosized materials from the macroscopic level whose frame is designed by lithographical and etching techniques [13–16]. The accomplishment of nanoparticle synthesis by the present approach involves abundant physical methods. Numerous physical and chemical methods are available and opted for the synthesis of nanomaterials. Some vital and highly advantageous methods are discussed as follows.

Lithography: Lithography is a well-known top-down method, which has been opted by the semiconductor industry to develop computer chips. Lithography is broadly categorized by two means:

1. A method that uses a physical mask, which is positioned nearby to resist exposure to photons/electron beam.
2. A computer-controlled software scans the surface of the resist by a scanning beam, termed scanning lithography.

The lithography fabrication technique involves the surface coating of a substrate with a photon/electron-beam-sensitive polymer coating, which is termed "resist." The resist placed at an appropriate orientation below the mask is exposed to light (photolithography) or electron beam, which causes chemical changes in the resist at appropriate sites corresponding to the mask pattern. The obtained pattern is transformed to the substrate by chemical etching.

Chemical etching technique: This technique designs the topographical features of a substrate by selective removal of molecules by physiochemical ways. Etching works in both directions, either isotropically or anisotropically. The physiochemical processes utilize liquid chemical and gaseous molecules, i.e., wet etching and dry etching. The rate of etching designs the crystal structure of materials.

Attrition-milling process: Simplicity and cost effectiveness are the famous merits of the attrition-milling methodology. Attrition is equipped with stationary jacketed tank, which is spherical in shape. The jacketed tank contains grinding media such as carbon steel, stainless steel, chrome steel, and tungsten carbide like materials. After feeding the tank with the reaction mixture and grinding by a shaft with arms at a very high speed, the obtained final particles are very fine [17]. The primary uses of the discussed method are to reduce particle size.

Sputtering of nanocrystal powder: It involves the glow discharge production of energetic ions followed by momentum transfer from these energetic ions to a solid or molten target, which results in the ejection of surface atoms or molecules to produce the sputtered species. Finally, the sputtered species are condensed on the substrate. This method is commonly used for the synthesis of good-quality nanomaterials. The rate of generation of nanofilms is very high using a 75 mm target and 1.2 kW [18, 19].

11.5.1.2 Bottom-up approach

Atoms and molecules are building blocks of the bottom-up approach. The key point of the present approach is to tune the electronic structure of nanoparticles. The strategy of the bottom-up approach is to manipulate and assemble materials on the nanoscale to explore newly originated physical as well as chemical properties, which will be different for clusters, molecules, metal complexes as well as bulk materials. The following are the commonly used bottom-up synthesis techniques:

Chemical precipitation method: This method is one of the extensively opted routes of synthesis. The main advantages of the chemical precipitation method are less time consuming, cost effectiveness, and ease of handling. Semiconductor and metallic nanoparticles can be prepared in different media depending on structural requirements and applications by the chemical precipitation method. The morphology of the synthesized materials depends on precursor metal salts, temperature, pH, medium of reaction, etc. Here, the nucleation and growth of particles can be decided by Ostwald ripening [20]. Also capping agents can be used to have good control over morphology. In these days, the stabilization of nanomaterials has been advanced through the use of caged and layer-like solid materials, including layered titanates, mesoporous silica, polymer films, etc. [21].

Sol–gel method: Sol–gel is a widely accepted route to synthesize oxides and sulfides. As the name indicates, the precursor solutions are mixed to form an irreversible gel followed by drying and densification to yield nanosized and microsized materials. The sol–gel process includes the hydrolysis and polymerization of metal alkoxide and metal sulfide precursors. The polymer forms a network-like skeleton, which gives porosity to the synthesized materials. It can have low-temperature sintering capability.

Hydrothermal/solvothermal method: It involves a closed-vessel reaction, which induces decomposition reaction between precursors with highly boiling solvents. The hydrothermal/solvothermal route of synthesis can be heterogeneous and homogeneous depending on the states of precursors. The hydrothermal route of synthesis is similar to the solvothermal method, but instead of an organic solvent, here water is used as the solvent media. This route of synthesis is opted when very harsh reaction conditions are involved, such as high temperature and high pressure. Here, the precursor materials are suspended in a high boiling solvent, which plays the dual role of capping agent and reaction media. The hydrothermal method is famously used for the synthesis of nanorods, nanowires, and nanobelts.

Microemulsion/inverse microemulsion: The word "emulsion" is used when the constituent precursors are liquid in state. Microemulsion is an oil-in-water (o/w) route of synthesis, which has attracted the interest of researchers because this method provides good control over morphology.

11.6 Characterization Tools and Techniques

11.6.1 X-Ray Diffraction

The words "unparallel" and "intangible" are not enough to describe the advantageous technique of X-ray diffraction (XRD). XRD is used to measure strain, grain size, epitaxy, crystallographic phase, and atomic spacing at an unmatched level of accuracy. In the electromagnetic spectrum, X-rays lie at the wavelength position of ~1 Å. X-rays can provide photon energy in the range of 10–100 eV. As the size of atom is comparable to the wavelength of X-rays, these can interact with matter, as matter is composed of atoms. Due to the similarity between X-ray wavelengths and atomic dimensions, X-rays undoubtedly befit the investigation of the structural features of crystals composed of atoms. On the interaction of X-ray photons with atoms, some photons get deflected back from the foremost direction. This is known as elastic scattering. The diffracted photons interfere constructively to give a diffraction pattern according to Bragg's law given by W. L. Bragg in 1912 as follow:

$$2d_{hkl}\sin\theta_{hkl} = n\lambda \qquad (10.8)$$

where θ_{hkl} (Bragg's angle) is the angle between the atomic planes and the incident (and diffracted) X-ray beam (as shown in Fig. 1.8), d is the interplanar spacing between the planes having miller indices (hkl), n is the order of diffraction, and λ is the incident X-ray wavelength.

11.6.2 Electron Microscopy

The topographical and morphological characterization of material at the atomic scale is normally done by electron microscopy. Optical microscopy is a conventional technique to play with spatial resolution. To achieve more depth of resolution/focus, sophistication of the instrument was needed, and this was achieved by the discovery of scanning electron microscope (SEM) and transmission electron microscope (TEM), which are the two classifications of electron microscopy [22].

11.6.2.1 Scanning Electron Microscope

An SEM reveals the topography, morphology, and surface composition of materials. Here, a highly energetic electron beam probes the surface of the material at about 10×–300,000× magnification. When the electron beam contacts the surface atoms by a number of interactions, the emission/scattering of electrons or photons occurs [23]. The scattered/emitted electrons are gathered by a detector, and an image is produced on the cathode ray tube (CRT). Each signal is mapped onto the screen.

11.6.2.2 Transmission Electron Microscope

The morphology and properties of matter, irrespective of whether it is inorganic, organic, and biological, can be measured and understood by TEM. Here, a fast-moving electron beam having a wavelength of ~0.0037 nm, much smaller than that of light, X-ray, and neutron, penetrates and passes through the thickness of the sample. The penetrating and focused electron beam acts as a probe, which investigates the sample. The signal is measured by recording both undeflected and deflected electrons. The signals are delivered to a detector, a fluorescent screen, which may be a film plate or camera. TEM constitutes high lateral spatial resolution because it employs a highly focused e⁻ beam probe. The samples under investigation are usually made very thin (<200 nm) because sample thickness relates to scattering events. These events of electron can be minimized by maintaining the sample thickness as small as possible, which binds the spreading of probe, i.e., retains coherency. The lateral spatial resolution varies with the operating voltage (300–400 keV) of TEM.

11.6.3 UV-Visible Absorption Spectroscopy

As the name indicates, the UV-visible radiations of electromagnetic spectrum are used for optical analysis. UV-visible absorption spectroscopy serves as fingerprint for the identification of molecules. It works in the 190–900 nm range of the electromagnetic spectrum. It is a useful technique to determine the band gap of synthesized nanomaterials. The radiation absorbance by nanomaterials can be determined by Lambert–Beer's law [24, 25]:

$$A = -\log_{10}(I/I_0) = \varepsilon c l \tag{11.9}$$

where A is the optical density, I is the intensity of light after interaction with sample, I_0 is the intensity of incident light, ε is the molar absorptivity, c is the concentration of sample, and l is the path length.

11.6.4 Energy Dispersive X-ray Spectroscopy

Energy dispersive X-ray spectroscopy (EDX) is an important tool used for elemental characterization. Being a type of spectroscopy, it involves the exploration of a sample through interactions between light and materials (organic, inorganic, and biological) and analysis of characteristic X-rays emitted in this particular case. Highly energetic electron beams hit the material to be analyzed, and X-rays are emitted followed by detection on a suitable detector, and the mapped signal is traced on a computer screen. Basically, an X-ray photon strikes a diode in the detector, producing a charge that is transformed into a positive voltage pulse.

11.7 Nanostructured Photocatalysts

Nanophotocatalysts can be classified into various categories: which is made possible by developments in synthesis routes. Pure semiconductors (binary and ternary), solid solutions, and nanocomposites are the major synthesized nanomaterials for photocatalytic application.

11.7.1 Binary Semiconductor Photocatalysts

Binary semiconductors are basically divided into three categories: nitrides (GaN, Ta_3N_5, g-C_3N_4), oxides (ZnO, TiO_2, ZrO_2, CeO_2), and chalcogenides (sulfide and selenides). The aforementioned semiconductors can be modified by doping as well as varying synthesis routes. Modifications in the nanomaterials give birth to diversity in topography, generate specific surface sites, variable crystal phases, and interesting photochemical reaction at a blistering rate.

Both binary oxide and chalcogenide semiconductor nanomaterials begged heterogeneous photocatalysis activity. The aforementioned nanomaterials are widely researched for vital and successful applications in environment detoxification, reduction in carbon dioxide level, aldehyde, water splitting, and reductive dehalogenation of benzene derivative. The nitride-based semiconductor nanomaterials have immense application in sensing. Reproducible results and chemical stability are two requirements fulfilled by GaN nanoparticles. Along with sensing, water splitting and photodegradation of dyes/drugs/pollutants are the other important application areas due to their energy band gap lying in the UV-visible region of the electromagnetic spectrum [26]. The visible light active Ta_3N_5 nanoparticles are highly suitable for water-splitting applications [27]. Similarly, the visible light active g-C_3N_4 nanoparticles give rise to applications such as bacterial disinfection, pollutant degradation, and fuel generation [28].

11.7.2 Ternary Oxide Photocatalyst

The general formula of ternary oxide nanoparticles is ABO_3. Ternary oxide nanoparticles are comprised of three different cations (alkali, alkaline, and rare earth metals), whereas cation B is transition metal with single anion. These photocatalysts are significantly studied in heterogeneous photocatalysis. $SrTaO_3$ possesses an energy band gap of 3.2 eV and UV light active photocatalyst. Its n-type semiconductor nanomaterial can easily tune its physical as well as chemical properties. It has enhanced its water-splitting capability by coupling with metal cocatalyst [29]. Furthermore, $LiTaO_3$ possesses a wide band gap of 4.6–4.7 eV; it can successfully perform water splitting without any cocatalyst [30]. $NaTaO_3$ is another ternary oxide nanophotocatalyst having a direct band gap of 4.1 eV [31]. It possesses an orthorhombic crystal structure.

11.7.2.1 AB_2O_4 type

The special AB_2O_4 ternary oxide is CaI_2O_4, which is active in visible light and catalyzes the photodegradation of methylene blue (MB) dye as well as organic toluene [32].

11.7.2.2 ABO_2 type

This is the third category of ternary oxide semiconductor nanomaterial. Specifically, Ag metal binds with group III oxide $AgMO_2$ in which M is Al, Ga, and In: α-$AgGaO_2$ (2.38 eV), α-$AgInO_2$ (1.90 eV), β-$AgGaO_2$ (2.18 eV), and β-$AgAlO_2$ (2.95 eV). Here, α and β reflect delafossite and cristobalite structures. The photocatalytic activity of α-$AgGaO_2$ is higher than that of β-$AgGaO_2$ (2.18 eV) and β-$AgAlO_2$ (2.95 eV) for the production of acetone from IPA [33].

11.7.2.3 ABO_4 type

Bismuth vandate ($BiVO_4$) is a visible light active photocatalyst. There are three crystal structures of $BiVO_4$: scheelite-monoclinic (SM), scheelite-tetragonal (ST), and zircon-tetragonal (ZT). Among all, SM is fascinating and marvelous in the electronic industry as well as in the photocatalytic area. The significant properties are increased due to the crystal structure because the bond lengths in Bi–O atoms are variable. Consequently, the overlapping in the orbitals of VB increases. As a result, charge separation on photoexcitation results, which uplifts its photocatalytic activity [34].

11.7.2.4 Ternary oxide photocatalyst (ABxCy)

This is the last category of ternary oxide photocatalyst. Here, A = Cu, Ag, Zn, and Cd; B = Ga, In; C = S, Se. These semiconductors possess a wide range of adsorption profiles in the UV-visible region. These semiconductors are employed in optoelectronic, LED, solar cells, and photocatalysis [35–37]. Figure 11.7 depicts the important categories of ternary semiconductor photocatalysts.

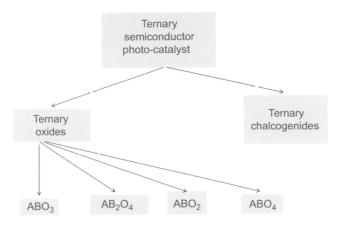

Figure 11.7 Classification of nanostructured materials.

11.7.3 Solid Solution Photocatalysts

The hybridization of a wide-band-gap semiconductor with a narrow-band-gap semiconductor leads to the modulation of energy band gaps. The modulation can be in either CB or VB or both CB and VB edges. Thus, optical properties can be tuned. Generally, the hybridized CB or VB promotes faster charge transfer due to its dispersed nature and increases the photocatalytic activity. According to IUPAC, solid solution is defined as "a crystal containing a second constituent, which fits into and is distributed in the lattice of the host crystal." The hybridization of CB and VB can be done by three ways:

1. Solid solution through continuous modulation of CB
2. Solid solution through continuous modulation of VB

3. Solid solution through continuous modulation of both CB and VB

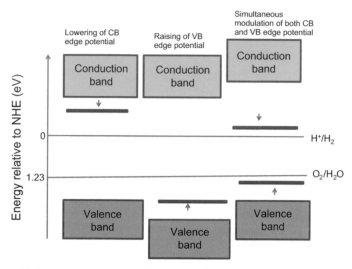

Figure 11.8 Energy band model in semiconductor nanomaterials.

Ternary semiconductors of the family $AgMO_2$ (M = Al, Ga, In) have noteworthy consideration in the photocatalytic field due to their discrete VB but modulation of CB edge. The visible light activity of β-AgAlO$_2$ (2.83 eV) for IPA decomposition through continuous fine-tuning of the CB edge of β-AgAlO$_2$, by combining it with narrow-band-gap β-AgGaO$_2$ (2.18 eV) semiconductor, results in a β-AgAl$_{1-x}$Ga$_x$O$_2$ (0 ≤ x ≤ 1) solid solution [38]. A similar CB modulation was achieved recently in the family of AM_2O_6 (A = Sn and M = Nb or Ta) ternary semiconductors by fabricating the series of $Sn(Nb_{1-x}Ta_x)_2O_6$ (0 ≤ x ≤ 1) solid solutions, which degrades MB dye in visible light very efficiently [39]. $Ag_xNa_{1-x}NbO_3$ (0 ≤ x ≤1) solid solutions were synthesized by continuous modulation of VB edge, and photocatalytic activity was performed by the decomposition of IPA [40]. Similarly, $(Ga_{1-x}Zn_x)(N_{1-x}O_x)$ solid solutions were synthesized for water-splitting activity in visible light. $Ag_{1-x}Sr_x(NbO_3)_{1-x}(TiO_3)_x$ (0 ≤ x ≤ 1) solid solutions have been explored for photocatalytic activity and synthesized by both modulation of CB and VB [41]. The widely studied sulfur solid solution is $Cd_xZn_{1-x}S$ (0 ≤ x ≤ 1), which possess a band gap of 2.3–3.1 eV. Here, the synthesized photocatalyst absorbs visible light for photocatalytic activity and generates hydrogen [42]. Multitudinous nanophotocatalysts are synthesized by the modulation of CB and VB edges. The complete modulation in band edges is shown in Fig. 11.8.

11.7.4 Nanocomposites

The coupling of two or three semiconductor–semiconductor nanomaterials at appropriate composition leads to the synthesis of nanocomposites. For the efficient charge separation on the absorption of visible light, only a narrow-band-gap material transfers its electron from CB2 to CB1, but holes remain in VB2, resulting in enhanced photocatalytic activity. On the other hand, when the materials are excited in UV light, the energy band gaps of both coupled nanomaterials are excited, i.e., electrons migrate from CB2 to CB1 and holes migrate from VB2 to VB1. Consequently, charges are separated for enhanced photocatalytic activity. In TiO_2/CdS type II heterojunction, upon illumination by visible light, photogenerated electrons from the CB of CdS are transferred to the CB of TiO_2 and holes remain within CdS, resulting in a more efficient charge separation as compared to that of the two individual semiconductors, which ultimately increases photocatalytic efficiency [43]. Similarly, TiO_2/ZnO is an example of such visible light active systems formed by combining two wide-band-gap UV-active semiconductors TiO_2 (3.32 eV) and ZnO (3.37 eV) and perform photodegradation of brilliant green dye and MB dye in visible light [44]. Both SnO_2/TiO_2 and SnO_2/ZnO nanocomposites exhibited high photocatalytic activity for the degradation of RhB under UV light as compared to the reduced activities of their individual constituent semiconductors [45]. SnO_2/Fe_2O_3 nanocomposite reveals exalted activity for the degradation of acid blue 62 dye under visible light [46]. SnO_2/SnS_2 is yet another visible light active nanocomposite system developed and studied especially for photocatalytically mediated environmental remediation applications [47].

11.8 Applications

The main focus of nanotechnology and nanophotocatalysts is in the direction of hydrogen production by using solar energy, photocatalytic disinfection, control over air pollution, and degradation/decomposition of contaminants in water to make it suitable for drinking.

11.8.1 Photocatalytic Hydrogen Production

The utility of nonrenewable sources such as petroleum and coal has been rising on earth. The incomplete burning of fossil fuels has caused many environmental evils. To overcome the environmental crisis, the scientific community has focused on engineering materials that can work in the direction of energy production. Hydrogen is considered the most promising clean energy source of the 21st century. Both fossil fuels and water are the primary sources of hydrogen energy production [48]. Enormous ways are available in the literature to produce hydrogen, and the most frequent mode is to decay fossil fuels to produce hydrogen. On the nanotechnology front, it has become very easy to generate hydrogen by the photocatalytic degradation of water. Subsequently, the produced hydrogen utilizes solar energy to yield hydrogen energy. The photocatalytic decomposition of water to generate hydrogen can convert solar energy into hydrogen energy and solve energy and environmental problems.

11.8.2 Wastewater Treatment

One of the main causes of water pollution is the discharge of industrial waste into water sources. The industrial waste includes oils, hazardous acids dyes, and drugs. The polluted water has adverse effects on both aquatic life and human health [49]. Nanotechnology has participated in wastewater treatment by its important branch called photocatalytic degradation of contaminated water by utilizing small sized nanoparticles by using photonic energy.

11.8.3 Photocatalytic Disinfection

The major cause of infection is pathogenic microorganisms. These pathogens enter in the environment by excess disposal of antibiotic drugs, exponential use of pesticides, and untreated household wastewater. After entering the environment, these harmful pathogens cause skin allergy, respiratory diseases, and many impairments in body functioning. However, various disinfection modes have been studied such as chlorination and ozonization. These disinfection techniques have major disadvantages. After chlorination, the generated byproducts are equally dangerous, i.e., chlorination destroys pathogens but creates harmful byproducts, which again enters the biological chain. Ozone oxidations demand highly sophisticated equipment, which are very costly. Nanotechnology has made it possible by designing photocatalytic antibacterial materials that utilize light to degrade harmful pathogens [50].

11.8.4 Air Purification

Air pollution surely causes respiratory diseases, lungs diseases, and eye infections. There are many causes of air pollution, such as automobile exhaust, industrial gases, and volatile organic compounds. Sulfur compounds such as sulfur dioxide, nitrogen oxides, halogen-containing compounds, and malodorous gases contained in industrial exhaust gases, as well as carbon monoxide, nitrogen oxides, and sulfur oxygen compounds in automotive exhaust gases, have a significant impact on people's health. Photocatalysis is a promising technology as it can adsorb as well as degrade air pollutants. This is a cost-effective technique. Photocatalysis is a technology with wide applications and great development potential. In addition to the above four applications, photocatalytic technology is also widely used in agriculture, construction, automobiles, roads, and household appliances. Semiconductor photocatalysis is mild, and the reaction process is relatively simple. In theory, photocatalysis can degrade almost all air pollutants. Therefore, compared with conventional methods such as filtration, adsorption, plasma, and ozone oxidation, photocatalytic technology can completely degrade the pollutants in the air under sunshine, thereby rapidly purifying the air [51, 52]. Some of the latest and highly cited published work is given in Table 11.2.

Table 11.2 Some important photocatalysts with corresponding photocatalytic potential applications

Sr. No	Type of nanoparticles	Photocatalytic application	References
1.	Ni-doped ZnS nanoparticles	MB dye degradation	[53]
2.	Cr-doped ZnS nanocomposites	Antimicrobial activity against MDR *S. aureus*	[54]
3.	ZnS-bipyridine nanoparticles	H_2 production	[55]
4.	Cu-doped TiO_2 nanoparticles	Degradation of diclofenac	[56]
5.	Ag-doped TiO_2/SiO_2 nanoparticles	Disinfection of sporulating ***Bacillus subtilis***	[57]
6.	TiO_2 nanoparticles	Hydrogen production	[58]
7.	Au-decorated CdS/CdO nanoparticles	Hydrogen evolution and dye degradation	[59]
8.	CdS nanoparticles	Photocatalytic degradation of erioglaucine and hydrogen generation	[60]
9.	ZnO rod decorated with Ag nanoparticles	MB dye degradation	[61]
10.	Ag nanoparticles loaded Ag_2SO_3	Organic pollutants degradation	[62]

11.9 Conclusion and Future Scope

Nanofabrication is the fabrication of materials in the nano regime. The remarkable progress in synthesis routes has opened a gateway toward magnificent properties. Size-tunable physical and chemical properties have channelized scientific application aspects. Here, a brief description of photocatalysis along with its mechanism has been provided. In this chapter, we have tried to explore the various categories of nanophotocatalysts employed in UV and visible lights.

It has been concluded that a strong competition exists between interfacial charge transfer and charge carrier recombination after suitable excitation of nanophotocatalysts, which strongly influences the photocatalytic behavior and acts as a detrimental factor for photocatalytic efficiency measurements. Further promotion of nanofabrication technology is highly demandable. The exploration of conceptual synthesis strategy for monodisperse nanoparticles is required for nanophotocatalyst engineering. The interpretation and design of devices based on quantum mechanics are necessary. A communal mindset from the level of basic science to the level of system architecture is vitally important in the future.

Acknowledgment

The author Dr. Balwinder Kaur is thankful to UGC-SAP and Chemistry Department, Punjabi University, Patiala, for providing necessary facilities. Dr. Balwinder Kaur is highly obliged to UGC-New Delhi for providing MANF.

References

1. Asiri, A. M. and Lichtfouse, E. (editors) (2019). *Nanophotocatalysis and Environmental Applications: Detoxification and Disinfection, Springer.*

2. Goh, S. N., Hamid, S. B. A., Juan, J. C., Ling, T. T., Samsudin, E. M., and Wu, T. Y. (2015). Evaluation on the photocatalytic degradation activity of reactive blue 4 using pure anatase nano-TiO_2, *Sains Malays.,* **44**, pp. 1011–1019.

3. Goei, R. and Lim, T. T. (2016). Combined photocatalysis-separation processes for water treatment using hybrid photocatalytic membrane reactors, In *Photocatalysis: Applications*, The Royal Society of Chemistry. pp. 130–156.

4. Matthews, R. W. (1988). An adsorption water purifier with in situ photocatalytic regeneration, *J. Catal.,* **113**, pp. 549–555.

5. Božić, A. M. L., Koprivanac, N., Kušić, H. M., and Peternel, I. T. (2007). Comparative study of UV/TiO_2, UV/ZnO and photo-Fenton processes for the organic reactive dye degradation in aqueous solution, *J. Hazardous Mater.,* **148**, pp. 477–484.

6. Saleh, B. E. and Teich, M. C. (1991). *Fundamentals of Photonics*, John Wiley & Sons, Hoboken, NJ.

7. Gaponenko, S. V. (1998). *Optical Properties of Semiconductor Nanocrystalls*, Vol. 23, Cambridge University Press.

8. Brus, L. E. (1983). A simple model for the ionization potential, electron affinity, and aqueous redox potentials of small semiconductor crystallites, *J. Chem. Phys.*, **79**, pp. 5566–5571.

9. Chemla, D. S. (1993). Nonlinear optics in quantum-confined structures, *Phys. Today*, **46**, pp. 46–52.

10. Chemla, D. S. and Miller, D. A. B. J. (1985). Room-temperature excitonic nonlinear-optical effects in semiconductor quantum-well structures, *J. Opt. Soc. Am. B*, **2**, pp. 1155–1173.

11. Capper, P. (1997). *Narrow-Gap II-VI Compounds for Optoelectronic and Electromagnetic Applications*, Springer Science & Business Media.

12. Data, Numerical (1982). *Functional Relationships in Science and Technology, Group III 17*.

13. Kittel, C. (1976). *Introduction to Solid State Physics*, Wiley New York.

14. Wise, F. W. (2000). Lead salt quantum dots: The limit of strong quantum confinement, *Accounts Chem. Res.*, **33**, pp. 773–780.

15. Brkic, S. (2018). Applicability of quantum dots in biomedical science. In *Ionizing Radiation Effects and Applications*, pp. 22–39. IntechOpen.

16. Burgess, J. D., Gokhale, R., and Sudhir, V. (2009). A comparative study of top-down and bottom-up approaches for the preparation of micro/nanosuspensions, *Inter. J. Pharma.*, **380**, pp. 216–222.

17. Corriu, R. and Nguyên, T. A. (2009). *Molecular Chemistry of Sol-Gel Derived Nanomaterials*, Wiley Online Library.

18. Reithmaier, J. (2009). *Nanostructured Materials for Advanced Technological Applications*, Springer Science & Business Media.

19. Barber, D. J., Conde, O., Gomes, M. J. M., Rolo A. G., and Ricolleau, C. (2003). HRTEM and GIXRD studies of CdS nanocrystals embedded in Al_2O_3 films produced by magnetron RF-sputtering, *J. Cry. Growth*, **247**, pp. 371–380.

20. Fendler, J. H. (2008). *Nanoparticles and Nanostructured Films: Preparation, Characterization, and Applications*. John Wiley & Sons.

21. Chen, J. S., Guo, Y., Liao, Z. L., Li, G. D., Wang, Y., and Zhang, H. (2005). Controlled growth and photocatalytic properties of CdS nanocrystals implanted in layered metal hydroxide matrixes, *J. Phy. Chem. B*, **109**, pp. 21602–21607.

22. Durach, D., Neudert, L., Schmidt, P. J., Oeckler, O., and Schnick, W. (2015). $La_3BaSi_5N_9O_2$: Ce^{3+} a yellow phosphor with an unprecedented

tetrahedra network structure investigated by combination of electron microscopy and synchrotron X-ray diffraction, *Chem. Mater.*, **27**, pp. 4832–4838.

23. Hornyak, G. L., Tibbals, H. F., Dutta, J., and Moore, J. J. (2008). *Introduction to Nanoscience*, CRC Press.

24. Park, J., Joo, J., Kwon, S. G., Jang, Y., and Hyeon, T. (2007). Large-scale nonhydrolytic sol–gel synthesis of uniform-sized ceria nanocrystals with spherical, wire, and tadpole shapes, *Angewandte Chemie Inter.*, **119**, pp. 4714–4745.

25. Wang, Y. (1995). Photophysical and photochemical processes of semiconductor nanoclusters, *Adv. Photochem.*, **19**, pp. 179–234.

26. Akiyama, M., Guan, G., Kida, T., Minami, Y., Nagano, M., and Yoshida, A. (2006). Photocatalytic activity of gallium nitride for producing hydrogen from water under light irradiation, *J. Mater. Sci.*, **41**, pp. 3527–3534.

27. Domen, K. Hisatomi, T., Ma, S. S. K., Maeda, K., and Moriya, Y. (2012). Enhanced water oxidation on Ta_3N_5 photocatalysts by modification with alkaline metal salts, *J. American Chem. Soc.*, **134**, pp. 19993–19996.

28. Ali, S., Chu, M., Hu, K., Jing, L., Liu, Y., Qin, C., and Wang, J. (2019). Synthesis of g-C_3N_4-based photocatalysts with recyclable feature for efficient 2,4-dichlorophenol degradation and mechanisms, *Appl. Catal. B: Environ.*, **243**, pp. 57–65.

29. Nakamura, Y., Ohno, T., Sayama, K., and Tsubota, T. (2005). Preparation of S, C cation-codoped $SrTiO_3$ and its photocatalytic activity under visible light, *Appl. Catal. A*, **288**, pp. 74–79.

30. Ahuja, S. and Kutty, T. R. N. (1996). Nanoparticles of $SrTiO_3$ prepared by gel to crystallite conversion and their photocatalytic activity in the mineralization of phenol, *J. Photochem. Photobio. A: Chem.*, **97**, pp. 99–107.

31. Kato, H. and Kudo, A. (2001). Water splitting into H_2 and O_2 on alkali tantalate photocatalysts $ATaO_3$ (A = Li, Na, and K), *J. Phys. Chem. B*, **105**, pp. 4285–4292.

32. Hu, C. C., Huang, H. H., and Huang, Y. C. (2017). N-doped $NaTaO_3$ synthesized from a hydrothermal method for photocatalytic water splitting under visible light irradiation, *J. Energy Chem.*, **26**, pp. 515–521.

33. Almeida, C. R. R., Araújo, V. D., Melo, M. M., Motta, F. V., Paskocimas, C. A., Tavares, M. T. S., and Tranquilin, R. L. (2016). Enhancement of the

photocatalytic activity and white emission of $CaIn_2O_4$ nanocrystals, *J. Alloys Comp.*, **658**, pp. 316–323.

34. Chen, D., Kikugawa, N., Ouyang, S., Ye, J., and Zou, Z. (2009). A systematical study on photocatalytic properties of $AgMO_2$ (M = Al, Ga, In): Effects of chemical compositions, crystal structures, and electronic structures, *J. Phys. Chem. C.*, **113**, pp. 1560–1566.

35. Kudo, A. and Yu, J. (2006). Effects of structural variation on the photocatalytic performance of hydrothermally synthesized $BiVO_4$, *Adv. Funct. Mater.*, **16**, pp. 2163–2169.

36. Kolny-Olesiak, J. and Weller, H. (2013). Synthesis and application of colloidal $CuInS_2$ semiconductor nanocrystals, *ACS Appl. Mater. Interfaces*, **5**, pp. 12221–12237.

37. Han, M. Y. and Regulacio, M. D. (2016). Multinary I-III-VI2 and I2-II-IV-VI4 semiconductor nanostructures for photocatalytic applications, *Accounts Chem. Res.*, **49**, pp. 511–519.

38. Kato, H., Kudo, A., and Tsuji, I. (2005). Visible-light-induced H_2 evolution from an aqueous solution containing sulfide and sulfite over a $ZnS–CuInS_2–AgInS_2$ solid-solution photocatalyst, *Angew. Chemie Inter. Ed,* **44**, pp. 3565–3568.

39. Ouyang, S. and Ye, J. (2011). $\beta\text{-}AgAl_{1-x}Ga_xO_2$ solid-solution photocatalysts: Continuous modulation of electronic structure toward high-performance visible-light photoactivity, *J. Am. Chem. Soc.*, **133**, pp. 7757–7763.

40. Chen, H., Lu, D., Ouyang, S., Ren, J., Umezawa, N., Wang, D., and Ye, J. (2015). Effective mineralization of organic dye under visible-light irradiation over electronic-structure-modulated $Sn\ (Nb_{1-x}Ta_x)_2O_6$ solid solutions, *Appl. Catal. B: Environ.*, **168**, pp. 243–249.

41. Kako, T., Li, G., Wang, D., Ye, J., and Zou, Z. (2007). Composition dependence of the photophysical and photocatalytic properties of $(AgNbO_3)_{1-x}(NaNbO_3)_x$ solid solutions, *J. Solid State Chem.*, **180**, pp. 2845–2850.

42. Kako, T., Wang, D., and Ye, J. (2009). New series of solid-solution semiconductors $(AgNbO_3)_{1-x}(SrTiO_3)_x$ with modulated band structure and enhanced visible-light photocatalytic activity, *J. Phys. Chem. C*, **113**, pp. 3785–3792.

43. Fu, J., Jaroniec, M., You, W., Yu, J., and Zhu, B. (2018). A flexible bio-inspired H_2-production photocatalyst, *Appl. Catal. B: Environ.*, **220**, pp. 148–160.

44. Bohorquez, M., Gopidas, K. R., and Kamat, P. V. (1990). Photophysical and photochemical aspects of coupled semiconductors: Charge-

transfer processes in colloidal cadmium sulfide-titania and cadmium sulfide-silver (I) iodide systems, *J. Phys. Chem.*, **94**, pp. 6435–6440.

45. Prasannalakshmi, P. and Shanmugam, N. (2017). Fabrication of TiO_2/ ZnO nanocomposites for solar energy driven photocatalysis, *Mater. Sci. Semicon. Proc.*, **61**, pp. 114–124.

46. Bakardjieva, S., Houšková, V., Lang, K., Murafa, N., and Štengl, V. (2008). Visible-light photocatalytic activity of TiO_2/ZnS nanocomposites prepared by homogeneous hydrolysis, *Micropor. Mesopor. Mater.*, **110**, pp. 370–378.

47. Xia, H., Xiao, D., Zhang, T., and Zhuang, H. (2008). Visible-light-activated nanocomposite photocatalyst of Fe_2O_3/SnO_2, *Mater. Lett.*, **62**, pp. 1126–1128.

48. Dionysiou, D. D., Du, Z. N., Li, K. W., Zhang, M., and Zhang, Y. C. (2011). High-performance visible-light-driven SnS_2/SnO_2 nanocomposite photocatalyst prepared via in situ hydrothermal oxidation of SnS_2 nanoparticles, *ACS Appl. Mater. Inter.*, **3**, pp. 1528 1537.

49. Chen, J., Hu, S., Jia, X., Li, Z., Lin, H., Lin, J. Li, T., Wu, X. L., Zhang, H., and Zhu, J. (2019). Organic dye doped graphitic carbon nitride with a tailored electronic structure for enhanced photocatalytic hydrogen production, *Catal. Sci. Technol.*, **9**, pp. 502–508.

50. Barraza, J. M., Betancourt-Buitrago, L. A., Machuca-Martínez, F., Marriaga, N., Ossa-Echeverry, O. E., and Rodriguez-Vallejo, J. C. (2019). Anoxic photocatalytic treatment of synthetic mining wastewater using TiO_2 and scavengers for complexed cyanide recovery, *Photochem. Photobiol. Sci.*, **18**, pp. 853–862.

51. Kai, C., Li, S., Na, H., Rong, W., Suo, Y., Wang, J., Yan, L., Zhang, W., and Zhu, W. (2018). Enhanced visible-light-driven photocatalytic sterilization of tungsten trioxide by surface-engineering oxygen vacancy and carbon matrix, *Chem. Eng. J.*, **348**, pp. 292–300.

52. Ichihara, F., Kako, T., Liu, G., Meng, X., and Ye, J. (2019). Study on the enhancement of photocatalytic environment purification through ubiquitous-red-clay loading, *Appl. Sci.*, **1**, pp. 138–145.

53. Othman, A. A., Osman, M. A., and Ali, M. A. (2020). Sonochemically synthesized Ni-doped ZnS nanoparticles: Structural, optical, and photocatalytic properties, *J. Mater. Sci: Mater. Electron.*, **31**, pp. 1752–1767.

54. Aqeel, M., Ikram, M., Asghar, A., Haider, A., Ul-Hamid, A., Naz, M., and Ali, S. (2020). Synthesis of capped Cr-doped ZnS nanoparticles with improved bactericidal and catalytic properties to treat polluted water, *Appl. Nanosci.*, pp. 1–11.

55. Ramírez-Rave, S., Rodríguez, A. A. R., Oros-Ruíz, S., García-Mendoza, C., Vengoechea, F. A., and Gómez, R. (2020). ZnS-Bipy hybrid materials for the photocatalytic generation of hydrogen from water, *Catalysis Today*, **341**, pp. 104–111.

56. Pedroza-Herrera, G., Medina-Ramírez, I. E., Lozano-Álvarez, J. A., and Rodil, S. E. (2020). Evaluation of the photocatalytic activity of copper doped TiO_2 nanoparticles for the purification and/or disinfection of industrial effluents, *Catalysis Today*, **341**, pp. 37–48.

57. Obuchi, E., Furusho, T., Katoh, K., Soejima, T., and Nakano, K. (2019). Photocatalytic disinfection of sporulating *Bacillus subtilis* using silver-doped TiO_2/SiO_2, *J. Water Proc. Eng.*, **30**, pp. 100511.

58. Patil, S. B., Basavarajappa, P. S., Ganganagappa, N., Jyothi, M. S., Raghu, A. V., and Reddy, K. R. (2019). Recent advances in non-metals-doped TiO_2 nanostructured photocatalysts for visible-light driven hydrogen production, CO_2 reduction and air purification, *Inter. J. Hydrogen Energy*, **44**, pp. 13022–13039.

59. Saha, M., Ghosh, S., and De, S. K. (2020). Nanoscale Kirkendall effect driven Au decorated CdS/CdO colloidal nanocomposites for efficient hydrogen evolution, photocatalytic dye degradation and Cr (VI) reduction, *Catalysis Today*, **340**, pp. 253–267.

60. Shenoy, S., Jang, E., Park, T. J., Gopinath, C. S., and Sridharan, K. (2019). Cadmium sulfide nanostructures: Influence of morphology on the photocatalytic degradation of erioglaucine and hydrogen generation, *Appl. Surface Sci.*, **483**, pp. 696–705.

61. Liu, H., Zhong, L., Govindaraju, S., and Yun, K. (2019). ZnO rod decorated with Ag nanoparticles for enhanced photocatalytic degradation of methylene blue, *J. Phys. Chem. Solids*, **129**, pp. 46–53.

62. Nahar, S., Hasan, M. R., Kadhum, A. A. H., Hasan, H. A., and Zain, M. F. M. (2019). Photocatalytic degradation of organic pollutants over visible light active plasmonic Ag nanoparticle loaded Ag_2SO_3 photocatalysts, *J. Photochem. Photobio. A: Chem*, **375**, pp. 191–200.

Multiple Choice Questions

1. What is the energy band gap value of CdS nanoparticles (in eV)?
   (a) 2.42
   (b) 3.42
   (c) 4.42
   (d) 5.0

2. What is the energy band gap value of TiO_2 nanoparticles (in eV)?
   - (a) 3.2
   - (b) 3.6
   - (c) 3.9
   - (d) 4.0

3. What are the important applications of nanoparticles?
   - (a) Air purification
   - (b) Photocatalysis
   - (c) Disinfection
   - (d) All of the above

4. Who gave the speech "There are plenty of rooms at the bottom"?
   - (a) Eric Drexler
   - (b) Richard Feynmann
   - (c) Harold Croto
   - (d) Albert Einstein

5. 10 nm = ____ m
   - (a) 10^{-8}
   - (b) 10^{-7}
   - (c) 10^{-9}
   - (d) 10^{-10}

6. The size of nanoparticles is between ____ nm.
   - (a) 100 to 1000
   - (b) 0.1 to 10
   - (c) 1 to 100
   - (d) 0.01 to 1

7. Nanoscience can be studied with the help of
   - (a) Quantum mechanics
   - (b) Newtonian mechanics
   - (c) Macro-dynamics
   - (d) Geophysics

8. The diameter of human hair is ___ nm.
   - (a) 50,000
   - (b) 75,000
   - (c) 90,000
   - (d) 100,000

9. The full form of STM is
   - (a) Scanning tunneling microscope
   - (b) Scientific technical microscope
   - (c) Systematic technical microscope
   - (d) Super tensile microscope

10. Which ratio decides the efficiency of nanosubstances?
    - (a) Weight/volume
    - (b) Surface area/volume
    - (c) Volume/weight
    - (d) Pressure/volume

11. Which is the correct form of XRD?
    - (a) X-ray diffraction
    - (b) X-ray diffractor
    - (c) X-ray difference
    - (d) X-ray diagram

12. Which is the correct form of TEM?
    - (a) Tunning electron microscopy

(b) Tunneling electron microscopy

(c) Transmission electron microscopy

(d) Trans electron microscopy

13. Which is the correct form of SEM?

(a) Scanning electron microscopy

(b) Surface electron microscopy

(c) Supra electron microscopy

(d) Super electron microscopy

14. Which is the full form of MB dye?

(a) Methylene blue   (b) Meth blue

(c) More blue   (d) Methine blue

15. Which is category of bottom-up approach?

(a) Chemical precipitation method

(b) Sol–gel method

(c) Hydrothermal/solvothermal method

(d) All three

16. Which is a category of top-down approach?

(a) Lithography   (b) Chemical etching technique

(c) Attrition-milling process   (d) All three

17. The prefix "nano" comes from a

(a) Greek word meaning dwarf

(b) French word meaning billion

(c) Spanish word meaning particle

(d) Latin word meaning invisible

18. Who first used the term nanotechnology and when?

(a) Richard Feynman, 1959   (b) Norio Taniguchi, 1974

(c) Eric Drexler, 1986   (d) Sumio Iijima, 1991

Answer Key

1. (a)   2. (a)   3. (d)   4. (d)   5. (a)   6. (c)   7. (a)

8. (a)   9. (a)   10. (b)   11. (b)   12. (c)   13. (a)   14. (a)

15. (d)   16. (d)   17. (a)   18. (a)

Short Answer Questions

1. Define nanotechnology.
2. What are the two approaches for the synthesis of nanomaterials?
3. What are the important applications of nanotechnology?
4. What is quantum confinement effect?
5. What is energy band gap?
6. What is photocatalysis?
7. Define exciton.
8. Write five examples of semiconductor photocatalysts?

Long Answer Questions

1. Explain the bottom-up approach with its various categories for the synthesis of nanoparticles.
2. Explain photocatalysis with its mechanism.
3. Explain the top-down approach with its various categories for the synthesis of nanoparticles.
4. Explain various categories of nanomaterials with suitable examples.
5. How nanotechnology is beneficial for society? Explain with suitable examples.

Chapter 12

Nanotechnology in Food Packaging: Current Uses and Future Applications

Sheetal Thakur,[a] Inderjeet Verma,[b] and Anjali Saharan[c]

[a]*Department of Food Science and Technology, M.M.I.C.T and B.M. (Hotel Management), MMDU, Mullana, Ambala, India*
[b]*MM College of Pharmacy, MMDU, Mullana, Ambala, India*
[c]*MM School of Pharmacy, Maharishi Markandeshwar University, Sadopur, Ambala, India*
indejeet.verma@mmumullana.org

12.1 Introduction

The birth of new technology with the inimitable term "nanotechnology" happened in the 21st century with the imaginary vision of Nobel Prize winner Richard Feynman in 1965. The study of nanoparticles in not new, but the spectacular rise in applications of nanoparticles was observed during the Industrial Revolution. The National Science and Technology Council describes the word nanotechnology as the knowledge and monitoring of the material at dimensions from 1 to 100 nm, where the exclusive phenomenon

Nanotechnology: Principles and Applications
Edited by Rakesh K. Sindhu, Mansi Chitkara, and Inderjeet Singh Sandhu
Copyright © 2021 Jenny Stanford Publishing Pte. Ltd.
ISBN 978-981-4877-43-5 (Hardcover), 978-1-003-12026-1 (eBook)
www.jennystanford.com

facilitates innovative appliances [1]. In 1959, at the American Physical Society meeting in Caltech, Richard presented a lecture titled "There's Plenty of Room at the Bottom," where he launched the thought of operating substances at the atomic level. This unique proposal paved the way for modern idea, and Feynman's supposition has since been proven truthful. This is the real reason that he is regarded as the father of modern nanotechnology. The word "nanotechnology" was first coined by N. Taniguchi in 1974 at the International Conference on Industrial Production in Tokyo with the vision of illustrating the ultrathin indulgence of materials with nanometer precision and the conception of nanosized means [2]. He explained that nanotechnology is made up of processing, severance, consolidation, and deformation of resources and structures by an atom or a molecule. The idea proposed by Richard Feynman was further developed by E. Drexler in his book *Vehicles of Creation: The Arrival of the Nanotechnology Era* published in 1986.

In the late 1980s and the first half of 1990s, many major findings and innovations were made with an influence on the auxiliary progression of nanotechnology. The first nanotechnology program of the National Scientific Fund started to operate in the United States in 1991. China, South Korea, and other countries have added their interest and mind for the development of nanotechnology. Along with this, countries in the Commonwealth of Independent States (CIS) have started concentrating in this field contained by state scientific programs. Therefore, the nanotechnology archetype was founded in the 1960s, whereas the development of nanotechnology started in the 1980s and 1990s. Nanoscience and nanotechnology (N&N) represent a swiftly growing research field. Distinctive nano-objects are molecules, atomic clusters, nanocrystallites, nanoparticles, nanowires, nanolayers, etc. [3].

Nanotechnology has an influence on quite a lot of phases of food science, from augmentation to processing and packaging. The food industry has developed nano-objects, which will put an important and beneficial expression in developing the taste of food as well as providing health benefits with respect to food safety. Nanoscience and nanotechnology are new boundaries of this century, and food nanotechnology is a rising technology, which has opened new prospects for the food industry. Improved quality of food and safety appraisal can be attained by means of nanotechnology. Further, food processing can be chiefly improved by applying nanotechnologies such as nanoliposomes, nanoemulsion, nanofibers, nanoencapsulation, and nanocoating [4]. In the recent years, nanotechnology will be effective in reinstating various areas with remarkable functions in the field of food processing and packaging and also be set to cover various progressions in the field of nanotechnology and their applicability to food and nutraceutical structure, together with recognizing the exceptional challenges.

Nanopackaging, in the recent years, has been raised to a good extent as the technology has various advantages from increasing shelf life to protecting the packed food. Nanopackaging is described as the route of intersecting, powering, cooling, and guarding the nanocompounds that are made up of nanomaterials to form electronic and bioelectronic structures in order to improve the cost as well as functionality [5]. The technology of nanomaterials has already been adding desirable impact in the growth of functional or interactive foods consumed by human to fulfill body's needs and transport nutrients more efficiently [6]. Nanomaterials, nanoparticles, and nanocomposites are good examples in the field of food packaging, which have been used effectively to provide improved mechanical strength, increased resistance to heat, reduction in weight of packaged material and improve hurdle against carbon dioxide, oxygen, UV radiation, humidity, and volatiles of food-packaging materials.

Nanofilters have already been used in brewing and dairy industry to filter microbes and viruses. The usage of nanopowder for the packaging of certain fruits is characterized with lower moisture rate and lower oxygen transmission along with high tensile strength of the final packaging material. Nanosensors are implemented to track and trace packaging starting from the packaging of food items, including pellets and containers involved in food packaging all the way through the food supply system. The utilization of nanotechnology in food packaging is already sustaining the advancement to enhance the color, taste, texture, and flavor as well as the stability of food along with many other important functions like, enhanced nutrient absorption and bioavailability of health supplements, improved mechanical, hurdle and antimicrobial properties of food package, and applications of nanosensors to monitor and trace the state of food during storage and transport [7].

12.2 Improved Packaging Using Nanotechnology

In conjunction with other innovative techniques, nanotechnology has made its place in the food sector and this technology can be validated in all facets of the food chain with the purpose of enhancement in quality control and food safety [8]. The principles and systems used in biology and biochemistry are same for administrating food mechanisms that are also complex biological systems. For that reason, the inventions made in nanotechnology will have an impact on the food industry. In order to compute, manage, and manipulate objects at the nanoscale level, nanotechnology helps to modify those functions and properties. Without modifying the food, it can be named "nanofood" when production, processing, or packaging include nanomaterials, nanotechniques, or tools [9].

Recent studies have notified that nanotechnology will prove undoubtedly the most promising technology in the food industry as narrative food-packaging technology in the coming years [10], where the agro-food sector is the best ever-growing field in nanoresearch and applications of nanotechnology. In order to provide quality products with safety and hygiene, packaging technologies have been segregated as active and intelligent packaging. The technologies have been aimed for the provision of providing safe food with enhanced shelf life. The active and intelligent packaging methods have a capable future by incorporation into packaging structures or techniques.

12.2.1 Active Packaging

Active packaging may be described as an approach where the final product, package, and the atmosphere interact with each other in a constructive manner to expand the shelf life or to accomplish a few characteristics [12]. Active packaging deals with the integration of certain additives in the package to maintain the quality of product as well as make the product shelf stable. In active packaging, nanomaterials interact with the food or the surrounding directly to provide superior security to the product. As an example, silver nanoparticles and coatings have the antimicrobial property when used in the presence of other materials such as oxygen or UV scavengers. Nanomagnesium and copper oxide, nanotitanium dioxide, and carbon nanotubes are also expected to be used for antimicrobial food packaging in the near future [13]. Here are certain types of active packaging used in the food industry (Fig. 12.1).

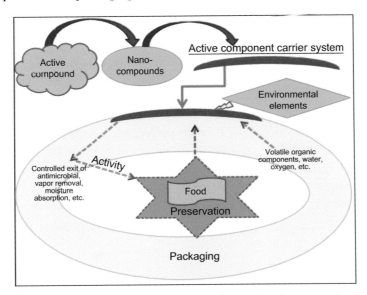

Figure 12.1 Active packaging model and its relationship with nanotechnology.

12.2.1.1 Flavor or odor scavengers

Volatile compounds such as aldehydes, amines, and sulfides gathered within the package due to food deterioration can be selectively scavenged [14]. During the transportation of multiple loads of products, these flavor scavengers inhibit the cross-contamination of overpowering odor. The development of odor-resistant packages came in light for the shipping of Durian fruit [15].

12.2.1.2 Ethylene forager

Ethylene (a natural plant growth hormone) works to hasten fruit and vegetable respiration, along with stimulating fruit ripening and softening [16]. Ethylene causes yellowing of vegetables and has injurious effect on the shelf life of various fruits and vegetables [17]. Activated carbon along with different metal catalysts is effective in removal of ethylene. Electron-deficit nitrogen-containing trienes introduced in ethylene permeable packages is also utilized to hinder ethylene from freshly yielded product.

12.2.2 Intelligent/Smart Packaging

In intelligent packaging, a system is maintained to examine the state of packaged food with the purpose of gaining the information concerning the quality of the packaged food for the period of storage and transport. Nanomaterials in such type of packaging are used for detecting microbiological/biochemical changes in food, for instance, discovering particular pathogens growing in food, or certain gases causing food spoilage [18]. In smart packaging, nanoparticles behave as reactive units in packaging structures to provide information regarding the situation of packaged item. In divergence to active components, intelligent components do not have the objective to liberate their components in the food (Fig. 12.2).

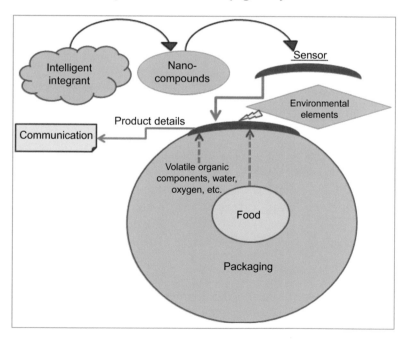

Figure 12.2 Intelligent packaging model and its relationship with nanotechnology.

The current discoveries in nanotechnology for polymer nanomaterials for intelligent packaging comprise spoilage indicators, oxygen pointers, product recognition, and traceability. Chiefly, there are three intelligent systems: radiofrequency identification (RFID) system, sensors, and indicators [19].

12.2.2.1 Radiofrequency identification systems (active tags)

Generally, active tags are electronic information-based systems that utilize radiofrequency to relocate information from a label stick to an entity to map out and recognize the entity mechanically. Radiofrequency detection is an augmentation to the preceding physical tracking methods/barcodes. The strength, longer reading range, and capacity to withstand extreme temperatures and different pressures are certain unique properties of active tags. Also it can be sensed at distances of higher than 100 m, and various tags can be read simultaneously [20]. Nanotechnology is also facilitating sensor packaging to integrate cheap, nanoenabled radiofrequency identification tags that are very tiny, elastic, and can be published on slender labels.

12.2.2.2 Smart sensors

Sensors are equipment used to identify the physical magnitude of material and translate into viewer legible signals. To adjust the interior atmosphere of food object and its properties, the signals are sensed frequently, which, in turn, are signified by sensors. A major chunk of market has been covered by oxygen scavengers, moisture absorbers, and obstacle packing product in the current smart-packaging system where the bakery and meat industry is using nanoenabled packaging technology so far. The major indicators for studying the food environment are oxygen content, temperature, and pathogens, which are to be used for accurate alarming. Besides, the shelf life of the produce can be tracked by using nanosensors. Few examples are gold nanoparticle-integrated enzymes for microbe identification, gas sensors (nanofibrils) associated with the condition of food products, and nanobarcodes tagging and security [21]. Moreover, the packages incorporated with nanosensors provide information regarding enzymes yielded during the breakdown of food molecules, hence making them insecure for human utilization.

Such packages help to let air and other enzymes release and not reside inside, hence extending shelf life and reducing artificial food preservatives [22]. Thus, smart sensors help end users in terms of high-quality recognition and manufacturers for fast circulation and endorsement of food items.

12.2.2.3 Time–temperature indicators

Temperature is a rate-limiting factor for establishing the kinetics of physical, chemical, and microbial spoilage of food objects. As mentioned by EC/450/2009, time–temperature indicators (TTIs) are designed to provide knowledge on whether a threshold temperature has been surpassed in due course or to have an approximation of the smallest period of time a produce has exhausted above the threshold temperature. Such labels give visual signals of temperature history during storage and transport. Three major types of TTIs are present in the market: critical temperature, partial history, and full history indicators [23]. On the whole, TTIs are small labels that retain calculation of time–temperature of a fragile product from fabrication to consumption [24].

12.2.3 Freshness Indicators

The information regarding the quality of products in the context of microbial expansion or chemical alterations in food objects can be collected by using freshness indicators. A visual response after the reaction between the microbial escalation metabolites and indicators inside the package gives the knowledge about the microbial quality of food, hence alerting the consumer about its freshness [25]. The researchers of Sejong University developed carbon dioxide indicators made up of aqueous solutions of chitosan or whey protein isolate. The presence of this gas was identified by the alteration in intelligibility by the pH-dependent whey [26].

The major drawback of a freshness indicator is that it gives the color change indications as contaminations even if the product is free from any sensory or quality decline. The existence of some target metabolite is not essentially a sign of deprived quality [27].

12.2.4 Integrity Indicators

To ensure the integrity of a package all the way through the fabrication and dispersal chain, a leak indicator is used. Visual oxygen markers in foods with less primary oxygen have been studied [28, 29]. The change in color with a change in the concentration of oxygen is a category of visual oxygen indicators with redox dyes. The major problem in this indicator is that the apparatus should be extremely responsive and the leftover oxygen in the container is vulnerable to indicators. Oxygen arriving via a leakage may also be utilized by the natural microbes available in the food [29].

12.3 Commercially Available Food-Packaging Systems

In terms of food safety, food packaging is one of the most crucial steps. As cellular respiration continues during the packaging of fresh fruits and vegetables, the complete obstruction of migration and permeability of gases is not desirable [30]. Whereas during the packaging of carbonated beverages, the elimination of O_2 and CO_2 is required to avoid oxidation and de-carbonation of beverages [30]. The type of food matrices and packaging material used decides the flow of CO_2, oxygen, and water vapors, which varies accordingly. Therefore, such type of difficulties in food packaging can be dealt with and defeated by making refined and extraordinarily varied packaging utilities, and the demands on the packaging business will only amplify when food is shipped over increasingly longer distances between producers and consumers. One such advanced improvement in the field of science and technology is the formulation of various nanocomposite materials [31].

12.3.1 Nanomaterials

In food packaging, the usage of nanomaterials has already become veracity. The bottles prepared by using nanocomposites have minimized the problem of carbon dioxide leakage from bottles. This innovation has helped in increasing the shelf life of carbonated beverages along with the replacement of expansive cans and heavy glass bottles. Moreover, the utilization of nanoparticles entrenched in plastic for the making of food storage bins is another example of improved packaging. The action of silver nanoparticles on bacteria helps in the eradication of harmful microbes from food stored in bins. Many such other techniques for improving food packaging are under progression. As an example, the use of nanosensors in plastic packaging helps in the identification of gases given off by food on spoilage and the package has the property of changing color itself, which alerts the consumer about the food gone rotten. The manufacturing of plastic films is under way that will help in maintaining the freshness of food for a longer time. Such films are embedded with silicate nanoparticles, which help in reducing the inward flow of oxygen to package and the escape of moisture from the package. The use of nanosensors can help in detecting various contaminants and other harmful bacteria on food surfaces at a packaging plant. This will help in saving the time utilized for lab testing and analysis of samples, and quick testing at a very low cost of analysis would be possible. The most important function of nanosensors in food packaging is to aware the consumer on whether the food item is safe to eat or not [32]. The application of nanotechnology in food science and technology has been shown in Fig. 12.3.

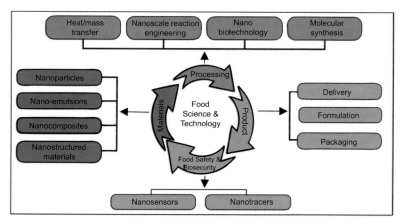

Figure 12.3 Application of nanotechnology in food science and technology [9].

12.3.2 Nanoparticles

In nanotechnology, the classification of particles has been done on the basis of size; fine particles range between 100 and 2500 nm, while ultrafine particles range between 1 and 100 nm. Nanoparticles range similar to the size of ultrafine particles. The rationale behind the double name is that during initial comprehensive studies conducted by the United States [33] and Japan [34] with "nanoparticles" in the 1970–80s, they were known "ultrafine particles" (UFP). Whereas during the 1990s, before the National Nanotechnology Initiative was launched in the United States, a new name "nanoparticle" became fashionable [35]. The usage of different types of nanoparticles along with their effects and applications has been examined in food-packaging systems [36, 37]. Nanoparticles having size in the range of 100 nm or less are named fine nanoparticulates and are used to incorporate into plastics to enhance the properties compared to those of conventional counterparts. A category of thermoplastic polymers called polymer nanocomposites has nanoscale additions of 2–8% by weight. Nanoscale enclosures consist of nanoclays, nanoscale metals, carbon nanoparticles, and oxides and polymeric resins.

Nanoclusters, nanopowders, and nanocrystals are distinct types of nanoparticles varied in size and are used in food packaging. The nanoparticles spread all over the plastic hinder O_2, CO_2, and humidity from approaching fresh foods. Nanoclay plays the role of creating plastic buoyant, tough and more heat defiant. By implanting nanocrystals in plastic, molecular obstacle has been constructed by scientists that facilitate the inhibition of oxygen escape.

12.3.3 Nanocomposites

The potential of nanotechnology to employ new types of food packaging has paved the way to a modern era of packaging. Nanocomposites, one of such inventions, maintains the capacity to enhance mechanical power, reduce weight, improve heat resistance, and expand obstruction against O_2, CO_2, UV rays, humidity, and volatiles of food-packaging matter. Extremely high surface-to-volume ratio of nanocomposites is their unique characteristic, which makes them highly reactive when compared with their macroscale counterparts, and this fundamental property of nanocomposites makes them different [38].

An additional property of nanocomposites is to provide added stability in the form of sustainable antimicrobial activity and help in reducing the migration of metal ions into stored foods. Nanocomposites are typically composed of a polymer matrix in a continuous or discontinuous phase [39]. It is a multiphase material resulting from the incorporation of matrix and a nanodimensional matter. Polymers such as low-density polyethylene (LDPE), gelatin, isostactic polypropylene, and polylactic acid are chiefly used to formulate nanocomposites with metal/metal oxide nanomaterials for food application [40]. On the basis of nanomaterial, the nanodimensional phase is usually exemplified into nanospheres or nanoparticles, nanowhiskers or nanorods, nanotubes and nanosheets, or nanoplatelets [41]. Furthermore, enhancing the mechanical and hurdle characters, nanoparticles add active or smart features to the packaging material [42]. A relatable manufacturing procedure was used to construct clay/poly (ethyleneimine) polymer nanocomposites that conserved the interweave integrity of cotton fabrics during increased burning times when they were used as a coating [43]. Ultimately, the benefit from polymer nanocomposites to the food-packaging industry is to provide better down gauging opportunities, as well as benefits in price savings and reducing waste. This is due to the fact that there is the use of very small amounts of polymer, which is required to achieve packaging materials with better mechanical features [42]. Nanocomposites may offer environmental benefits over conventional plastics: When a nanofiller is dispersed within the biocompatible polymer PLA, the PLA bionanocomposite has a faster rate of biodegradation than PLA having no such additives [44].

In the recent times, the usage of nanobiocomposites in food packaging has improved the capacity of food-packaging system to work as a hurdle against gases [45]. Current trends in food packaging are promoting the use of biodegradable polymers fortified with nanofillers, which are ecofriendly [46].

12.3.4 Chemical-Release Nanopackaging

This is a type of packaging that interacts with the foodstuff it encloses, and the switching can be kept in dual directions. The benefit of such a packaging system is that there can be release of nanorange antimicrobials, essence, aroma, antioxidants, or nutraceuticals in the food or beverage, which helps in enhancing the shelf stability of food or to enhance its flavor or smell [47, 48]. In certain cases, such type of packaging also encompasses scrutiny elements, which means that the liberation of nanochemicals will take place in reaction to a specific trigger event [49]. On the other hand, nanopackaging that includes carbon nanotubes has been created with the capacity to "pump out" O_2 or CO_2, which could cause food/beverage deterioration [50]. Nanopackaging having the capacity to soak up unwanted aromas is also under progress. Nanomagnesium and copper oxide, nanotitanium dioxide and carbon nanotubes are certain resources that are also assumed to be used so far in antimicrobial food packaging [51].

The British Airways, Nestlé, MonoPrix Supermarkets, and many more are utilizing the packaging prepared with chemical sensors. Also nanotechnique is providing recent and extra-refined devices to enhance such potential and to cut down price [52].

12.3.5   Nano-Based Antimicrobial Packaging

Food-packaging systems including packaging containers are prepared by adding antimicrobial nanomaterials, to hinder or stop the deterioration of food occurring because of microbes. Nanoparticles of silver, zinc oxide, and chlorine dioxide have been used for the production of such products. Packaging materials using nanomagnesium oxide and titanium dioxide and carbon nanotubes are also being built for usage in antimicrobial food packaging [53].

Another widely used category under antimicrobial packaging is the use of silver nanoparticles, which are used in a large range of consumer goods [54]. Silver nanoparticulate liberates ions more proficiently than the mass metal, and silver ions have a bactericidal effect as they can inhibit a broad range of biological procedures within the bacteria [22]. As the toxic effects of liberation of silver ions on humans are very low, it is probable that nanoparticulate silver will be incorporated in additional amalgamated materials. Nonetheless, the discharge of a large amount of silver ions into the environment and the accumulation in the ecosystem may have some adverse effects as silver ions are said to be harmful to the marine world. Antimicrobial action is exhibited by zinc oxide, which increases with decrease in particle size [55]. This movement is accelerated by visible light and does not need the availability of

UV light [56]. Otherwise, the precise mechanism of action is still not known. Nanomaterials containing zinc oxide have been impregnated in a variety of distinct polymers together with polypropylene [56]. Also, zinc oxide absorbs UV light efficiently without re-emitting heat and hence enhance the steadiness of polymer compounds. Chitosan is a biopolymer resulting from chitin that exhibits antimicrobial properties, hence utilized as a packaging material [57]. This property of chitosan has developed the interest of researchers to explore its introduction into divergent composite materials, which may possibly be used in human well-being and food packaging as well as its usage as a "green reagent" to lessen and alleviate silver ions [58], in conjunction with clays such as rectorite, which may perhaps then be utilized in polymer composites [59].

12.3.6 Antimicrobial Packing

In the recent years, antimicrobial packing has received considerable attention. It acts to trim down, restrain, or hinder the escalation of microbes that are possibly found in the packaged foodstuffs or packaging system itself [60]. Antimicrobial agents can be integrated into or coated onto the food package material in order to control unwanted microbes on foods [61].

An antimicrobial film has an acceptable structural integrity, so it is desirable. Moreover, the hurdle characteristics revealed by nanoparticles and the antimicrobial behavior added by the antimicrobial mediators incorporated inside the film make the system more acceptable [62]. The film permits nanomaterials to be able to stick more replicates of biological particles that give higher efficiency [63].

12.3.7 Nanosensors

Another advanced type of commercial food-packaging system is the packages prepared with nanosensors that help to track internal and external situation of food objects and containers from production to packaging. This type of packaging helps to monitor temperature or humidity. Nanosensors used in plastic packaging can perceive gases liberated by food on spoilage, and the color change of packaging alerts the consumers. Hence, nanosensors can act in response to ecological alterations, degradation products, or microbial contamination [64]. On the incorporation of nanosensors in food packaging, evident chemical composites, contaminants, and pathogens in food can be easily detected and the requirement for imprecise expiry dates can be removed, providing freshness to food stuffs [65].

The use of nanoparticles made from magnetic materials or different colors in nanosensors helps in selectively binding themselves to food pathogens. This helps in detecting the detrimental pathogens. The benefit of such coordination is that numerous nanoparticles can be kept on one nanosensor for accurate and speedy detection of the presence of distinct bacteria and pathogens [66].

12.3.8 Nanobiodegradable Packaging

The most concerning factor affecting the environment and surroundings is pollution. The use of nondegradable plastics affects the properties of soil in a negative way and plays a role in accumulating harmful gases in the environment that cause global warming. To overcome these defects, biodegradable plastics have been put to action, but they have the deficiency of mechanical power and more permeability to gases and water. Such drawbacks have been covered by the usage of nanotech-based packaging materials constructed of natural or synthetic nanoparticles with distinct properties such as more mechanical power and biodegradability. Nanoparticles are found as carbohydrates, proteins, and lipids obtained from living sources [67].

The usage of nanoparticles to empower bioplastics (based on plant sources) may facilitate such plastics to be utilized in place of fossil-fuel-based plastics for the packaging of foods and carry bags [51]. On the mixing of bioplastics together with nanoclay particles, the resultant nanocomposites attain enhanced hurdle properties as compared to the unadulterated bioplastic, and on the completion of functional lifespan, nanocomposites can be manured and sent back to the soil. Other nanomaterials can be used, such as nanoparticles, nanofibers, and nanowhiskers [68].

Various biopolymers such as chitosan, collagen, cellulose, and zinc have been manufactured as nanofibers from distinct biopolymers by means of electrospinning technology. In certain cases, they have better characteristics when compared to the traditionally cast polymer, such as amplified heat resistance, antibacterial action, and as biocatalyst. Additionally, sheets of such nanofibers acquire an extremely nanoporous arrangement and may be utilized as supporting mediums for more functionality as they perform the minimum thermal conductance of the entire solids [69].

12.3.9 Nanocoatings

In order to improve the physical property of nanoparticles, only a few researchers have described the use of nanoparticles toward coating films. With the purpose of decreasing the dispersal of oxygen, nanoclay particles have been inserted into pectins [70]. Similarly, for the substantial enhancement of physical properties, nanocomposites prepared by montmorillonite and gelatin have been used [71].

12.3.10 Nanolaminates

Food scientists have generated various ideas from nanotechnology to construct novel nanolaminate films that are appropriate for usage in food business. Normally, a nanolaminate is made up of multiple coatings of matters having nanometer dimensions attached physically or chemically with each other. A highly powerful method of preparing nanolaminates has been derived from the layer-by-layer deposition method, where charged surfaces are varnished with interfacial coatings made up of numerous layers of varied nanomaterials [72]. Various adsorbing materials may possibly be utilized to prepare the divergent coatings, which include charged lipids, natural polyelectrolytes, and colloidal particles [73].

12.3.11 Clay Nanoparticles and Nanocrystals

By adding and entrenching nanoclays within food-packaging substances, the hurdle characteristics of food package can be enhanced. In nanocomposites, layered silicates made up of two-dimensional coatings have usually been used, which are 1 nm broad and numerous microns long depending on the specific silicate. The occurrence of silicates in the formulation of polymers provides excellent barrier properties [74].

Oxygen scavenger coatings have been productively made by incorporating titania nanoparticles into distinct polymers in order to be utilized for packaging of oxygen-sensitive products. The photocatalytic action of nanocrystalline titania under UV radiation is the major area of interest. The existence of O_2 in a package speeds up the oxidative decline of food. The escalation of aerobic microbes, development of bad flavor and odor, color alterations, and nutritional deficiency along with adverse effects on shelf stability of foodstuffs are certain effects of oxygen availability in foods [75]. Hence, it is essential to restrict the levels of oxygen in food packages to control the speed of such deterioration effects in food. The utilization of oxygen scavengers, which play a role in absorbing the residual oxygen after packaging, helps in minimizing the alteration of quality in oxygen-sensitive foods [76].

12.4 Nanotechnology Applications in Processing and Packaging of Foods

In the area of food science and research, there have been incredible improvements within nanotechnology applications. Nanotechnology plays a role in the detection of insect killers [77], pathogens [78], and contaminants [79], which ultimately facilitates the preservation of food quality. Packaging of foodstuffs has a significant function in ensuring the security of food materials. Nanotechnology is playing a critical part to make better packaging materials to enhance functionality and, as a result, ensure food safety and security for end users. The integration of carbon nanotubes into packaging materials with the purpose of detecting microbes, lethal proteins, and food spoilage is of prime interest in the nanotechnology market [80]. Moreover, carbon nanotubes have the capability to shift the forthcoming food-packaging materials to smart and active packaging structures [81]. The packing of food has an enormous implication in food preservation to make it profitable. The prime applications of food-packaging systems are to provide an innovative system that maintains food quality, provides biodegradable packaging and end-user satisfaction, and many more. Keeping in mind these applications of food packaging, nanotechnology has been classified based on the rationale of functions [82].

According to researchers in the United Kingdom, nanomaterials of zinc are certainly extremely efficient in killing microbes and, therefore, have relevance in the packaging of foodstuffs [83]. Consumers would be able to "read" the situation of packaged food upon implanting nanosensors in the packaging system [84].

In order to inhibit the microbial expansion and spoilage, food items are protected by keeping the food in a static and low O_2 environment; therefore, the matter employed should be resistant to gases. Nanocomposites are integrated in the polymer medium of the materials because of their big surface area, which supports the interactions of the filler medium and its presentation. Many researches have achieved advancement in nanocomposites and carbon nanotubes since packaging plays a backbone for profitability of products.

Nanosensors check for the microbial load and spoilage-causing organisms in the packaged food and focus on food safety. They check for pesticide residues on fruits and vegetables to maintain process parameters through carbon nanotubes, chemical nanosensors, and nanobiosensors. Nanosensors in plastic film packages are employed to autodetect odors from spoiled food, specific to organisms, trace temperature fluctuations, detect pathogen, and ensure preservation. Silicate nanoparticles encrusted in polymeric matrix are used for detecting bacteria in meat, fish packs, and fungi affecting fruit packs. Nanofibers are used in the form of micro-emulsions composed of nanoparticles of SiO_2 with diverse functions viz. action as catalysts, role in increasing the surface-to-mass ratio, for increased kinetics, helps in increasing gelation, and as viscosifying agents. They are applied in DNA microarray, microfluidics, and microelectromechanical systems [85].

Color-changing labels detect ripeness, fluctuations in temperature, and period of storage and trace either interior or exterior situations through silver and gold nanoparticles, carbon nanotubes as biosensor, and carbon black nanoparticles. Nanocochleates/nanodroplets are for more specific delivery of nutrients to cells without affecting taste and color. This also helps in increasing the bioavailability of product lycopene, beta-carotenes, and phytosterols as coiled nanoparticles and vitamin sprays. Nanofilms help to develop edible films that protect food from gases and lipid moisture, maintain texture, strengthen bioplastics, and provide thermal insulation and corrosion protection in metal containers [85].

A desired packaging matter must have permeability for gases and humidity in combination with potency and biodegradability [86]. In order to improve food-packaging system, usage of nanocomposites as a functioning material for packaging and coating can also be used [87]. Utilization of inert nanoscale fillers such as clay and silicate nanoplatelets and silica nanoparticles in the polymer medium provides it with light, strong, and better thermal properties [42]. Antimicrobial nanocomposite coatings, which are prepared by permeating the fillers into the polymers, offer duel advantage due to their structural reliability and hindrance properties [88].

12.5 Safety Issues and Regulations

Although there are massive advantages of nanotechnology to the food-packaging industry, problems of food safety related to nanomaterials cannot be ignored. Various scientists have talked about the safety issues related to nanomaterials and focused on the likelihood of migration of nanoparticles from the packaging system into the food and their influence on the health of end users [89, 90]. The toxicity level or effect of nanomaterials will probably be determined on a case-by-case basis because each nanomaterial has its exclusive characteristics [91]. The entry points for different nanoparticles originated from other nanomaterials can be by inhalation, ingestion, or dental dissemination [8]. The other sources of entry of nanoparticles could be nanotechnique-based medical equipment, drugs and vaccines, and discharge from implants [64]. From the viewpoint of the food industry, breathing and skin contact are specifically linked with the people working inside nanomaterials-manufacturing units, but the major matter of concern is end users where problem occurs due to ingestion [92]. The existence of nanoparticles in food is mainly due to the direct association of packaging material and food along with the relocation of nanoparticles from nanopackaging [64].

Liver and spleen are the two main organs that cause the dispersal of nanoparticles after intake and spread from intestine till distribution [93]. Only a few studies have inspected the effect of nanomaterials within the alimentary canal, and the reports produced from this research illustrated that the nanostructured particles go through intestines and are abolished quickly [94]. Investigations demonstrated that gulped nanoparticles may penetrate the olfactory system under the forebrain via the axons of olfactory nerves within the nose via which they arrive to certain sections of brain through inhalation [95]. The consequences of nanomaterials on human body also depend on their characteristics. The entrance of nanoparticles in the blood stream may have an effect on the vessel lining or its function or could also be linked to cardiovascular effects associated with inhaling ambient nanoparticles [96]. Certain studies demonstrate that a few nanoparticles are competent in crossing the blood–brain stream and go into cells and organs and interfere with body's metabolism or pass through the fetus [93]. The lethal effect of nanomaterials depends on their path of entry in the body as well as properties, intensity and length of contact with nanoparticles, along with individual receptiveness and condition of person [97, 98]. Carcinogenic consequences of importunate elements such as asbestos have been proposed to migrate from the local generation of granulomas and fibrosis in lungs [99, 100]. Evidences related to the health hazards linked with the inhalation of nanomaterials are increasing day by day [101]. Additionally, bio-accretion of nanoparticles such as nanosilver drawn from either nanopackaging or plants and animals has been corroborated in food and humans [102].

The first authorized meaning of nanomaterials was given in 2011 within the European Union in the class of a nonbinding official instrument. Subsequently, the description of "engineered nano-materials" has been added within Reg. 1169/2011 on the delivery of food information to end users, where the labeling of food items having nanomaterials is obligatory. Extra comprehensive regulations have been put in place whose diligence depends on the anticipated utilization of the nanoparticles. When nanomaterials are used as chief components such as nanoemulsions, these are added in the span of "Novel Food" Regulation (258/97) as "foods and food components with a novel or deliberately altered primary molecular structure" and they are subjected to a risk appraisal protocol before market endorsement. On the usage of nanomaterials as food additives, a particular module has been followed (Reg. 1333/2008) and the nanomaterials are anticipated to be placed in the EU register prior to use. In the section of additives, the description of nanoparticles seems contentious, as a few permitted additives, such as silica (E551), may have certain primary constituents fulfilling the necessities of meaning of nanomaterial. The major queries in such cases are whether previously endorsed products exposed to be nanomaterials should be reconsidered or not and whether the occurrence of nanoparticles as a result of fabrication procedure and not to the readiness of the manufacturer should be excused from advanced assessment. Furthermore, before adding the food contact materials within the EU catalogue, an endorsement process is mandatory to follow. At present, certain materials, for example titanium nitride nanoparticles used in plastic materials to hinder CO_2 seepage from carbonated beverages, have been permitted because these are known to have no effects as the relocation has not been confirmed [103]. In so far as the products to be used in the agricultural zone, the impact on ecosystem should be assessed along with the probability for human contact.

In Europe, the legislation at present restricts an overall migration limit of 10 mg ingredient/dm^2 area for all or any matter that will move from food contact materials to foods. On the other hand, excluding certain substances precisely scheduled in Annexure 1 of the legislation, risk assessment related with nanomaterial must be done on a case-by-case basis [104] (Commission Regulation (EU) No. 10/2011). The migration of silver nanoparticles from three distinct types of nanocomposites into food, together with an investigation of the kind of silver transferring (ions or atoms), has been examined [105]. It was observed that silver moved into food stimulants having acidic food represented the exceptionally finest level of migration. Furthermore, the effect of heat on migration increases on heating where microwave heating results in more migration when compared with a classical oven. The writers proposed that silver migration could occur via two distinct means: disengagement of silver nanoparticles from the compounds or the oxidative dissolution of silver ions.

A research organization established in the United Kingdom considers that antimicrobial nanoparticles are found secure and efficient for food packaging, an invention that would modernize food packaging in the near future. The researchers reported that nanoparticles of zinc have been found to be efficient in eradicating microbes. Improvement has been made in developing novel nanocomposites that are clusters of functional nanoparticles, hundreds of micrometers in dimension, competent in splitting liquid into nanoparticles that can get affixed to and destroy microbes [106]. The chemical-release nanopackaging technique is an additional inclination that facilitates food packaging to intermingle with food. This swapping can be processed in both ways. The role of packaging in liberating nanoscale antimicrobials, antioxidants, flavors, fragrances, or nutraceuticals into food/beverages can be considered to improve the shelf life of food or boost its taste or aroma [107, 108].

The food-processing techniques involved in producing nanomaterials and nanosized emulsions have various major health insinuations that alert food regulations. The investigation is needed in order to determine whether or not the related food safety principles are required with respect to the potential for such foods to pose new health hazards [109]. The regulatory structures competent of supervising any threats associated with nanofoods are urgently needed and hence the use of nanotechnologies in the food industry. The broader communal, financial, social, and moral challenges related to nanotechnology have to be responded by the governments. The developed new technologies in the significant section of agriculture and food need community participation in executing nanotechnology-related decisions just to guarantee democratic control [110]. The U.K.RS/RAE2004 has suggested that chemicals inside certain nanomaterials or nanotubes must be considered new materials. The components in such nanoparticles have to go through a complete safety evaluation to be conducted by the pertinent panel of scientists prior to giving permission to be consumed in food products. Also it is mandatory to reveal in the ingredients lists that a particular nanomaterial has been added in the food product. Moreover, the liberation of nanomaterials and nanotubes in the atmosphere must be evaded as much as possible [111].

A statement released by the USFDA said that if there is already an approval regarding the commercial usage of chemicals in large particle form, nanoparticles of such chemicals do not lawfully need any further endorsement or prompt novel safety testing [111].

Moreover, the food components that are categorized as "Generally Recognized as Safe" (GRAS) do not necessitate any premarket approval from the USFDA. The GRAS system fails to discriminate between matter in larger particle or nanoparticle form.

The Food Standards Australia and New Zealand (FSANZ) regulates additives and ingredients added in nanofoods in Australia under the Food Standards Code [112]. Besides, Australian laws seem to be struggling to keep on adding latest techniques for the speedy growth of nanotechnology in the field of food and agriculture.

Overall, the safety of nanoproducts has turned out to be the major objective of accelerating awareness. Regardless of the swift commercialization of nanotechnology, there are no nanospecific policies to be present anywhere in the world. The majority of regulatory organizations remain in an information-collecting approach and are deficient in authorized and scientific tools, information, and possessions to sufficiently supervise the exponential growth of the nanotechnology market. On the other hand, the FDA anticipates that numerous nanotechnology products will come underneath its authority; thus, the Office of Combination Products will probably grasp any related responsibilities [113]. The FDA has legalized a good array of manufactured goods, counting foods, cosmetics, medicines, apparatus, and veterinary items in which a variety of produced items are making use of nanotechnology or they have nanomaterials. With the introduction and extension of nanotechnology in the food sector, common people should be aware of the allied physical wellness, safety, and environmental consequences of nanotechnology.

12.6 Conclusion

Nowadays, perishability decides the quality as well as shelf life of food products. Ambient temperature is required to maintain the quality and freshness of perishable foods during storage and transport. By supervising the level to which perishable foods come across deterioration-endorsing causes, mainly light, oxygen, and ethylene, the perishability can be controlled. Food must be freed from any kind of infectivity, whether chemical, physical, or biological during production, handling, and delivery. Food packages and the packaging materials play a primary and crucial function in maintaining the quality of food as well as its shelf life.

Recent advancements in nanotechnology have modernized the food business with its distinct functions in food processing, protection, and security, along with broadening shelf life and lessening the packaging waste. Nanomaterials, including carbon nanotubes, metal nanoparticles, quantum dots, and other active nanomaterials, are frequently used to construct biosensors for the quantification of microorganisms and other tests for food safety.

The hurdle characteristics are primarily persuaded by packaging to make an indisputable food environment. The beginning of the nanotechnology era has additionally unlocked the innovative opportunities and technical progressions in the food-packaging region. Connecting nanoparticles to polymers with an aim to formulate nanomaterial packaging promotes regular packaging with improved blockade properties, elasticity and steadiness, and mechanical and thermal strength. Innovative nanopackaging methods (active/smart/intelligent packaging) have prospective to function as a critical device to overcome obtainable packaging challenges with respect to customer and manufacturer satisfaction. It is expected that the traditional packaging system is going to be comprehensively substituted with multipurpose smart or active packaging system. Nanostructured materials help to control microbial attack, hence promising microbiological food safety. Furthermore, nanosensors help to aware and alert end users related to the protection and precise nutritional eminence of packaged food substances. Numerous companies have participated in this region with the establishment of most modern packaging techniques with restructured expertise. On the other hand, interruption in acquaintance is still present as this is an undeveloped division due to which the research society has been obstructed with plentiful questions principally related to its toxicity and hazardous effect on the environment. Concerns on the migration of nanomaterials to packaged food materials have been raised; however, migration analysis and risk evaluations are still not decisive. Indeterminate poisonous, insufficiency of supportive clinical experimental data, and hazard measurement investigations bound the applicability of nanomaterials in the food-packaging system. Finally, for the victorious accomplishment of nanotechniques at the colossal level, the consent of end user is obligatory for which the advantages and risk review should be recognized unquestionably. Qualified scientific bodies should come forward with suitable cataloging, and they should also set aside the usual set of laws that enhance the purchaser's satisfaction. Government organizations should come forward and put their efforts in cooperation with one another to undertake this matter prudently and graciously and make a path for the emerging world. Policies and legislation should be structured wisely for the sake of society related to the administration and application of nanomaterials in the food-packaging structure.

12.7   Future Scope

Nanotechnology is a budding area that connects almost every technical field from chemical science to computer technology. In the future, the development of biosensors for the detection of pollutants and pathogens in agricultural produce will help guarantee the security of food supply. In order to improve the quality of food and meat industry, the new inventions in nanomaterials will play a promising role in promoting the trends of food packaging with respect to reduction in waste and enhancement of food quality and shelf life. For the delivery of bioactive compounds, nanotechniques will establish a certainty in health enhancement activities through food. The production of components with improved functionality, which could be used to manufacture foods with new and exclusive taste and texture, is a potentially hopeful future outlook of nanotechnology. Moreover, in the area of food packaging, the manufacturing of nanostructured packaging objects with improved hurdle or self-stamping/sealing distinctiveness will pave the way for the modern era of the food-packaging industry.

References

1. Roco, M. C. (2000). National Nanotechnology Initiative (NNI), National Science Foundation.

2. Tolochko, N. K. (2009). History of nanotechnology. *Encyclopedia of Life Support Systems* (EOLSS). UNESCO.

3. Stix, G. (2001). Little big science. *Sci. Am.*, **285**(3), 32–37.

4. Raj, A. A. S., Ragavi, J., Rubila, S., Tirouthchelvamae, D., and Ranganathan, T. V. (2013). Recent trends in nanotechnology applications in foods. *Inter. J. Edu. Res. Tech.*, **2**(10), 956–961.

5. Tummala, R., Wong, C. P., and Raj, P. M. (2009). Nanopackaging research at Georgia Tech. IEEE *Nanotech. Magaz.*, **3**(4), 20–25.

6. Sujithra, S. and Manikkandan, T. R. (2019). Application of nanotechnology in packaging of foods: A review. *Int. J. Chem. Tech Res.*, **12**(4), 7–14.

7. Chaudhry, Q. and Castle, L. (2011). Food applications of nanotechnologies: An overview of opportunities and challenges for developing countries. *Trends Food Sci. Tech.*, **22**(11), 595–603.

8. Baltic, Z. M., Boskovic, M., Ivanovic, J., Dokmanovic, M., Janjic, J., Loncina, J., and Baltic, T. (2013). Nanotechnology and its potential applications in meat industry. *Tehnol. Mesa*, **54**(2), 168–175.

9. Ozimek, L., Pospiech, E., and Narine, S. (2010). Nanotechnologies in food and meat processing. *ACTA Sci. Polon. Technol. Aliment.*, **9**(4), 401–412.

10. Dasgupta, N., Ranjan, S., Mundekkad, D., Ramalingam, C., Shanker, R., and Kumar, A. (2015). Nanotechnology in agro-food: From field to plate. *Food Res. Inter.*, **69**, 381–400.

11. Mihindukulasuriya, S. D. and Lim, L. T. (2014). Nanotechnology development in food packaging: A review. *Trends Food Sci. Technol.*, **40**(2), 149–167.

12. Miltz, J., Passy, N., and Mannheim, C. H. (1995). Trends and applications of active packaging systems. In: *Food and Packaging Materials: Chemical Interactions* (Ackerman, P., Jagerstad, M., and Ohlsson, M., Eds.), The Royal Society of Chemistry, London, England, pp. 201–210.

13. Chaudhry, Q., Scotter, M., Blackburn, J., Ross, B., Boxall, A., Castle, L., Aitken, R., and Watkins, R. (2008). Applications and implications of nanotechnologies for the food sector, *Food Additi. Contam.*, **25**(3), 241–258.

14. Day, B. P. F. (1998). Active packaging of foods. In: *Smart Packaging Technologies for Fast Moving Consumer Goods* (Kerry, J. and Butler, P., Eds.), John Wiley & Sons, pp. 1–18.

15. Morris, S. C. (1999). Odour proof package, Patent No WO1999025625A1.

16. Abeles, F. B., Morgan, P. W., and Saltveit Jr., M. E. (2012). *Ethylene in Plant Biology*, Academic Press Inc., San Diego, California.

17. Zagory, D. (1995). Ethylene-removing packaging. In: *Active Food Packaging* (Rooney, M. L., Ed.), Springer, Boston, MA, pp. 38–54.

18. Kuswandi, B., Wicaksono, Y., Abdullah, A., Heng, L. Y., and Ahmad, M. (2011). Smart packaging: Sensors for monitoring of food quality and safety, *Sens. Instrum. Food Qual Safe*, **5**(3–4), 137–146.

19. Vanderroost, M., Ragaert, P., Devlieghere, F., and De Meulenaer, B. (2014). Intelligent food packaging: The next generation. *Trends Food Sci. Technol.*, **39**(1), 47–62.

20. Abad, E., Zampolli, S., Marco, S., Scorzoni, A., Mazzolai, B., Juarros, A., Gómez, D., Elmi, I., Cardinali, G. C., Gómez, J. M., and Palacio, F. (2007). Flexible tag microlab development: Gas sensors integration in RFID flexible tags for food logistic, *Sensors Actuators B: Chem*, **127**(1), 2–7.

21. Jans, H. and Huo, Q. (2012). Gold nanoparticle-enabled biological and chemical detection and analysis, *Chem. Soc. Rev.*, **41**(7), 2849–2866.

22. Sondi, I. and Salopek-Sondi, B. (2004). Silver nanoparticles as antimicrobial agent: A case study on *E. coli* as a model for Gram-negative bacteria, *J. Colloid. Interface Sci.*, **275**, 177–182.

23. Singh, R. P. (2000). Scientific principles of shelf-life evaluation. In: *Shelf Life Evaluation of Foods* (Man, C. M. D. and Jones, A. A., Eds.), Springer, Boston, MA, pp. 3–26.

24. Fu, B. and Labuza, T. P. (1995). Potential use of time-temperature indicators as an indicator of temperature abuse of MAP products. In: *Principles of Modified-Atmosphere and Sous Vide Product Packaging* (Farber, J. M. and Dodds, K. L., Eds), CRC Press, pp. 385-423.

25. Jung, J., Puligundla, P., and Ko, S. (2012). Proof-of-concept study of chitosan-based carbon dioxide indicator for food packaging applications, *Food Chem.*, **135**(4), 2170–2174.

26. Kuswandi, B., Maryska, C., Abdullah, A., and Heng, L. Y. (2013). Real time on-package freshness indicator for guavas packaging, *J. Food Meas. Charact.*, **7**(1), 29–39.

27. Hogan, S. A. and Kerry, J. P. (2008). Smart packaging of meat and poultry products. In: *Smart Packaging Technologies for Fast Moving Consumer Goods* (Kerry, J. and Butler, P., Eds.), John Wiley & Sons, pp. 33–54.

28. Davies, E. S. and Gardner, C. D. (1996). Oxygen indicating composition, British Patent. 2298273.

29. Mattila-Sandholm, T., Ahvenainen, R., Hurme, E., and Järvi-Kääriänen, T. (1998). Oxygen sensitive colour indicator for detecting leaks in gas-protected food packages, *European Patent* EP.666977.

30. Robertson, G. L. (2016). *Food Packaging: Principles and Practice*, 2nd edn, CRC Press, New York.

31. Abbaspour, A., Norouz-Sarvestani, F., Noori, A., and Soltani, N. (2015). Aptamer-conjugated silver nanoparticles for electrochemical dual-aptamer-based sandwich detection of staphylococcus aureus, *Biosens. Bioelectron.*, **68**, 149–155.

32. Sorrentino, A., Gorrasi, G., and Vittoria, V. (2007). Potential perspectives of bio-nanocomposites for food packaging applications, *Trends Food Sci. Technol.*, **18**(2), 84–95.

33. Sujithra, S. and Manikkandan, T. R. (2019). Application of nanotechnology in packaging of food: A review, *Int. J. Chem. Tech. Res.*, **12**(4), 7–14.

34. Granqvist, C., Buhrman, R., Wyns, J., and Sievers, A. (1976). Far-infrared absorption in ultrafine Al particles, *Phys. Rev. Lett.*, **37**(10), 625–629.

35. Uyeda, T., Hayashi, C., and Tasaki, A. (1995). *Ultra-fine Particles: Exploratory Science and Technology*, Elsevier.

36. Costa, C. (2010). Nanocomposites to extend the shelf life of ready to use fruit and vegetables. In: *14th Workshop on the Developments in the Italian PhD Research on Food Science Technology and Biotechnology*, University of Sassari, Oristano, 16–18.

37. De Azeredo, H. M. C. (2009). Nanocomposites for food packaging applications, *Food Res. Intern.*, **42**(9), 1240–1253.

38. Brody, A. L. (2006). Nano and food packaging technologies converge, *Food Technol. (Chicago)*, **60**(3), 92–94.

39. Arora, A. and Padua, G. W. (2010). Review: Nanocomposites in food packaging, *J. Food Sci.*, **75**(1), R43–R49.

40. He, X. and Hwang, H. M. (2016). Nanotechnology in food science: Functionality, applicability, and safety assessment, *J. Food Drug Anal.*, **24**(4), 671–681.

41. Bratovcic, A., Odobašić, A., Ćatić, S., and Šestan, I. (2015). Application of polymer nanocomposite materials in food packaging, *Croatian J. Food Sci. Technol.*, **7**(2), 86–94.

42. Duncan, T. V. (2011). Applications of nanotechnology in food packaging and food safety: Barrier materials, antimicrobials and sensors, *J. Colloid Interface Sci.*, **363**(1), 1–24.

43. Li, Y. C., Schultz, J., Mannen, S., Delhom, C., Condon, B., Chang, S., Zammarano, M., and Grunlan, J. C. (2010). Flame retardant behavior of polyelectrolyte-clay thin film assemblies on cotton fabric, *ACS Nano*, **3**, 3325.

44. Sinha Ray, S., Yamada, K., Okamoto, M., and Ueda, K. (2002). Polylactide-layered silicate nanocomposite: A novel biodegradable material. *Nano Lett.*, **2**(10), 1093–1096.

45. Ghanbarzadeh, B., Oleyaei, S. A., and Almasi, H. (2015). Nanostructured materials utilized in biopolymer-based plastics for food packaging applications, *Critic Rev. Food Sci. Nutri.*, **55**(12), 1699–1723.

46. Abdollahi, M., Rezaei, M., and Farzi, G. (2012). A novel active bionanocomposite film incorporating rosemary essential oil and nanoclay into chitosan, *J. Food Eng.*, **111**(2), 343–350.

47. Nobile, M. D., Cannarsi, M., Altieri, C., Sinigaglia, M., Favia, P., Iacoviello, G., and D'agostino, R. (2004). Effect of Ag-containing nano-composite active packaging system on survival of *Alicyclobacillus acidoterrestris*, *J. Food Sci.*, **69**(8), E379–E383.

48. Lopez-Rubio, A., Gavara, R., and Lagaron, J. M. (2006). Bioactive packaging: Turning foods into healthier foods through biomaterials, *Trends Food Sci. Technol.*, **17**(10), 567–575.

49. Nachay, K. (2007). Analyzing nanotechnology, *Food Technol. (Chicago)*, **61**(1), 34–36.

50. Gander, P. (2007). The smart money is on intelligent design. *Food Manufacture*, 82, 1.

51. LeGood, P. A. U. L. and Clarke, A. (2006). Smart and active packaging to reduce food waste. *Smart Materials, Surfaces and Structures Network (SMART. mat)*.

52. Alfadul, S. M. and Elneshwy, A. A. (2010). Use of nanotechnology in food processing, packaging and safety: Review. *Af. J. Food, Agric. Nutri. Dev.*, **10**(6).

53. ElAmin, A. (2007). Carbon nanotubes could be new pathogen weapon, Available at http://www foodproductiondaily com/news/ng asp.

54. Berube, D. M., Searson, E. M., Morton, T. S., and Cummings, C. L. (2010). Project on emerging nanotechnologies: Consumer product inventory evaluated, *Nanotechnol. Law Business*, **7**, 152.

55. Yamamoto, O. (2001). Influence of particle size on the antibacterial activity of zinc oxide, *Int. J. Inorg. Mater.*, **3**(7), 643–646.

56. Jones, N., Ray, B., Ranjit, K. T., and Manna, A. C. (2008). Antibacterial activity of ZnO nanoparticle suspensions on a broad spectrum of microorganisms, *FEMS Microbiol. Lett.*, **279**(1), 71–76.

57. Chandramouleeswaran, S., Mhaske, S., Kathe, A., Varadarajan, P., Prasad, V., and Vigneshwaran, N. (2007). Functional behaviour of

polypropylene/ZnO–soluble starch nanocomposites, *Nanotechnology*, **18**(38), 385702.

58. Qi, L. F., Xu, Z. R., Jiang, X., Hu, C. H., and Zou, X. F. (2004). Preparation and antibacterial activity of chitosan nanoparticles, *Carbohydr. Res.*, **339**(16), 2693–2700.

59. Sanpui, P., Murugadoss, A., Prasad, P. V. D., Ghosh, S. S., and Chattopadhyay, A. (2008). The antibacterial properties of a novel chitosan-Ag-nanoparticle composite, *Int. J. Food. Microbiol.*, **124**(2), 142–146.

60. Appendini, P. and Hotchkiss, J. H. (2002). Review of antimicrobial food packaging, *Innov. Food Sci. Emerg. Technol.*, **3**(2), 13–26.

61. Labuza, T. P. (1996). An introduction to active packaging for foods, *Food Technol.*, **50**, 68–71.

62. Rhim, J. W. and Ng, P. K. W. (2007). Natural biopolymer-based nanocomposite films for packaging applications, *Crit. Rev. Food Sci. Nutr.*, **47**(4), 411–433.

63. Luo, P. G. and Stutzenberger, F. J. (2008). Nanotechnology in the detection and control of microorganisms, *Adv. Appl. Microbiol.*, **63**, 145–181.

64. Bouwmeester, H., Dekkers, S., Noordam, M. Y., Hagens, W. I., Bulder, A. S., De Heer, C., ten Voorde, S. E. C. G., Wijnhoven, S. W. P., Marvin, H. J. P., and Sips, A. J. A. M. (2009). Review of health safety aspects of nanotechnologies in food production, *Regul. Toxicol. Pharmacol.*, **53**(1), 52–62.

65. Liao, F., Chen, C., and Subramanian, V. (2005). Organic TFTs as gas sensors for electronic nose applications, *Sens. Actuat. B Chem.*, **107**(2), 849–855.

66. Verma, A. K., Singh, V. P., Vikas, P., Andersen, H. J., Assadi, P., Bastani, B., Fernandez, D., Brody, A. L., Cagri, A., and Ustunol, Z. (2007). Nanotechnology to play important and prominent role in food safety, *Am. J. Food Technol.*, **7**(8), 240–254.

67. Momin, J. K., Jayakumar, C., and Prajapati, J. B. (2013). Potential of nanotechnology in functional foods, *Emir. J. Food Agr.*, **25**(1), 10–19.

68. Fahlman, B. D. (2007). *Materials Chemistry*, Vol. 1, Springer Mount Pleasant, pp. 282–283.

69. Robinson, D. K. R. and Salejova-Zadrazilova, G. (2010), *Nanotechnology for Biodegradable and Edible Food Packaging*. Working paper version 1.

70. Mangiacapra, P., Gorrasi, G., Sorrentino, A., and Vittoria, V. (2006). Biodegradable nanocomposites obtained by ball milling of pectin and montmorillonites, *Carb. Polym.*, **64**(4), 516–523.

71. Zheng, J. P., Li, P., Ma, Y. L., and Yao, K. D. (2002). Gelatin/montmorillonite hybrid nanocomposite. I. Preparation and properties, *J. Appl. Polym. Sci.*, **86**(5), 1189–1194.

72. Decher, G. and Schlenoffm, J. B. (2003). *Multilayer Thin Films: Sequential Assembly of Nanocomposite Materials*, John Wiley & Sons-VCH, Weinheim, 543.

73. Dasgupta, N., Ranjan, S., Patra, D., Srivastava, P., Kumar, A., and Ramalingam, C. (2016). Bovine serum albumin interacts with silver nanoparticles with a "side-on" or "end on" conformation, *Chem. Biol. Interact.*, **253**, 100–111.

74. Bharadwaj, R., Mehrabi, A., Hamilton, C., Trujillo, C., Murga, M., Fan, R., Chavira, A., and Thompson, A. K. (2002). Structure–property relationships in cross-linked polyester clay nanocomposites, *Polymer*, **43**(13), 3699–3705.

75. Hogan, S. A. and Kerry, J. P. (2008). Smart packaging of meat and poultry products. In: *Smart Packaging Technologies for Fast Moving Consumer Goods* (Kerry, J. and Butler, P., Eds.), John Wiley & Sons, pp. 33–54.

76. Kerry, J. P., O'grady, M. N., and Hogan, S. A. (2006). Past, current and potential utilisation of active and intelligent packaging systems for meat and muscle-based products: A review, *Meat Sci.*, **74**(1), 113–130.

77. Fu, J., Park, B., Siragusa, G., Jones, L., Tripp, R., Zhao, Y., and Cho, Y. J. (2008). An Au/Si hetero-nanorod-based biosensor for Salmonella detection, *Nanotechnology*, **19**(15), 155502.

78. Hahn, M. A., Keng, P. C., and Krauss, T. D. (2008). Flow cytometric analysis to detect pathogens in bacterial cell mixtures using semiconductor quantum dots, *Anal. Chem.*, **80**(3), 864–872.

79. Yang, L. and Li, Y. (2005). Quantum dots as fluorescent labels for quantitative detection of *Salmonella typhimurium* in chicken carcass wash water, *J. Food. Protect.*, **68**(6), 1241–1245.

80. Tully, E., Hearty, S., Leonard, P., and O'Kennedy, R. (2006). The development of rapid fluorescence-based immunoassays, using quantum dot-labelled antibodies for the detection of Listeria monocytogenes cell surface proteins, *Int. J. Biol. Macromol.*, **39**(1–3), 127–134.

81. Wang, C. and Irudayaraj, J. (2008). Gold nanorod probes for the detection of multiple pathogens, *Small*, **4**(12), 2204–2208.

82. Chellaram, C., Murugaboopathi, G., John, A., Sivakumar, R., Ganesan, S., Krithika, S., and Priya, G. (2014). Significance of nanotechnology in food industry, *APCBEE Procedia,* **8**, 109–113.

83. Farhang, B. (2009). Nanotechnology and applications in food safety. In: *Global Issues in Food Science and Technology*, Academic Press, pp. 401–410.

84. Asadi, G. and Mousavi, S. M. (2006). Application of nanotechnology in food packaging. In: *13th World Congress of Food Sciences and Technology,* 2006, pp. 739–739.

85. Jaiswal, S. (2016). Applications of nanotechnology in food processing and packaging, *INROADS: Int. J. Jaipur Nat. Univ.,* **5**(1s), 45–48.

86. Couch, L. M., Wien, M., Brown, J. L., and Davidson, P. (2016). Food nanotechnology: Proposed uses, safety concerns and regulations, *Agro Food Industry Hi-Tech,* **27**(1), 36–39.

87. Pinto, R. J. B., Daina, S., Sadocco, P., Neto, C. P., and Trindade, T. (2013). Antibacterial activity of nanocomposites of copper and cellulose, *BioMed. Res. Int.,* **6**, 280512.

88. Rhim, J. W. and Ng, P. K. (2007). Natural biopolymer-based nanocomposite films for packaging applications, *Crit. Rev. Food Sci. Nutr.,* **47**(4), 411–433.

89. Bradley, E. L., Castle, L., and Chaudhry, Q. (2011). Applications of nanomaterials in food packaging with a consideration of opportunities for developing countries, *Trends Food Sci. Technol.,* **22**(11), 604–610.

90. Jain, A., Ranjan, S., Dasgupta, N., and Ramalingam, C. (2018). Nanomaterials in food and agriculture: An overview on their safety concerns and regulatory issues, *Crit. Rev. Food Sci. Nutr.,* **58**(2), 297–317.

91. Mahler, G. J., Esch, M. B., Tako, E., Southard, T. L., Archer, S. D., Glahn, R. P., and Shuler, M. L. (2012). Oral exposure to polystyrene nanoparticles affects iron absorption, *Nat. Nanotech.,* **7**(4), 264–271.

92. Aschberger, K., Micheletti, C., Sokull-Klüttgen, B., and Christensen, F. M. (2011). Analysis of currently available data for characterizing the risk of engineered nanomaterials to the environment and human health: Lessons learned from four case studies, *Env. Int.,* **37**(6), 1143–1156.

93. Silvestre, C., Duraccio, D., and Cimmino, S. (2011). Food packaging based on polymer nanomaterials, *Prog. Polym. Sci.,* **36**(12), 1766–1782.

94. Oberdorster, G., Sharp, Z., Atudorei, V., Elder, A., Gelein, R., Kreyling, W., and Cox. C. (2004). Translocation of inhaled ultrafine particles to the brain, *Inhal. Toxicol.,* **16**(6–7), 437–445.

95. Elder, A., Lynch, I., Grieger, K., Chan-Remillard, S., Gatti, A., Gnewuch, H., Kenawy, E., Korenstein, R., Kuhlbusch, T., Linker, F., and Matias, S. (2009). Human health risks of engineered nanomaterials. In: *Nanomaterials: Risks and Benefits* (Linkov, I. and Steevens, J., Eds.), Springer, Dordrecht, pp. 3–29.

96. Pekkanen, J., Peters, A., Hoek, G., Tiittanen, P., Brunekreef, B., de Hartog, J., Heinrich, J., Ibald-Mulli, A., Kreyling, W. G., Lanki, T., and Timonen, K. L. (2002). Particulate air pollution and risk of ST-segment depression during repeated submaximal exercise tests among subjects with coronary heart disease: The exposure and risk assessment for fine and ultrafine particles in ambient air (ULTRA) study, *Circulation*, **106**(8), 933–938.

97. Maynard, A. D. (2006). Nanotechnology: Assessing the risks, *Nano Today*, **1**(2), 22–33.

98. Rim, K. T., Song, S. W., and Kim, H. Y. (2013). Oxidative DNA damage from nanoparticle exposure and its application to workers' health: A literature review, *Safe Health Work*, **4**(4), 177–186.

99. Li, J. G., Li, W. X., Xu, J. Y., Cai, X. Q., Liu, R. L., Li, Y. J., Zhao, Q. F., and Li, Q. N. (2007). Comparative study of pathological lesions induced by multiwalled carbon nanotubes in lungs of mice by intratracheal instillation and inhalation, *Env. Toxicol. Int. J.*, **22**(4), 415–421.

100. Muller, J., Decordier, I., Hoet, P. H., Lombaert, N., Thomassen, L., Huaux, F., Lison, D., and Kirsch-Volders, M. (2008). Clastogenic and aneugenic effects of multi-wall carbon nanotubes in epithelial cells, *Carcinogenesis*, **29**(2), 427–433.

101. Yang, F. M., Li, H. M., Li, F., Xin, Z. H., Zhao, L. Y., Zheng, Y. H., and Hu, Q. H. (2010). Effect of nano-packing on preservation quality of fresh strawberry (*Fragaria ananassu* Duch. cv Fengxiang) during storage at 4°C, *J. Food Sci.*, **75**(3), C236–240.

102. Jovanovic, B. (2015). Critical review of public health regulations of titanium dioxide, a human food additive, *Integr. Env. Assess Manag.*, **11**(1), 10–20.

103. EFSA Panel on food contact materials, enzymes, flavourings and processing aids (CEF). (2012). Scientific opinion on the safety evaluation of the substance, titanium nitride, nanoparticles, for use in food contact materials, *EFSA J.*, **10**(3), 2641.

104. Silvestre, C., Pezzuto, M., Cimmino, S., and Duraccio, D. (2013). Polymer nanomaterials for food packaging: Current issues and future trends. In: *Ecosustainable Polymer Nanomaterials for Food Packaging: Innovative Solutions, Characterization Needs, Safety and Environmental Issues* (Silvestre, C. and Cimmino, S., Eds.), CRC Press, pp. 2–28.

105. Echegoyen, Y. and Nerín, C. (2013). Nanoparticle release from nano-silver antimicrobial food containers, *Food Chem. Toxicol.*, **62**, 16–22.

106. Ding, Y. and Povey, M. (2005). Nanotech discovery promises safer food packaging. http://www.foodprductiondaily.com/news/ng.asp?n=59980- discovery-pro.

107. LaCoste, A., Schaich, K. M., Zumbrunnen, D., and Yam, K. L. (2005). Advancing controlled release packaging through smart blending, *Pack. Technol. Sci. Int. J.*, **18**(2), 77–87.

108. Nachay, K. (2007). Analyzing nanotechnology, *Food Technol. (Chicago)*, **61**(1), 34–36.

109. Bowman, D. M. and Hodge, G. A. (2007). A small matter of regulation: An international review of nanotechnology regulation, *Columbia Sci. Technol. Law Rev.*, **8**(1), 1–36.

110. U.K.RS/RAE. (2004). Nanoscience and nanotechnologies: Opportunities and uncertainties, Available at http://www.nanotec.org.uk/finalReeport.html (accessed 17 January 2008).

111. U.S. FDA. (2007). Nanotechnology: A Report of the U.S. Food and Drug Administration Nanotechnology Task Force. Available at http://www.fda.gov/nanotechnology/taskforce/report2007.html (accessed 15 January 2008).

112. Bowman, D. M. and Hodge, G. A. (2006). Nanotechnology: Mapping the wild regulatory frontier, *Future*, **38**(9), 1060–1073.

113. Taver, T. (2006). Food nanotechnology. *Food Technol. Nov.*, Available at http://members.ift.org/NR/rdonlyres/E725D811-3620-4CC1-8AAD-40E2BF66CE7E/0/1106 Nano.pdf.

Multiple Choice Questions

1. Who first used the term nanotechnology and when?

    (a) Richard Feynman, 1959

    (b) Norio Taniguchi, 1974

    (c) Eric Drexler, 1986

    (d) Sumio Iijima, 1991

2. What exactly is a quantum dot?

   (a) A semiconductor nanostructure that confines the motion of conduction band electrons, valence band holes, or excitons in all three spatial directions.

   (b) The sharpest possible tip of an atomic force microscope.

   (c) A fictional term used in science fiction for the endpoints of wormholes.

   (d) Unexplained spots that appear in electron microscopy images of nanostructures smaller than 1 nm.

3. What is the 2017 budget for the US National Nanotechnology Initiative?

   (a) $587 million
   (b) $917 million
   (c) $1.4 billion
   (d) $2.1 billion

4. Nanorobots (nanobots)

   (a) Do not exist yet

   (b) Exist in experimental form in laboratories

   (c) Are already used in nanomedicine to remove plaque from the walls of arteries

   (d) Will be used by NASA in the next unmanned mission to Mars

5. Which one of these condiments is unique due to the nanoscale interactions between its ingredients?

   (a) Ketchup
   (b) Mustard
   (c) Mayonnaise
   (d) All of the above

6. Which of these consumer products is already being made using nanotechnology methods?

   (a) Fishing lure
   (b) Golf ball
   (c) Sunscreen lotion
   (d) All of the above

7. Which one of these statements is NOT true?

   (a) Gold at the nanoscale is red.

   (b) Copper at the nanoscale is transparent.

   (c) Silicon at the nanoscale is an insulator.

   (d) Aluminum at the nanoscale is highly combustible.

8. How many oxygen atoms lined up in a row would fit in a 1 nm space?

   (a) None; an oxygen atom is bigger than 1 nm

   (b) One

(c) Seven

(d) Seventy

9. Nanoscience is the study of

    (a) Phenomena on the scale of 1–100 nm

    (b) Phenomena on the scale of electrons

    (c) Phenomena on the scale of single atoms

    (d) Phenomena on the scale of molecules

10. Which of the following examples are in the nanometer scale?

    (a) A red blood cell   (b) Pollen grain

    (c) DNA molecules   (d) Human hair

11. At the nanoscale, a material can have different properties compared to its bulk form.

    (a) True   (b) False

12. Nanomaterials are not the only human-made materials. Can you remember some examples of "non-intentionally made" nanomaterials, i.e., a nanomaterial that you can find in nature?

    (a) Bacteria   (b) Nanotransistors

    (c) Gold nanoparticles   (d) Proteins

13. Nanoparticles are now being used in some commercial products.

    (a) No   (b) Yes

14. Nanomaterials are

    (a) Very promising in many applications and toxicology; studies are now under way to assess their safety

    (b) Have same properties than normal materials and do not require toxicology studies

    (c) Technology development should take into consideration social and ethical questions

    (d) None of above

15. How many nanometers are there in a meter?

    (a) One million

    (b) One billion

    (c) One trillion

    (d) This is not a real unit of measurement

16. In what year did Richard Feynman give his lecture "There's Plenty of Room at the Bottom," which helped inaugurate the study of nanotechnology?

    (a) 1949   (b) 1959

    (c) 1969   (d) 1979

17. Some research has raised concerns that carbon nanotubes used in products such as high-tech spray paint might share qualities with this carcinogen.

    (a) Asbestos   (b) Coal tar pitch

    (c) Wood dust   (d) Formaldehyde

18. In this 1986 book, K. Eric Drexler helped popularize the promise of nanotechnology.

    (a) Radical abundance   (b) Engines of creation

    (c) The diamond age   (d) Too small for school

19. Which of these terms describes an apocalyptic nanotechnological scenario?

    (a) Ochre jelly   (b) Gray goo

    (c) Gelatinous   (d) Green slime

20. Which of these fields has not drawn on nanotechnology?

    (a) Visual art

    (b) Computing

    (c) Textile manufacturing

    (d) They have all drawn on nanotechnology

21. Which of these is not currently a goal of nanomedicine?

    (a) Targeting tumors

    (b) Re-growing bones

    (c) Increasing lung capacity

    (d) Connecting brain with computers

22. Nanomaterials are of the following category?

    (a) Tubes   (b) Rods

    (c) Fibers   (d) All of above

23. Antimicrobial activity in packaging is done by which of following type?

    (a) Nanocomposites   (b) Nanosensors

    (c) Both of above   (d) None of these

24. Carbon nanotubes are

    (a) Carbon molecules with a unique structure and ability to produce energy

    (b) Carbon molecules with a unique structure and ability to create new molecules

    (c) Carbon molecules with a unique structure and ability to bond with each other

    (d) None of above

25. A nanometer is

    (a) A unit of measure equal to one millionth of a meter

    (b) A unit of measure equal to one billionth of a meter

    (c) A unit of measure equal to one trillionth of a meter

    (d) A unit of measure equal to one billionth of a millimeter

Answer Key

1. (b)   2. (a)   3. (c)   4. (a)   5. (c)   6. (d)   7. (d)

8. (c)   9. (a)   10. (c)   11. (a)   12. (d)   13. (b)   14. (c)

15. (b)   16. (b)   17. (a)   18. (b)   19. (b)   20. (d)   21. (c)

22. (d)   23. (c)   24. (c)   25. (b)

Short Answer Questions

1. What type of nanotechnique is used to provide mechanical strength to the food packaging?

2. What are the health effects of using nanotechnology in food production?

3. What do you understand by active packaging?

4. What is the role of ethylene in fruits and vegetables?

5. Define the RFID system.

6. What is the role of freshness indicator?

7. Are there specific health risks from nanoproducts?

8. How nanotechnology helps in antimicrobial packaging?

9. What are the applications of silver nanoparticles in food technology?

10. Write a note on oxygen scavengers.

Long Answer Questions

1. Write a note on commercially available food packaging systems.

2. What are the regulations in different countries for the usage of nanotechnology in food packaging?

3. Differentiate between active packaging and intelligent packaging? Explain giving examples.

4. Write a detailed note on improved food packaging using nanotechnology.

5. How nanotechniques help in providing good shelf life and microbe free food.

Chapter 13

Biomedical Diagnostics through Nanocomputing

Varun Sapra,[a] Luxmi Sapra,[b] Jasminder Kaur Sandhu,[b] and Gunjan Chhabra[a]

[a]University of Petroleum and Energy Studies, Dehradun,
Uttarakhand, India
[b]Chitkara University Institute of Engineering and Technology,
Chitkara University, Punjab, India
luxmi.sapra@chitkara.edu.in

A decade ago, communicable diseases were a major threat in the developing countries, but there is a paradigm shift in the last few years and now noncommunicable diseases such as cancer and cardiac diseases account for 80% of deaths worldwide. The palpable reason for this change is due to the increased affordability of people for medical resources. These communicable diseases can be cured only if they can be diagnosed in their early stages. Although there are multiple methods, both invasive and noninvasive, for disease diagnostics and some of them are really accurate, such as angiography, which is an invasive technique for diagnosing

Nanotechnology: Principles and Applications
Edited by Rakesh K. Sindhu, Mansi Chitkara, and Inderjeet Singh Sandhu
Copyright © 2021 Jenny Stanford Publishing Pte. Ltd.
ISBN 978-981-4877-43-5 (Hardcover), 978-1-003-12026-1 (eBook)
www.jennystanford.com

cardiovascular diseases, yet such tests are expensive and require high clinical expertise and infrastructure. In the last few years, researchers have worked extensively for formulating noninvasive methods using artificial intelligence. It is an umbrella term, which contains many computationally intelligent methods such as machine learning and deep learning. The espousal of artificial intelligence with nanocomputing or Internet of Things opened another chapter in the ever-growing technology industry in the form of Industry 4.0. This new tide has opened gates for researchers in many fields, such as optimization of production and monitoring processes in almost all major sectors such as manufacturing, mechanical engineering, and chemical industry, to name a few. The medical field is still a growing industry, especially in the case of nanocomputing. The amalgamation of the two dominant technologies—machine learning, which has the ability to find complex relationships within the huge data, and nanocomputing, which is capable of encompassing the physical processes to computational methods—can be considered another novel approach for the early detection of medical diseases and data communication.

13.1 Introduction

Early-stage diagnosis of medical diseases is still a challenge in modern medicine. Noncommunicable chronic diseases such as cardiovascular diseases or cancer are spreading at a rapid pace. As per the report published by the WHO, 71% deaths occur due to noncommunicable diseases [1]. The projected global deaths till 2030 predicted by the WHO show a rise in noncommunicable diseases [2]. Figure 13.1 shows the global projections of diseases from 2004 to 2030.

Early diagnosis is required in these diseases for the survival of the patient and successful prediction of the disease. A number of invasive and noninvasive techniques are available for the successful identification of these diseases, such as angiography for cardiovascular diseases; it is a type of invasive technique that requires good medical expertise as well as a laboratory setup to perform. Moreover, the technique is costly, painful, and not easily available in rural areas. Similarly, for the successful detection of cancer, there are methods such as mammography and fine needle aspiration cytology (FNAC). Between the two frequently used methods, mammography is a preferred method for the detection of breast cancer, but it also needs medical expertise and analysis skills of the clinical practitioner and the results are not 100% accurate [3]. Such complications encouraged researchers to develop noninvasive methods for the early and accurate diagnosis of diseases.

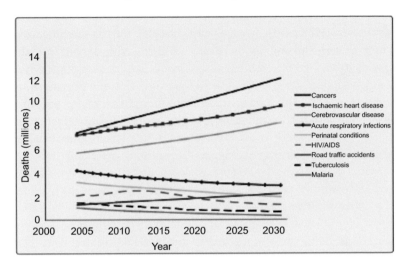

Figure 13.1 Major causes of mortality by 2030 [2].

In the past one decade, many researchers have worked on a number of possibilities and developed many noninvasive models for the early identification of diseases. Liu et al. proposed a data-mining-based method for the prediction of the survivability of patients positive with breast cancer. The authors applied decision table (DT)-based predictive models with the implementation of C5 and bagging techniques for classification, and their proposed model could achieve an accuracy of 86.52% [4]. Karabatak et al. suggested association rules and neural-network-based hybrid model for the successful detection of breast cancer. They used the Wisconsin dataset for the experiment. In the first phase, the authors implemented association rules for dimensionality reduction and then carried out classification using neural network. The proposed model was much more efficient for the diagnosis of breast cancer [5].

In another prognostic study, Marateb and Goudarzi [6] proposed a neuro-fuzzy-based classifier for the detection of coronary artery disease (CAD). They implemented their model on a data acquired from the UCI repository. The authors reduced the dimensionality of the dataset using sequential feature selection and multiple logistic regression techniques. The selected features were age, induced angina, thallium, exercise, slope, and a number of major vessels colored by fluoroscopy. They further implemented multiple logistic regression (MLR) with neuro-fuzzy classifier for the classification. Their model showed an accuracy of 84% with sensitivity and specificity of 79.4% and 89%, respectively. Banerjee et al. proposed another noninvasive novel technique for the identification of CAD using the two-phase classification approach. In the first stage, the clinical parameters and the demographic information of the patients were used to produce a rule-based engine. After generating the rule-based engine, the authors extracted two sets of signals, i.e., phonocardiogram and photoplethysmogram signals, and the support vector machine technique was used to perform classification. Their model achieved a sensitivity of 92% and a specificity of 90% [7]. Arabasadi et al., in 2017, proposed another method of using genetic algorithm for determining initial weights and then implemented a neural-network-based approach to present a basis for the detection of CAD. They implemented the Z-Alizadeh Sani dataset having records of 303 subjects to compute information gain, principal component analysis. The proposed method achieved higher accuracy [8]. Kora et al., in their work, produced a system for the early detection of CAD. Their proposed clinical support system was based on fuzzy approach. They identified significant features by reducing the dimensionality of the datasets retrieved from the Hungarian Institute of Cardiology, Budapest, and the Cleveland clinical database. The datasets contained details of 597 subjects and have 76 attributes. The author implemented a neuro-fuzzy hybrid approach to compute the prediction value for the occurrence of the disease using linguistic variables and membership function. They further compared their model with other techniques such as K-nearest neighbor (KNN) and support vector machine (SVM). The integrated neuro-fuzzy logic achieved the highest accuracy of 99.3% in comparison to KNN (73.1%) and SVM (80.9%) [9]. Verma et al. (2016) proposed another method for coronary disease diagnostic using noninvasive clinical parameters. They proposed a blended approach where the authors reduced the feature space using correlation feature subset (CFS) for reducing the dimensionality of the biomarkers with particle swarm optimization (PSO) search. Further, they implemented classification techniques, and their method improved the prediction accuracy of diagnostic models [10].

Despite of the development of all these methods, we are just able to diagnose these diseases to a certain extent, but many diseases, including cancer and cardiovascular diseases, originate at the molecular level from alterations and changes to normal cellular arrangement and metabolic pathways [11]. Accurate and early diagnosis in such cases is still complicated due to the limited knowledge and usage of biosensors and molecular probes capable of rapidly recognizing the distinct molecular features of these diseases [12]. In the last few years, the demand for disposable devices with fast response times, user friendliness, and cost efficiency for the successful diagnosis of these diseases has grown. The capability of nanomaterials and nanobiosensors or nanorobots to interact with biomolecules and to convert those interactions into signals has opened a new gateway of early-stage diagnostic techniques.

13.2   Nanotechnology in Medical Science

Nanotechnology, the term coined by Eric Drexler, is used nowadays in the medical domain not only for the diagnosis of diseases but also for the successful treatment of the diseases. Nanotechnology can be defined as a hybrid science at the level of atoms and molecules with the amalgamation of biotechnology and nanotechnology to provide a productive platform for the development of systems exhibiting physical, biological, and chemical properties with clinical relevance. "Nanomedicine" can be considered a branch of science that has the ability to deal with diagnosing, repairing of specific diseased cells, and preventing diseases with the help of genetic engineering and nanomaterials like the natural healing process.

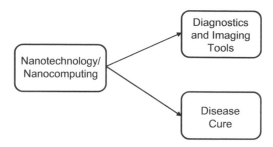

Figure 13.2 Classification of nanodiagnostics.

The recent advances in nanotechnology have revolutionized the medical domain. Figure 13.2 shows the classification of nanotechnology as diagnostics and imaging tools and for disease cure. Nanodiagnostics can be further classified into in vitro and in vivo diagnostics. In the in vitro diagnosis, disease identification is carried out using selective human tissues or fluids. These tissues can be extracted using specialized nanodevices for multiple analysis, thus increases the efficiency and performance of nanomedicines. However, in vivo deals with nanodevices that can be ingested into the human body to diagnose the early presence of diseases [13].

13.2.1 Nanorobots

Nanorobots are nanoelectromechanical systems constructed of nanoscale components. They contain self-propelled nanomotors ranging from 0.1 to 10 μm for performing a specific task of delivering a drug to the diseased cells at the nanoscale dimensions of 1–100 nm in the human body for providing treatment against pathogens. With the development of this new technology, not only the medical field is benefitted but also this technology can be applied for energy conservation, food processing, oil and gas industry, and optical displays, to name a few. Figure 13.3 displays the various application areas that are using nanotechnology. They are designed to work efficiently at atomic or molecular levels in different industrial fields [14, 15].

Further, nanobots can be explained as any structure that can perform any of the following tasks: sensing, signaling, manipulation, actuation, processing of the information, and propulsion at the nanoscale [16]. The application of nanorobots in the domain of healthcare has become popular in the last few years because of their ability to identify diseases with least preconception to the healthy cells and curing with reconstructive treatment at the molecular level. Figure 13.4 gives an overview of different applications of nanorobotics in the field of medicine.

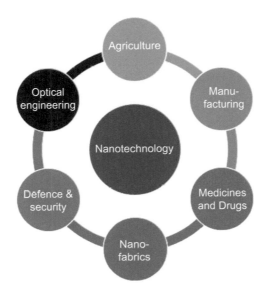

Figure 13.3   Applications of nanocomputing/nanorobots.

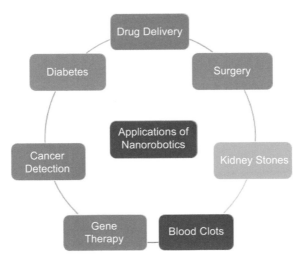

Figure 13.4   Applications of nanorobotics in the field of medicine.

13.2.2   Structure and Design of Nanorobots

Generally, we have two types of nanorobots: The first category of nanorobots are known as organic, also called bionanorobots, and the other one is called inorganic nanorobots. The bionanorobots consist of proteins, viruses, and polynucleotides and are less toxic in nature. They are easy to produce with the use of genetic programming and can easily communicate with other organic systems. Inorganic nanorobots are made up of metals or synthesized proteins. Inorganic nanorobots are relatively less toxic in nature with good component behavior and are easy to program. Some of the disadvantages with this type of inorganic robots are difficulty in self-reproduction and high expense when compared to organic nanorobots.

South Korean scientists from the Chonnam National University were successful in developing an organic nanorobot called Bacteriobot for detecting toxicity. They used salmonella bacteria, which was genetically changed to hide toxicity. This nanorobot is attracted to molecules released by cancer cells and releases the drug [17]. Juul et al. presented their research idea of nanorobots that can carry medicine and can be controlled for opening and closing of bots based on the atmospheric temperature [18]. A nanorobot has been developed by a group of researchers at the Max Plank Institute of Germany to detect protozoa, which is light activated and made up of liquid crystal elastomers [17]. A Professor from the University of Toronto in Canada is working on a project for the development of robots to collect tissue biopsies or carry drug capsules inside the body.

13.2.2.1   Components of nanorobots

Some of the common components in nanorobots include

1. **Power supply:** Source of energy for operating and performing function.

2. **Sensors:** Tiny devices to receive the signal for movement.

3. **Propeller:** Devices used to drive in the case of opposite blood flow direction.

4. **Payload:** A small section that has the ability of holding the drug or medicine and can release it at the location of injury.

5. **Microcamera:** A miniature version of the camera for navigating the nanorobot through the human body.

6. **Lasers:** It is used to burn harmful materials such as cancer cells or blood clots.

13.2.3 Types of Nanorobots

Some of the nanorobots that have been developed for medical purposes are as follows:

Microbivore nanorobots: Microbivore nanorobots, also known as nanorobotic phagocytes, behave just like artificial white blood cells in the human body. It is a spheroid-shaped device with 3.4 μm and 2.0 μm diameters along its major and minor axes, respectively. The main purpose or use of these nanobots is to check the bloodstream for unwanted pathogens, including bacteria and viruses, at a very fast pace and thus eliminate infections within minutes. It uses the process called phagocytosis to absorb the pathogens in the bloodstream [19].

Respirocyte nanorobots: These nanorobots have been designed to serve as artificial mechanical red blood cells in our bodies. They are spherical-shaped devices with 1 μm diameter. They are used as a carrier for carrying oxygen and carbon dioxide in the human body. One such nanorobot consists of 18 billion atoms, which are arranged in diamond-shaped pressure tanks that have the capability of storing 3 billion oxygen and carbon dioxide molecules [20]. It has three different types of propellers: the first to store oxygen, the second to extract all the carbon dioxide, and the third to take glucose from the bloodstream.

Clottocyte nanorobots: Hemostasis is the process of stopping blood from a cut or a damage to the endothelium cells by platelets. These platelets create a clot and stick to the wound. This whole process takes up to 2–5 min, and depending on the size of the cut, a significant blood loss can occur. The theoretically designed clottocyte is a spherical nanorobot of approximately 2 μm diameter, which would store and release fibers at the infected place where they come together to form a clot to stop the bleeding.

Cellular repair nanorobots: These nanorobots have been created to aid medical practitioners in performing surgeries more accurately and with fewer complications. These bots work at the cellular level and can be helpful in controlling the damage caused by scalpel [21].

Chromallocyte nanorobots: The chramallocyte bots are used for replacing chromosomes in the human body and are effective in minimizing the effects of genetic diseases. They are also helpful in fighting or preventing ageing in the human body.

13.2.4 Applications of *Nanorobots* in Medicine

1. **Cancer detection and treatment (oncology):** In the last one decade, cancer has become one of the biggest reasons for mortality throughout the world. It is important to diagnose it accurately at the earliest stage. Many different approaches have been followed by medicos to diagnose cancer in its early stages, but still more efficient methods are being searched for reducing the mortality and morbidity associated with oncological conditions and their treatment. Nanorobots with implanted chemical biosensors have been proved as an adjunct tool for detecting, diagnosing, and treatment of cancer cells in early stages in the human body. One most widely used therapy to treat cancer is chemotherapy, but we cannot control its effects on other body parts. It can damage healthy tissues as well, but nanorobots can be sent to the targeted tissues where the robots can release their drugs and target only the diseased tissues. Figure 13.5 shows the usage of nanobots in cancer treatment.

2. **Detection and treatment of CAD:** CAD is one of the foremost reasons of deaths in India as well as throughout the world. It happens due to the accumulation of plaque in the arteries, and over the time, these arteries get harden and narrow, which restricts the flow of blood to vital organs and hence increases the chances of heart attack or stroke.

   Theoretically, these blockages can be cleared by using nanorobots. In the Drexel University, researchers have developed a microrobotic technology for drilling through the clogged arteries. In the worst situation if these bots are not enough to clear the entire blockage, they can still reduce the severity or the chances of getting heart attack or stroke in the human body. Figure 13.6 shows the process of drilling through the blocked arteries using nanobots.

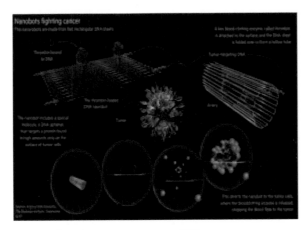

Figure 13.5   Nanobots in cancer treatment [22].

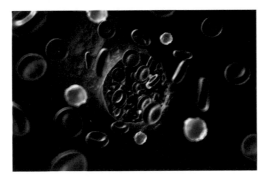

Figure 13.6   Nanobots drill through blocked arteries [23].

3. **Dentistry:** Nanorobots have been used significantly in dentistry. Virtually all aspects of dentistry can make use of nanorobots to improve the treatment quality. These nanorobots are helpful to medical practitioners in straightening of irregular set of teeth, teeth durability, desensitizing tooth, and oral anesthesia. In the process of root canal, they play a crucial role as they are equipped with tiny cameras, which can help in visualization and can also release drugs at the targeted

location. Further, they can be used in toothpaste for improving daily dental care [24]. Figure 13.7 shows the applications of nanobots in dentistry.

Figure 13.7   Applications of nanorobotics in dentistry [25].

4. **Nanorobots as antibodies:** Nanobots are also used to boost the immune systems of individuals who have weak immunity by destroying the dangerous foreign materials in the human body.

Figure 13.8   Nanobots monitoring blood glucose levels [26].

5. **Bots to monitor blood glucose level:** These bots are used to monitor the blood sugar level by using embedded nanobiosensors. These are tiny sensors consisting of tiny chips, which use electrical signal. These sensors are connected

to the mobile phones of patients and transmit a special signal to take necessary action or prescription regarding the disease. In some of the cases, these nanobots not only transmit signals but also carry drugs, and when they detect the sign of a disease, drug is released to handle the disease. The merit of these bots is that patients with diabetes do not need to prick their finger several times for blood samples to monitor glucose level. The glucose can be monitored constantly by using medical nanobots. Figure 13.8 shows nanobots in action for monitoring blood glucose levels in a patient.

13.3 Conclusion

Although the healthcare sector is one of the fastest evolving sectors with a plenty of automation, it still needs a lot of attention. Diagnosing a disease in its early stages is still a challenge and is a key requirement for saving patients and improving the quality of treatment. Various invasive and noninvasive methods are used by researchers for the early diagnosis of chronic diseases such as cancer and CAD. Some of the researches have been very useful and helped medical practitioners for taking early decisions regarding the treatment of patients, but we still need some more mechanisms that can be used as tools for the diagnosis and treatment of diseases at a very fast pace. The espousal of artificial intelligence methods with nanocomputing provides enormous advantages over conventional medicine, such as low cost, fast recovery actions, and low or almost no invasion. With nanobots inside a human body, we can expect a better life expectancy.

References

1. Bennett, J. E., Stevens, G., Bonita, R., Rehm, J., Kruk, M., Riley, L., Dain, K., Kengne, A., Chalkidou, K., Beagley, J., Kishore, S., Chen, W., Saxena, S., Bettcher, D., Grove, J., Beaglehole, R., and Ezzati, M. (2018). NCD countdown 2030: Worldwide trends in non-communicable disease mortality and progress towards Sustainable Development Goal target 3.4. *The Lancet*, **392**, 1072–1088.

2. World Health Organization. (2008). *The Global Burden of Disease: 2004 Update*. World Health Organization.

3. Elmore, J. G., Wells, C. K., Lee, C. H., Howard, D. H., and Feinstein, A. R. (1994). Variability in radiologists' interpretations of mammograms. *New England Journal of Medicine*, **331**(22), 1493–1499.

4. Liu, Y. Q., Wang, C., and Zhang, L. (2009). Decision tree based predictive models for breast cancer survivability on imbalanced data. In *2009 3rd International Conference on Bioinformatics and Biomedical Engineering*, Beijing, 2009, pp. 1–4, IEEE.

5. Karabatak, M. and Ince, M. C. (2009). An expert system for detection of breast cancer based on association rules and neural network. *Expert Systems with Applications*, **36**(2), 3465–3469.

6. Marateb, H. R. and Goudarzi, S. (2015). A noninvasive method for coronary artery diseases diagnosis using a clinically-interpretable fuzzy rule-based system. *Journal of Research in Medical Sciences*, **20**(3), 214.

7. Banerjee, R., Bhattacharya, S., Bandyopadhyay, S., Pal, A., and Mandana, K. M. (2018). Non-invasive detection of coronary artery disease based on clinical information and cardiovascular signals: A two-stage classification approach. In *2018 IEEE 31st International Symposium on Computer-Based Medical Systems (CBMS)*, Karlstad, 2018, pp. 205–210, IEEE.

8. Arabasadi, Z., Alizadehsani, R., Roshanzamir, M., Moosaei, H., and Yarifard, A. A. (2017). Computer aided decision making for heart disease detection using hybrid neural network-Genetic algorithm. *Computer Methods and Programs in Biomedicine*, **141**, 19–26.

9. Kora, P., Meenakshi, K., Swaraja, K., Rajani, A., and Islam, M. K. (2019). Detection of cardiac arrhythmia using fuzzy logic. *Informatics in Medicine Unlocked*, **17**, 100257.

10. Verma, L., Srivastava, S., and Negi, P. C. (2016). A hybrid data mining model to predict coronary artery disease cases using non-invasive clinical data. *Journal of Medical Systems*, **40**(7), 178.

11. DeBerardinis, R. J., Lum, J. J., Hatzivassiliou, G., and Thompson, C. B. (2008). The biology of cancer: Metabolic reprogramming fuels cell growth and proliferation. *Cell Metabolism*, **7**, 11–20.

12. Hu, Y., Fine, D. H., Tasciotti, E., Bouamrani, A., and Ferrari, M. (2011). Nanodevices in diagnostics. *Wiley Interdisciplinary Reviews: Nanomedicine and Nanobiotechnology*, **3**(1), 11–32.

13. Mallanagouda, P., Dhoom, S. M., and Sowjanya, G. (2008). Future impact of nanotechnology on medicine and dentistry, *Journal of Indian Society of Periodontology*, **12**(2), 34–40.

14. Meena, K., Monika, N., and Sheela, M. (2013). Nanorobots: A future medical device in diagnosis and treatment. *Research Journal of Pharmaceutical, Biological and Chemical Sciences*, **4**(2), 1229–1307.

15. Sarath, K. S., Nasim, B. P., and Abraham, E. (2018). Nanorobots: A future device for diagnosis and treatment. *Journal of Pharmacy and Pharmaceutics*, **5**, 44–49.

16. Mazumder, S. (2014). Nanorobots: Current state and future perspectives. *International Journal of Development Research*, **4**(6).

17. da Silva Luz, G. V., Barros, K. V. G., de Araújo, F. V. C., da Silva, G. B., da Silva, P. A. F., Condori, R. C. I., and Mattos, L. (2016). Nanorobotics in drug delivery systems for treatment of cancer: A review. *Journal of Materials Science and Engineering A*, **6**, 167–180.

18. Juul, S., Iacovelli, F., Falconi, M., Kragh, S. L., Christensen, B., Frøhlich, R., Franch, O., Kristoffersen, E. L., Stougaard, M., Leong, K. W., Ho, Y. P., Sørensen, E. S., Birkedal, V., Desideri, A., and Knudsen, B. R. (2013). Temperature-controlled encapsulation and release of an active enzyme in the cavity of a self-assembled DNA nanocage. *ACS Nano*, **7**(11), 9724–9734.

19. Manjunath, A. and Kishore, V. (2014). The promising future in medicine: Nanorobots. *Biomedical Science and Engineering*, **2**(2), 42–47.

20. Freitas, R. A. (2005). Current status of nanomedicine and medical nanorobotics. *Journal of Computational and Theoretical Nanoscience*, **2**(1), 1–25.

21. https://www.dummies.com/education/science/nanotechnology/types-of-nanorobots-being-developed-for-use-in-healthcare/ [Last accessed on 15-03-2020].

22. Vishwanathan, S. N. and Anitha, S. (2019). Nanobots in medical field: A critical overview. *International Journal of Engineering Research & Technology*, **8**(12).

23. https://www.seeker.com/corkscrew-nanobots-drill-though-blocked-arteries-1769963310.html.

24. Saadeh, Y. and Vyas, D. (2014). Nanorobotic applications in medicine: Current proposals and designs. *American Journal of Robotic Surgery*, **1**(1), 4–11.

25. Verma, S. K. and Chauhan, R. (2014). Nanorobotics in dentistry: A review. *Indian Journal of Dentistry*, **5**, 62–70.

26. Zehe, A., Thomas, A., and Ramírez, A. (2016). Nanotechnology tackles problems with noninvasive glucose monitoring. *European Pharmaceutical Review*, **21**(2), 68–71.

Multiple Choice Questions

1. The size of nanoparticles is between _____ nm.
   - (a) 1 to 10
   - (b) 0.1 to 100
   - (c) 0.01 to 1
   - (d) 1 to 100

2. The term nanotechnology was coined by whom?
   - (a) Richard Feynman
   - (b) Norio Taniguchi
   - (c) Eric Drexler
   - (d) Dennis Heymann

3. Pick up the correct answer.
   Nanotechnology is not used in
   - (a) Agriculture
   - (b) Defense and security
   - (c) Optical engineering
   - (d) Education business

4. Nanobots are capable of performing the following tasks:
   - (a) Sensing, actuation, smelling
   - (b) Signaling, information processing, sensing
   - (c) Manipulation, propulsion, prediction
   - (d) All of the above

5. The nanorobots can be classified as
   - (i) Organic
   - (ii) Semi detached
   - (iii) Inorganic
   - (iv) Embedded
   - (a) (i) and (ii)
   - (b) (i) and (iii)
   - (c) (ii), (iii), and (iv)
   - (d) All of the above

6. Which nanobots are less toxic in nature?
   - (a) Organic
   - (b) Semi detached
   - (c) Inorganic
   - (d) Embedded

7. Bacteriobots were discovered by
   - (a) Canadian scientists
   - (b) South Korean scientists
   - (c) European scientists
   - (d) British Research Lab

8. Liquid crystal elastomers were used to create nanobots for the diagnosis of which disease?
   - (a) Cancer
   - (b) Protozoa
   - (c) Heart disease
   - (d) Diabetes

9. The component responsible for the movement of nanorobots inside the human body is

(a) Sensor   (b) Payload

(c) Propeller   (d) Lasers

10. The nanorobots that work in enhancing the immunity of human body are

(a) Respirocyte   (b) Microbivore

(c) Clottocyte   (d) Monocyte

11. The process of blocking blood from a cut is known as

(a) Hemostasis   (b) Prognostics

(c) Oncology   (d) Dentistry

12. The clottocyte nano robots are

(a) 1 μm in diameter   (b) 2 μm in diameter

(c) 3 μm in diameter   (d) 0.5 μm in diameter

13. The nanorobots that can help medicos in surgical procedures are

(a) Chromallocyte   (b) Clottocyte

(c) Cellular repair nanorobots   (d) Respirocyte

14. Which of the diagnostic methods is invasive for the detection of heart disease?

(a) Angiography   (b) Colorography

(c) ECG   (d) MRI

15. Which of the following statements do not relate to nanotechnology?

(a) General purpose

(b) Also known as Green technology

(c) Rearrangement of atoms

(d) Described by Newtonian Mechanics

16. _____ is the most significant property of nanomaterials.

(a) Temperature   (b) Force

(c) Friction   (d) Pressure

17. _____ is the hardest material originating in nature.

(a) Iron   (b) Diamond

(c) Steel   (d) Quartz

18. _____ is the general name for a class of structures made up of rolled carbon lattices.

(a) Nanorods   (b) Nanotubes

(c) Nanosheets   (d) Nanoscale

19. Nano signifies the magnitude of
    (a) 10^{-6}
    (b) 10^{-3}
    (c) 10^{-9}
    (d) 10^{-12}
20. Nanorobots are not used in dentistry for
    (a) Teeth durability
    (b) Sensitizing tooth
    (c) Straightening of irregular teeth
    (d) Cutting of stitches

Answer Key

1. (d)   2. (c)   3. (d)   4. (b)   5. (b)   6. (c)   7. (b)
8. (b)   9. (c)   10. (b)   11. (a)   12. (b)   13. (c)   14. (a)
15. (a)   16. (c)   17. (b)   18. (b)   19. (c)   20. (d)

Short Answer Questions

1. What is the need of nanocomputing in medicines?
2. What are the different noninvasive methods researchers followed to diagnose cancer and coronary artery disease?
3. Discuss the classification of nanodiagnostic methods.
4. Discuss the functioning of "Bacteriobots" with their merits and demerits.
5. What are the design considerations for nonorobotics?
6. How machine learning methods are useful in nanotechnology?

Long Answer Questions

1. Discuss the applications of nanorobotics in the field of medicine.
2. Illustrate the components of nanorobots. Elaborate the functioning of each component.
3. What are the applications of nanobots in healthcare?
4. Justify the statement "early stage diagnosis of medical diseases is still a challenge in modern medicine."

Chapter 14

Nanofluids: Current Applications and Future Challenges

Balwinder Kaur,[a] Subhash Chand,[b] and Balraj Saini[c]
[a]*Department of Chemistry, Punjabi University, Patiala, Punjab, India*
[b]*Lajpat Rai DAV College, Jagraon, Punjab, India*
[c]*Chitkara College of Pharmacy, Chitkara University, Punjab, India*
balraj.saini@chitkara.edu.in

14.1   Introduction

Nanomaterials include those materials whose crystalline size lies below 100 nm [1]. The term nanomaterial includes a great range of materials such as nanocrystals, quantum dots, carbon nanotubes, nanowires, nanosheets, and nanocomposites. The beauty of nanosized materials is that their thermal, physical, mechanical, chemical, optical, and other properties get changed to the nanosize as compared to bulk particles [2, 3]. When these materials (dispersion phase) are added in some fluids (dispersion medium) such as ethylene glycol (EG), distilled water (DW), oil, and ethanol, they form a colloidal solution. This colloidal solution (nanoparticles

Nanotechnology: Principles and Applications
Edited by Rakesh K. Sindhu, Mansi Chitkara, and Inderjeet Singh Sandhu
Copyright © 2021 Jenny Stanford Publishing Pte. Ltd.
ISBN 978-981-4877-43-5 (Hardcover), 978-1-003-12026-1 (eBook)
www.jennystanford.com

+ base fluid) is called "nanofluids" (Fig. 14.1). The ions present in the dispersion medium (base fluid) have an impact on the properties of the base fluid. Due to the presence of nanoparticles in fluid, the properties of the fluid (thermal conductivity, density, viscosity) get changed and these properties of nanofluids can be altered by changing in nanomaterials, etc. Nanoparticles present in the base fluid also contribute in maintaining the small size; however, this is very challenging because particles generally come in contact with each other and form bunch-like morphologies called agglomerates, which settle down under the influence of gravity [4, 5].

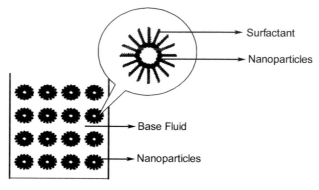

Figure 14.1 Schematic diagram of nanofluids solution.

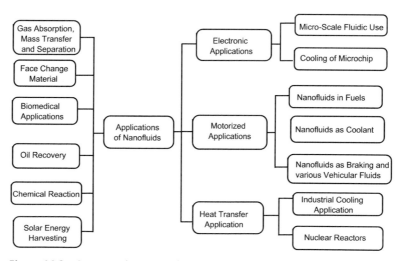

Figure 14.2 Some applications of nanofluids.

Nowadays nanofluids are playing a major role in our daily life [6, 7]. Some of these applications are listed in Fig. 14.2.

14.2 Types of Nanofluids

The nanomaterials in nanofluids can be a single element (Fe, Cu, and Ag), alloys (Fe–Ni, Cu–Zn, and Ag–Cu), oxides of single elements (Al_2O_3, CuO, Cu_2O, TiO_2), metal oxides (B_4C, ZrC, and SiC), multielement oxides ($ZnFe_2O_4$, $CuZnFe_4O_4$, $NiFe_2O_4$), and carbon nanomaterials (CNT, graphite, and diamond could be dispersed in the base fluid (EG, oil, water, ethanol, etc.)) [8–10]. Nanofluids can be classified into hybrid nanofluids and single material nanofluids.

- **Hybrid nanofluids:** Hybrid nanofluids are made up of a combination of more than one type of nanomaterial in the base fluids [11].
- **Single material nanofluids:** In these nanofluids, a single type of nanomaterial is used and mixed in various base fluids. These nanomaterials could be metal, metal oxide, metal carbides, etc. [12, 13].

14.3 Preparation

Nanofluids cannot be synthesized by just mixing nanoparticles and base fluids because nanoparticles are very reactive and can react with each other or with the base fluids or in some cases (usually nanoparticles of active metals) with the environment if they are exposed to air. So the uniformity and thermophysical properties of nanofluids depend on the methods that are opted for them. It shows that if the same nanofluid is made by two different methods, they tend to agglomerate and their thermophysical properties could be different. So the best method for the preparation of nanofluids depends on the kind of material, base fluid, and their interactions, i.e., chemical stability, homogencity, and durability with each other. There are mainly two ways of preparation of nanofluids [8, 14] (Fig. 14.3): one-step method and two-step method.

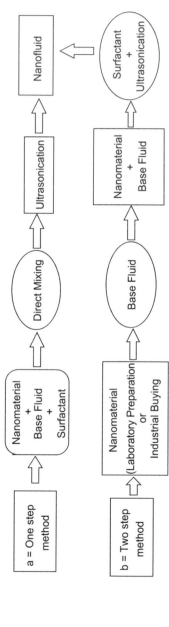

Figure 14.3 Schematic procedure for (a) one-step method and (b) two-step method.

14.3.1   One-Step Method

The one-step method is mainly used for highly reactive nanomaterials. In this approach, the synthesis of nanoparticles, addition of surfactant, and their dispersion into the base fluid are achieved in a single step. So all the reactions in this approach have no external handling, no external storage of nanoparticles, and no unwanted expose to the environment. There are many techniques to achieve nanofluid synthesis in a single step, and in many techniques, there may be multiple steps, but all steps are completed in a rapid sequence from starting (synthesis of nanoparticles) to end (complete preparation of nanofluid) [15, 16]. The one-step approach is always challenging due to its complexity. The methods in this approach may involve many steps of heating, cooling, and mixing to form stable nanofluids. There are some that follow one-step techniques, which are used for the synthesis of nanofluids, and out of them vacuum evaporation onto a running oil substrate (VEROS) (Fig. 14.4) is the most common one [8, 17]. Some of the one-step methods are as follows:

- Phase change materials
- Laser and electric arc ablation in liquids
- Dual plasma source method
- One-step approach through chemical reaction
- Vacuum evaporation onto a running oil substrate (VEROS) (Fig. 14.4)

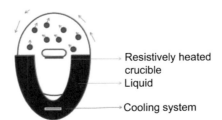

Figure 14.4   Preparation of nanofluids using vacuum evaporation onto a running oil substrate (VEROS) [17].

14.3.2 Two-Step Method

The two-step method is more popular for the synthesis of nanofluids as compared to the one-step approach due to lower processing cost, easy handling, and easy availability of nanoparticles from the industry. In this method, nanoparticles are initially purchased and then mixed/dispersed in the base fluid. The dispersion of nanoparticles is achieved by ultrasonic sonication, magnetic stirrer, homogenizer, beads mill, and high shear mixtures [8, 18–20]. Lee et al., Wang et al., and Easmax et al. used this method for the preparation of Al_2O_3 nanofluids. Xuan and Lee dispersed Cu nanoparticles in water and oil base fluids. Similarly, by using the same approach, Murshed et al. prepared TiO_2 nanofluid in water. Some authors (Table 14.1) also synthesized multi-walled or single-walled nanotubes without adding surfactant (for the stabilization of nanoparticles) and prepared nanofluids using the two-step method. Despite many advantages, there are some disadvantages also. In this method, aggregation of nanoparticles is more as compared to the single-step approach.

Table 14.1 Summary of various nanofluids

S. No.	Categories of nanoparticles	Nanoparticles	Base liquid	References
		ZnO	PEG, deionized water, EG	[21, 22]
		CuO	Deionized water, oil, oleic acid	[23, 24]
1.	Metal oxide nanoparticles	TiO_2	Deionized water, double distilled water, EG-TNT, oil	[25, 26]
		Al_2O_3	Deionized water, CMC, WEG50, EG	[27, 28]
		Fe_2O_3	FG/water, kerosene	[29]

S. No.	Categories of nanoparticles	Nanoparticles	Base liquid	References
		Ag	Deionized water	[30]
2.	Metallic nanoparticles	Cu	Deionized water, transfer oil, R113, EG	[31, 32]

14.4 Stability of Nanofluids

The properties and durability of nanofluids depend on their stability. The more the stability of nanofluids, the longer their life. The major challenge in the field of nanofluids is their stability. The nanoparticles of nanofluids can agglomerate due to interaction with each other, or in some cases, they can also react with the base fluid. This type of behavior is due to the van der Waals forces of attractions on the surface of nanoparticles. So if the van der Waals forces of attraction are high, these nanoparticles can attract toward each other during their Brownian motion and get agglomerated [8, 33, 34]. As a result of agglomeration, nanoparticles in base fluids/dispersion medium separate out and settle down at the bottom due to gravity. These agglomerates may be soft or hard. If the agglomerates are soft, they may be broken by stirring, sonication, or other methods of fluid agitation [8]. However, hard agglomerates cannot be broken up by normal methods because of the sintering of nanoparticles. So according to the Derjaguin, Landau, Vervey, and Overbeek (DLVO) theory, two forces of attraction play an important role in the stability of nanofluids: (i) van der Waals force of attraction and (ii) electric double-layer repulsion. The attractive potential energy [17, 33] and the repulsive potential energy can be calculated by using the Hamaker formula:

$$V_A = \frac{-Ar}{12H} \qquad \text{.......equation 1}$$

$$V_R = \frac{64\pi r n_o kT \gamma_o^2}{k^2} \exp(-kH) \qquad \text{.......equation 2}$$

where V_A is the attractive potential energy, V_R is the repulsive potential energy, and A is the Hamaker constant.

The stability of nanofluids depends on the balance of attractive and repulsive forces. For good stability of nanofluids, the opposing force should be greater than the attractive force between nanoparticles. The repulsive forces in the nanoparticles of nanofluids are generated through steric repulsion (Fig. 14.5a) and electrostatic repulsion (Fig. 14.5b) [17].

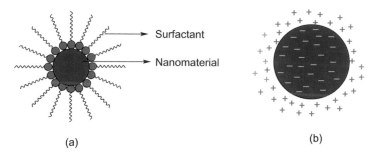

(a) (b)

Figure 14.5 Types of stabilization: (a) steric stabilization and (b) electrostatic stabilization [17].

Through steric repulsion, various surfactants (stabilizing agents) are used (Table 14.2) to prevent the aggregation of nanoparticles in nanofluids. The surfactants have two ends: one is called head and the other is called tail. The head is hydrophilic, and the tail is hydrophobic in nature. Surfactants are added during the synthesis of nanofluids. The surfactant gets adsorbed on the surface of nanoparticles and causes stearic hindrance to each other, but the hydrophilic end of these compounds is compatible with the dispersion medium (Fig. 14.5a). In the stability of nanofluids, surfactants have a great role, but they also have the following limitations:

- Surfactants can change the optical properties of nanofluids.
- Surfactants can change the viscoelastic properties of nanofluids.
- Surfactants can deteriorate the surface of nanoparticles.
- Surfactants can also affect the thermal conducting power of nanofluids because at high temperatures, these surfactants may desorb/detach from the surface of nanoparticles in nanofluids and get agglomerated. During this process, at high

Table 14.2 List of various surfactants with their chemical structure

S. No.	Name of surfactant	Chemical structure of surfactant	Reference
1.	Sodium octanoate (SOCT)		[35]
2.	Dodecyltrimethylammonium bromide (DTAB)		[36]
3.	Sodium dodecyl benzene sulfonate (SDBS)		[37]
4.	Polyvinylpyrrolidone (PVP)		[38]
5.	Sodium dodecyl sulfate (SDS)		[39]
6.	Oleic acid (OA)		[40]
7.	Polyacrylic acid sodium salt (SAAS)		[41]

temperature, the deterioration of nanoparticle surface may increase. Typically, the degradation of surface increases above 60°C; as a result, the use of nanofluids becomes limited in high-temperature processes.

In the absence of surfactants, some nanofluids are also synthesized. The stability of these types of nanofluids is based on the electrostatic interaction theory (Fig. 14.5b). According to this theory, when nanoparticles come in contact with the base fluid, they preferentially adsorb one particular type of ions on their surface and form a fixed layer or stern layer. To counterbalance the charge of the stern layer, an addition movable layer is also formed next to it, and this layer is called diffused layer (present in the body of the solution) (Fig. 14.6). As a result, a diffused electrical double layer is formed around the nanoparticles.

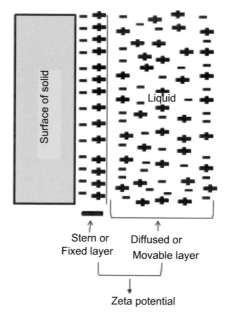

Figure 14.6 Formation of diffused electrical double layer.

A potential difference also exits between the stern layer and the diffused layer, and this potential difference is called zeta potential, which is measured in millivolts. This layer is responsible for the repulsive columbic force between the nanoparticles and keeps them apart from each other or prevent them from agglomeration. In nanofluid if zeta potential is greater than −30 and less than +30 have high probability of agglomeration (less stable). Nanofluid consider stable when zeta potential is between ± 30mV to ± 40 mV and good stability between ±40 mV to ±60 mV. Nanofluids have excellent stability when zeta potential is less than −60 mV and greater than +60 mV.

Table 14.3 Correlation between zeta potential values and stability of nanofluids

Serial No.	Zeta Potential (mV)	Stability of Nanofluids
1.	0 to ±5	Highly unstable (Coagulation)
2.	±10 to ± 30	Very low stability
3.	±30 to ± 40	Modest stability
4.	±40 to ± 60	Good stability
5.	Greater than ±60	Exceptional stability

14.5 Applications of Nanofluids

Nanofluids have combined properties of nanoparticles and fluids in which these particles are dispersed. These fluids are primarily designed for the use of heat transfer [1, 8] because of their excellent heat transfer properties and stability. Nowadays nanofluids are also used in a variety of new areas, which are expanding day by day (Fig. 14.2). Some applications of nanofluids are as follows:

14.5.1 Electronic Applications

In the electronic industry, many nanofluids are used as cooling agents. Some of these applications are as follows:

14.5.1.1 Microscale fluidic use

The change in liquid volume is an important factor in microelectronic mechanical systems (MEMS), optical devices, and fluidic digital display devices. This can be achieved by the electro-wetting technique in which with a change in volume, the solid–electrolyte angle gets changed by the application of a potential difference between the solid and the electrolyte (Fig. 14.7).

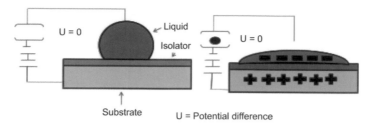

Figure 14.7 Electro-wetting technique.

The electro-wetting technique is one of the crucial techniques of microscale liquid manipulation. Vafaei et al. [42] found that nanofluids are helpful in the wettability of surface engineering. They found that the addition of bismuth telluride nanofluid even at a very low concentration astonishingly changed the wetting characteristics of the surface. The contact angle increases up to 40° even at a very low concentration of 3×10^{-6}. It was also observed that with an increase in the quantity of nanoparticles in the nanofluids, the contact angle increases up to a certain concentration and then decreases [43]. The contact angle of the droplet also changes with the size of the nanoparticles. Small-sized nanoparticles increase the contact angle for the same concentration as compared to larger ones because of their large surface-to-volume ratio.

14.5.1.2   Cooling of microchips

One of the major drawbacks during the development of microchips is that quick heat loss cannot be achieved during their working. It has been anticipated that the new-generation computer chips will produce 10 MW/m^2 localized heat flux with a total power exceeding 300 W. So to remove this heat, nanofluids can be used as coolant due to their high thermal conductivity. It has been observed that the nanofluids-based oscillating heat pipe cooling system with thin-film evaporation can remove heat flux up to 10 MW/m^2. Thus, nanofluids can be used as coolant in next-generation devices [44, 45]. Nguyen et al. [46] examined the Al_2O_3–H_2O nanofluid heat transfer behavior in closed cooling systems designed for electronic devices such as microprocessors. The experimental data show that water-containing nanoparticles have 40% more cooling convective heat transfer coefficient compared to 68% base fluid water. It has also been observed that the smaller the size of Al_2O_3 nanoparticles in nanofluids, the higher the convective heat transfer coefficient value

[46a]. Lin et al. examined silver nanoparticles in pulsating heat pipes and found that Ag particles improved the heat transfer property of the heat pipes. Further research on using nanofluids as cooling agents in electronic devices will help in making next-generation devices [46b].

14.5.2 Motorized Applications

For cooling, lubrication, and automatic transmission, various high-energy transfer fluids are used. These fluids are used in thermal system engine, radiators, air conditioners, etc. But these conventional fluids are less stable at high temperature and have low heat transfer capacity. These properties can be improved by the use of nanofluids. Some automotive applications of nanofluids are as follows:

14.5.2.1 Nanofluids in fuels

The addition of Al nanofluids in fuel increases the total combustion and reduces smoke and NO_2 emission [47, 48]. An Al_2O_3 layer is formed on the Al nanoparticles when they are formed through the plasma arc system. This layer increases the oxidation property of pure Al and forms a large surface area with H_2O, which improves the decomposition of H_2 gas from water. In this process, Al_2O_3 acts as a catalyst and Al nanoparticles help in the decomposition of water to produce H_2 gas.

14.5.2.2 Nanofluids as coolant

Aerodynamics is very important in the design of vehicles because it directly affects their performance. So manufacturers try to decrease air resistance by making vehicles more aerodynamic. Approximately 65% of the total amount of energy is wasted in overcoming the aerodynamic drag in large vehicles such as buses and trucks. In large vehicles, the front area is large and the engine is positioned toward the front to achieve maximum cooling from the coming air. Due to the high capacity of conducting heat, nanofluids make it possible to reduce the size and position of radiator. With the use of such fluids, the size of truck engines can be reduced without reducing their performance. In the radiator, Routbort et al. [49] used a high-thermal-conductivity fluid and found that the size of the front radiator can be reduced by up to 10% and the design can be made more aerodynamic, which can save fuel up to 5%. These nanofluids also help in reducing friction in the radiator and parasitic loses, which helps to save fuel by up to 6%. All these properties can be improved in the future.

14.5.2.3 Nanofluids as braking and various vehicular fluids

Day by day, the speed of vehicles is increasing. Thus, it is crucial to develop very good braking systems. In the hydraulic braking system, the kinetic energy of the vehicle is transferred to the brake fluid. When the break is applied repeatedly, the brake fluid reaches its boiling point and then the heat dissipation power of the fluid gets reduced, which causes dis-functioning of the brakes. This creates serious concern toward safety. Because of having better dissipating power than base fluids, nanofluids can be used in braking systems. CuO (CBN) and Al_2O_3 brake nanofluids have higher conducting heat capacity, boiling point, and viscosity as compared to the commonly used brake fluid (DOBT3) as a result of their high boiling point. The risk of vapor lock decreases even in high temperature. Thus, it increases the safety of vehicles [47, 48].

14.5.3 Gas Absorption, Mass Transfer, and Separation

One of the promising applications of nanofluids is found in gas absorption, separation, and mass transfer [50]. When a gas is passed through a nanofluid, its nanoparticles break larger gas bubbles into smaller ones, which increases the dissolution of the gas. Moreover, nanofluids can play the same role of chemical solvents by increasing the diffusion process similar to the role played by hemoglobin and myoglobin during the transportation of gases in blood.

14.5.4 Nanofluid Phase Change Materials

The area of phase change materials nanofluids has emerged in current time. The most of work focus on addition of nanoparticles to PCMs. For example, paraffin and aqueous solution of salt hydrates (potassium carbonate, lithium carbonate, and barium chloride) added to base PCM to increase its heat capacity or thermal conductivity. In this process nanoparticles provide their higher thermal conductivity power as seeding nuclei to speed up the freezing process. During repetitive cooling/heating process the nanoparticles form cluster and network which increase the thermal conductivity also. Han et al. used emulsification based synthesis method in disperse phase-changeable medium nanoparticles into polyalphaolefin (PAO) which increases heat capacity significantly [88]. This area will grow significantly in near future.

14.5.5   Biomedical Applications of Nanofluids

Thermal properties, optical properties, magnetic field interaction, and suspension stability of nanofluids depend on the surrounding medium. Nanofluids are extensibility used in the medical field, such as in drug delivery, magnetic cell separation, nanocryosurgery, cryopreservation, and sensing and imaging (Table 14.4) [51–58]. In cancer treatment, magnetic iron-based nanoparticles are used for drug delivery in cancer patients. These magnetic nanoparticles are more adhesive to cancer cells as compared to healthy cells. These nanoparticles can then be heated up under controlled conditions, which destroys cancer cells.

Optical properties such as scattering, extinction coefficient, absorption, and transmittance are based on particle size, shape, volume fraction, and path length. In general, the absorption of nanofluid increases, or transmittance decreases with an increase in the concentration of nanoparticles. With an increase in the optical path length, transmittance decreases.

The transmittance is always less than that of the base fluid and increases with the size of nanoparticles. Scattering light and extinction coefficient also increase with an increase in the size of particles and their concentration in the base fluid. All these nanofluid properties change on changing their vicinity. So nanofluids are used in the field of sensing and imaging to detect various diseases.

14.5.6   Nanofluid for Oil Recovery

The adhesion and spread of nanofluids on a solid surface are different compared to simple fluids [59]. Due to this difference, nanofluids can be used in lubrication, soil remediation, and oil recovery. Nikolov and Wasan [60] found that when an oil drop encounters polystyrene nanoparticles, these nanoparticles rearrange and concentrate around the oil drop and form a wedge-like area between the oil drop and the surface of nanoparticles. These nanoparticles diffuse into this wedge film; as a result, pressure around the oil drop increases with an increase in the concentration of nanoparticles, which helps the oil drop to detach from the surface [60].

Table 14.4 List of some nanomedicine studies

S. No.	References	Particle size	Treatment type	Result
1.	Huang et al. [51]	Au	Laser photo thermal therapy	70–80°C required for the destruction of cancer cell
2.	Johannsen et al. [52]	Silica-core Fe_2O_3 nanoshells	Magnetic hyperthermia	This technique is effective but not selective action
3	Dombrovsky et al. [53]	Au silica-core nanoshells	Laser photo thermal therapy	Established parameterized model
4.	Parveen and Sahoo [54]	Polymeric nanoparticles	Drug delivery	Successful targeted delivery
5.	Aryal et al. [55]	Nanocarrier as for hydrophobic–hydrophilic drug conjugate	Drug delivery	Successful targeted delivery
6.	Kikumori et al. [56]	Magnetic Fe_2O_3	Nanoparticle hyperthermia	Specific and effective hyperthermia
7.	Salloum et al. [57]	Polymeric encapsulated Fe_2O_3	Nanoparticle hyperthermia	Nanoparticles distribute nonuniformly near injection site
8.	Bergey et al. [58]	Silica-core Fe_2O_3 nanoshells	Magnetocytolysis	Selectively lyse cell

14.5.7 Heat Transfer Applications

Fluids are extensively used as heat transfer agents in various fields such as hedonic heating and cooling systems, buildings, transportation industry, pulp and paper industry, chemical industry, and food-processing industry. Nanofluids have exceptionally high thermal conductivity. In the following fields, nanofluids are used as heat transfer agents.

14.5.7.1 Industrial cooling applications

Routbort et al. [61] reported that replacement of heating and cooling water with nanofluids has conserved approximately 1 trillion Btu energy and reduced approximately 5–6 metric tons of CO_2, 21,000 metric tons of SO_2, and 8600 metric tons of NO_2. So nanofluids will play a major role in the field of energy conservation in the 21st century [62–71]. Some nanofluids are listed in Table 14.5, which show their increased thermal conducting power. This enhanced thermal conductivity is very helpful in energy conservation.

14.5.7.2 Nuclear Reactors

In nuclear reactors, nanofluids play a dual role in the waste heat removal system and the emergency core cooling system [72]. Nanofluids have gained this power due to their high critical heat flux, which can be increased up to 50% from the base fluid [72]. Nanofluids have defined their role in accident mitigation strategies also. Nanofluids can remove the extra heat from the nuclear reactor and can work at very high temperatures with great stability for a long time as compared to other fluids.

14.5.8 Uses of Nanofluids in Chemical Reactions

Many nanoparticles act as catalysts in various important reactions [73–79]. Nanoparticles of WO_3, TiO_2, ZnO, GaP, CdS, and SiC in a base fluid (usually water) have been used for many organic reactions (Table 14.6) in less time and ecofriendly manner.

Table 14.5 List of various nanofluids with their enhanced thermal conductivity

S. No.	References	Material type	Concentration	Thermal conductivity enhancement
1.	Wei et al. [62]	Cu_2O + DW	0.01 to 0.05 vol.%	Up to 24%
2.	Karthikeyan et al. [63]	CuO (8 nm) + DW + EG	1 vol.%	Up to 31.6%
3.	Zhu et al. [64]	Graphite + DW	2.0 vol.%	Up to 34%
4.	Kwak et al. [65]	CuO + EG	< 0.002 vol.%	Up to 54%
5.	Li et al. [66]	Cu + DW	0.1 wt.%	Up to 10.7%
6.	Lee et al. [67]	CuO + DW	0.3 vol.%	Three times increase
7.	Wang et al. [68]	Al_2O_3 + DW, Cu + DW	0.4 wt.%	Up to 13%
8.	Sunder et al. [69]	Nanodiamond-Ni + EG	3.01 wt.%	Up to 13%
9.	Trinh et al. [70]	Gr-CNT/Cu hybrid + EG	0.005–0.035%	Up to 10–41%
10.	Selvam et al. [71]	Graphene nanoplatelets + EG	0.5 vol.%	Up to 21%

Table 14.6 List of some selected chemical reactions performed by using nanoparticles

S. No.	References	Particle size	Base fluid	Desired product	Result
1.	Tao et al. [73]	Rh–Pd, Pt–Pd (core–shell)	Colloidal fab. in organic solvents	"Smart" catalysts adaptable to environment	Reversible chemical rearrangement when changing reactions
2.	Sınağ et al. [74]	SnO_2, ZnO	Cellulose (aqueous)	CO, H_2, CH_4, advanced organic compounds	ZnO helps to improve hydrogen formation
3.	Trepanier et al. [75]	CNT w/Co	Microemulsion w/H_2O in oil	Fischer–Tropsch process	Rapid rate of reaction, 15% enhancement in CO conversion
4.	Nutt et al. [76]	Pd-on-Au particles	H_2O	Hydro-chlorinated chlorine contaminates	Very useful for remediation of ground H_2O
5.	Hou et al. [77]	Au–Pd and Au, Pd	H_2O with alcohols	Formation of aldehydes and ketones from alcohol	Au–Pd nanoparticles act as an oxidizing agent
6.	Shi et al. [78]	Nanoparticles diamond C/Cu	Prepared in hydrazine sol.	Extremely oxidizing propellant	Decrease in oxidization temperature up to 6–15%

14.5.9 Solar Energy Harvesting

As mentioned earlier, the optical properties of nanofluids depend on the size of nanoparticles and the nature of the base fluid. Any small change in these areas can bring a large change in the properties of nanofluids. Hunt [79] and Anderson [80] in 1970 studies of collect of solar energy by using particles how mixed particles in gas working fluid. Recent research in this field shows that inappropriate design or unstable nanofluids can be harmful to the solar collector (Fig. 14.8) [81].

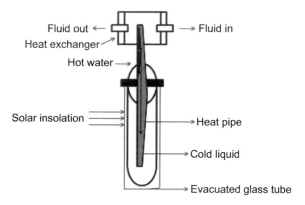

Figure 14.8 Outline diagram of a solar collector [81].

If the nanoparticles are large in size, then maximum light will be absorbed only at the surface. As a result, light will be lost in the environment and will not be stored. On the other hand, if the particle size is small, these particles will not properly absorb the incoming light. So an appropriate size of nanoparticles in nanofluids is important for solar heat collectors. The size of nanoparticles in nanofluids that are used in solar collectors is small. The nanoparticles are larger than smoke particles but smaller than pigments and paints particles. It has been found that nanofluids are better collectors of solar radiation as compared to the base fluids such as glycol and water (Table 14.7) [82–91].

Table 14.7 List of some selected nanofluids having solar applications

S. No.	References	Particle size	Base fluid	Application
1.	Taylor et al. [82]	Ag, Au, Cu	Water/VP-1 oil	Dish/power towers
2.	Taylor et al. [83]	Ag, Al, Au, Cu, TiO$_2$	Water/VP-1 oil	Lab testing, optical props. for solar absorption column
3.	Lenert et al. [84]	Graphite shell, cobalt core	VP-1 oil	Lab-testing solar absorption column
4.	Otanicar et al. [85]	Ag, CNT	Graph water	Micro-collector (flat plate)
5.	Sani et al. [86]	Carbon nanohorns	E: glycol, water lab testing	Nanofluid optimization
6.	Tyagi et al. [87]	Al	Water	Flat plate
7.	Han et al. [88]	Carbon black	Water	Test tube testing E: Good thermal and optical properties
8.	Otanicar and Golden [89]	Graphite	Water	Flat plate environmental comparison
9.	Lu et al. [90]	CuO	Water	Evacuated tube
10.	Veeraragavan et al. [91]	Graphite	VP-1 oil	Optimum geometry/nanofluid study

14.6   Merits of Nanofluids

Nanofluids have the following merits:

1. Fluids are used as a heat transfer agents and this capacity can be enhance by adding metallic nanoparticles, i.e., conversion of fluids into nanofluids = nanoparticles + fluids. As metals are good conductor of heat and electricity so these nanoparticles also contribute in the heat transfer capacity of fluid. So nanofluids become better heat transfer agents as compare to normal fluids.

2. Solar energy harvesting can be improved by using nanofluids in place of normal fluids in solar power harvesting systems and it can be maximized by changing the shape, size, and volume fraction of nanoparticles.

3. The biggest advantage of nanofluids is their heat-carrying capacity, due to which nanofluids improve the safety and efficiency of heat transfer systems.

4. Nanofluids can be easily made compatible to various systems by simply changing the base fluid material and the concentration of the suspending nanoparticles in the base fluid.

5. Nanofluids can be used to develop high-speed braking systems, to improve fuel efficacy, and to reduce the emission of harmful gases during the burning of fuels in modern vehicles.

6. Fluid turbulence mixing fluctuation can be increased.

7. Nanofluids can be used as a green reaction medium in environment friendly organic reactions.

8. Nanofluids can be used in drug-delivery systems, sensing and imaging, and cancer therapy.

14.7   Demerits of Nanofluids

Along with merits, there are still some demerits of nanofluids:

1. For the synthesis of nanofluids, highly purified chemicals and sophisticated laboratories are required. As a result, this requirement increases the overall cost of nanofluids.

2. Stability of nanofluids is one of the challenges during their synthesis. Despite various methods for increasing stability, such as addition of surfactants and surface modifications, the stability of nanofluids is still not long lasting due to their high reactive surface. After some time, they start to coagulate and settle down under the influence the gravity.

3. Nanoparticles deposit on the heat transfer surfaces of various devices in which the nanofluids are used. This deposition of nanofluids is called fouling effect. Fouling reduces the mechanical and thermal efficiency of heat exchangers, hinder fluid flow, and increase pressure drop and corrosion.

4. The high viscosity and density of nanofluids lead to enhancement in pressure drop. Many studies have shown that nanofluids show significant pressure drop as compared to the base fluid.

14.8 Future Outlook and Upcoming Challenges in Nanofluids

Further research is required for the better use and understanding of promising nanofluids. These fluids can play a very important role in the energy transfer mechanism, which will lead to the development of more efficient, sensitive, accurate, and energy-saving devices. The thermal conductivity of nanofluids depends on particle size, material, shape, pH, surfactant nature of the base fluid, interaction of nanoparticles with the base fluid, etc. So there are further miles to go to find out various parameters on which thermal conductivity depends. Achieving long stability of nanofluids is also a major challenge in the future because without long stability, nanofluids have limited scope. The presently available nanofluids are limited, and their specificity is also not accurate. So further research is required. In the future, we should encourage engineers working with nanofluids to opt for more green, time-saving, and energy-saving methods to produce environment friendly biodegradable materials. Further research is still required in the field of synthesis and commercialization of nanofluids and their application for the benefit of humankind.

14.9   Conclusion

Nanofluids are a novel class of fluids that have very exciting application in the fields of energy conservation, heat transfer, biomedicine, catalysts, compact and more efficient electronic devices, etc. Due to their various unique characteristics, nanofluids can play a crucial role in the 21st century in various fields after their full potential is understood. Hence, it is recommended that further research in the field of nanofluid synthesis and applications should be carried out.

Acknowledgment

The authors would like to thank the Punjabi University Patiala, Punjab, India; Lajpat Rai DAV College Jagraon, Ludhiana, Punjab, India; and Chitkara College of Pharmacy, Chitkara University, Punjab, India for providing necessary facilities.

References

1. Taylor, R., Coulombe, S., Otanicar, T., Phelan, P., Gunawan, A., Lv, W., Rosengarten, G., Prasher, R., and Tyagi, H. (2013). Small particles, big impacts: A review of the diverse applications of nanofluids. *Journal of Applied Physics*, **113**(1), 011301.

2. Bellos, E., Said, Z., and Tzivanidis, C. (2018). The use of nanofluids in solar concentrating technologies: A comprehensive review. *Journal of Cleaner Production*, 196, pp. 84–99.

3. Ghadimi, A., Saidur, R., and Metselaar, H. S. C. (2011). A review of nanofluid stability properties and characterization in stationary conditions. *International Journal of Heat and Mass Transfer*, **54**(17–18), pp. 4051–4068.

4. Memon, A. G. and Memon, R. A. (2017). Thermodynamic analysis of a trigeneration system proposed for residential application. *Energy Conversion and Management*, **145**, pp. 182–203.

5. Sajid, M. U. and Ali, H. M. (2019). Recent advances in application of nanofluids in heat transfer devices: A critical review. *Renewable and Sustainable Energy Reviews*, **103**, pp. 556–592.

6. Bahiraei, M. and Heshmatian, S. (2019). Graphene family nanofluids: A critical review and future research directions. *Energy Conversion and Management*, **196**, pp. 1222–1256.

7. Pordanjani, A. H., Aghakhani, S., Afrand, M., Mahmoudi, B., Mahian, O., and Wongwises, S. (2019). An updated review on application of nanofluids in heat exchangers for saving energy. *Energy Conversion and Management*, **198**, 111886.

8. Ali, N., Teixeira, J. A., and Addali, A. (2018). A review on nanofluids: Fabrication, stability, and thermophysical properties. *Journal of Nanomaterials*, **2018**. https://doi.org/10.1155/2018/6978130.

9. Gupta, M., Singh, V., Kumar, R., and Said, Z. (2017). A review on thermophysical properties of nanofluids and heat transfer applications. *Renewable and Sustainable Energy Reviews*, **74**, pp. 638–670.

10. Yang, L. and Du, K. (2017). A comprehensive review on heat transfer characteristics of TiO_2 nanofluids. *International Journal of Heat and Mass Transfer*, **108**, pp. 11–31.

11. Jana, S., Salehi-Khojin, A., and Zhong, W. H. (2007). Enhancement of fluid thermal conductivity by the addition of single and hybrid nano-additives. *Thermochimica Acta*, **462**(1–2), pp. 45–55.

12. Yang, L. and Du, K. (2017). A comprehensive review on heat transfer characteristics of TiO_2 nanofluids. *International Journal of Heat and Mass Transfer*, **108**, pp. 11–31.

13. Azmi, W. H., Sharif, M. Z., Yusof, T. M., Mamat, R., and Redhwan, A. A. M. (2017). Potential of nanorefrigerant and nanolubricant on energy saving in refrigeration system: A review. *Renewable and Sustainable Energy Reviews*, **69**, pp. 415–428.

14. Chamsa-ard, W., Brundavanam, S., Fung, C. C., Fawcett, D., and Poinern, G. (2017). Nanofluid types, their synthesis, properties and incorporation in direct solar thermal collectors: A review. *Nanomaterials*, 7(6), p. 131.

15. Eastman, J. A., Choi, U. S., Li, S., Thompson, L. J., and Lee, S. (1996). Enhanced thermal conductivity through the development of nanofluids. *MRS Online Proceedings Library Archive*, **457**, 3, doi:10.1557/PROC-457-3.

16. Zhu, H. T., Lin, Y. S., and Yin, Y. S. (2004). A novel one-step chemical method for preparation of copper nanofluids. *Journal of Colloid and Interface Science*, **277**(1), pp. 100–103.

17. Kong, L., Sun, J., and Bao, Y. (2017). Preparation, characterization and tribological mechanism of nanofluids. *RSC Advances*, 7(21), pp. 12599–12609.

18. Wang, X., Xu, X., and Choi, S. U. (1999). Thermal conductivity of nanoparticle-fluid mixture. *Journal of Thermophysics and Heat Transfer*, **13**(4), pp. 474–480.

19. Xuan, Y. and Li, Q. (2000). Heat transfer enhancement of nanofluids. *International Journal of Heat and Fluid Flow*, **21**(1), pp. 58–64.

20. Murshed, S. M. S., Leong, K. C., and Yang, C. (2005). Enhanced thermal conductivity of TiO_2 water based nanofluids. *International Journal of Thermal Sciences*, **44**(4), pp. 367–373.

21. Zafarani-Moattar, M. T. and Majdan-Cegincara, R. (2012). Effect of temperature on volumetric and transport properties of nanofluids containing ZnO nanoparticles poly (ethylene glycol) and water. *The Journal of Chemical Thermodynamics*, **54**, pp. 55–67.

22. Raykar, V. S. and Singh, A. K. (2010). Thermal and rheological behavior of acetylacetone stabilized ZnO nanofluids. *Thermochimica Acta*, **502**(1–2), pp. 60–65.

23. Akhavan-Behabadi, M. A., Hekmatipour, F., Mirhabibi, S. M., and Sajadi, B. (2014). An empirical study on heat transfer and pressure drop properties of heat transfer oil-copper oxide nanofluid in microfin tubes. *International Communications in Heat and Mass Transfer*, **57**, pp. 150–156.

24. Harikrishnan, S. and Kalaiselvam, S. (2012). Preparation and thermal characteristics of CuO–oleic acid nanofluids as a phase change material. *Thermochimica Acta*, **533**, pp. 46–55.

25. Wei, B., Zou, C., and Li, X. (2017). Experimental investigation on stability and thermal conductivity of diathermic oil based TiO_2 nanofluids. *International Journal of Heat and Mass Transfer*, **104**, pp. 537–543.

26. Chen, H., Ding, Y., and Lapkin, A. (2009). Rheological behaviour of nanofluids containing tube/rod-like nanoparticles. *Powder Technology*, **194**(1–2), pp. 132–141.

27. Mojarrad, M. S., Keshavarz, A., Ziabasharhagh, M., and Raznahan, M. M. (2014). Experimental investigation on heat transfer enhancement of alumina/water and alumina/water–ethylene glycol nanofluids in thermally developing laminar flow. *Experimental Thermal and Fluid Science*, **53**, pp. 111–118.

28. Beck, M. P., Sun, T., and Teja, A. S. (2007). The thermal conductivity of alumina nanoparticles dispersed in ethylene glycol. *Fluid Phase Equilibria*, **260**(2), pp. 275–278.

29. Sundar, L. S., Ramana, E. V., Singh, M. K., and De Sousa, A. C. M. (2012). Viscosity of low volume concentrations of magnetic Fe_3O_4 nanoparticles dispersed in ethylene glycol and water mixture. *Chemical Physics Letters*, **554**, pp. 236–242.

30. Phuoc, T. X., Soong, Y., and Chyu, M. K. (2007). Synthesis of Ag-deionized water nanofluids using multi-beam laser ablation in liquids. *Optics and Lasers in Engineering*, **45**(12), pp. 1099–1106.

31. Hu, H., Peng, H., and Ding, G. (2013). Nucleate pool boiling heat transfer characteristics of refrigerant/nanolubricant mixture with surfactant. *International Journal of Refrigeration*, **36**(3), pp. 1045–1055.

32. Xuan, Y. and Li, Q. (2000). Heat transfer enhancement of nanofluids. *International Journal of Heat and Fluid Flow*, **21**(1), pp. 58–64.

33. Popa, I., Gillies, G., Papastavrou, G., and Borkovec, M. (2010). Attractive and repulsive electrostatic forces between positively charged latex particles in the presence of anionic linear polyelectrolytes. *The Journal of Physical Chemistry B*, **114**(9), pp. 3170–3177.

34. Bar-Hen, A., Bounioux, C., Yerushalmi-Rozen, R., Solveyra, E. G., and Szleifer, I. (2015). The role of steric interactions in dispersion of carbon nanotubes by poly (3-alkyl thiophenes) in organic solvents. *Journal of Colloid and Interface Science*, **452**, pp. 62–68.

35. Dey, D., Kumar, P., and Samantaray, S. (2017). A review of nanofluid preparation, stability, and thermo-physical properties. *Heat Transfer—Asian Research*, **46**(8), pp. 1413–1442.

36. Li, X. F., Zhu, D. S., Wang, X. J., Wang, N., Gao, J. W., and Li, H. (2008). Thermal conductivity enhancement dependent pH and chemical surfactant for Cu-H_2O nanofluids. *Thermochimica Acta*, **469**(1–2), pp. 98–103.

37. Kong, L., Sun, J., and Bao, Y. (2017). Preparation, characterization and tribological mechanism of nanofluids. *RSC Advances*, **7**(21), pp. 12599–12609.

38. Pantzali, M. N., Mouza, A. A., and Paras, S. V. (2009). Investigating the efficacy of nanofluids as coolants in plate heat exchangers (PHE). *Chemical Engineering Science*, **64**(14), pp. 3290–3300.

39. Chandrasekar, M., Suresh, S., and Bose, A. C. (2010). Experimental investigations and theoretical determination of thermal conductivity and viscosity of Al_2O_3/water nanofluid. *Experimental Thermal and Fluid Science*, **34**(2), pp. 210–216.

40. Hwang, Y., Lee, J. K., Lee, J. K., Jeong, Y. M., Cheong, S. I., Ahn, Y. C., and Kim, S. H. (2008). Production and dispersion stability of nanoparticles in nanofluids. *Powder Technology*, **186**(2), pp. 145–153.

41. Kong, L., Sun, J., and Bao, Y. (2017). Preparation, characterization and tribological mechanism of nanofluids. *RSC Advances*, **7**(21), pp. 12599–12609.

42. Vafaei, S., Borca-Tasciuc, T., Podowski, M. Z., Purkayastha, A., Ramanath, G., and Ajayan, P. M. (2006). Effect of nanoparticles on sessile droplet contact angle. *Nanotechnology*, **17**(10), 2523.

43. Dash, R. K., Borca-Tasciuc, T., Purkayastha, A., and Ramanath, G. (2007). Electrowetting on dielectric-actuation of microdroplets of aqueous bismuth telluride nanoparticle suspensions. *Nanotechnology*, **18**(47), 475711.

44. Ma, H. B., Wilson, C., Yu, Q., Park, K., Choi, U. S., and Tirumala, M. (2006). An experimental investigation of heat transport capability in a nanofluid oscillating heat pipe. *Journal of Heat Transfer*, **128**(11), pp. 1213–1216.

45. Ma, H. B., Wilson, C., Borgmeyer, B., Park, K., Yu, Q., Choi, S. U. S., and Tirumala, M. (2006). Effect of nanofluid on the heat transport capability in an oscillating heat pipe. *Applied Physics Letters*, **88**(14), 143116.

46. (a) Nguyen, C. T., Roy, G., Gauthier, C., and Galanis, N. (2007). Heat transfer enhancement using Al_2O_3–water nanofluid for an electronic liquid cooling system. *Applied Thermal Engineering*, **27**(8–9), pp. 1501–1506.

    (b) Lin, Y. H., Kang, S. W., and Chen, H. L., (2008). Effect of silver nanofluid on pulsating heat pipe thermal performance. *Applied Thermal Engineering*, **28**, pp. 1312–1317.

47. Kao, M. J., Chang, H., Wu, Y. Y., Tsung, T. T., and Lin, H. M. (2007). Producing aluminum-oxide brake nanofluids using plasma charging system. *Journal of the Chinese Society of Mechanical Engineers*, **28**(2), pp. 123–131.

48. Kao, M. J., Lo, C. H., Tsung, T. T., Wu, Y. Y., Jwo, C. S., and Lin, H. M. (2007). Copper-oxide brake nanofluid manufactured using arc-submerged nanoparticle synthesis system. *Journal of Alloys and Compounds*, **434**, pp. 672–674.

49. Routbort, J. L., Singh, D., and Chen, G. (2006). Heavy vehicle systems optimization merit review and peer evaluation. *Annual Report*, Argonne National Laboratory, Chicago, Illinois, USA.

50. Cussler, E. L. and Cussler, E. L. (2009). *Diffusion: Mass Transfer in Fluid Systems*. Cambridge University Press.

51. Huang, X., Jain, P. K., El-Sayed, I. H., and El-Sayed, M. A. (2006). Determination of the minimum temperature required for selective photothermal destruction of cancer cells with the use of immunotargeted gold nanoparticles. *Photochemistry and Photobiology*, **82**(2), pp. 412–417.

52. Johannsen, M., Thiesen, B., Wust, P., and Jordan, A. (2010). Magnetic nanoparticle hyperthermia for prostate cancer. *International Journal of Hyperthermia*, **26**(8), pp. 790–795.

53. Dombrovsky, L. A., Timchenko, V., Jackson, M., and Yeoh, G. H. (2011). A combined transient thermal model for laser hyperthermia of tumors with embedded gold nanoshells. *International Journal of Heat and Mass Transfer*, **54**(25–26), pp. 5459–5469.

54. Parveen, S. and Sahoo, S. K. (2011). Long circulating chitosan/PEG blended PLGA nanoparticle for tumor drug delivery. *European Journal of Pharmacology*, **670**(2–3), pp. 372–383.

55. Aryal, S., Hu, C. M. J., Fu, V., and Zhang, L. (2012). Nanoparticle drug delivery enhances the cytotoxicity of hydrophobic–hydrophilic drug conjugates. *Journal of Materials Chemistry*, **22**(3), pp. 994–999.

56. Kikumori, T., Kobayashi, T., Sawaki, M., and Imai, T. (2009). Anti-cancer effect of hyperthermia on breast cancer by magnetite nanoparticle-loaded anti-HER2 immunoliposomes. *Breast Cancer Research and Treatment*, **113**(3), p. 435.

57. Salloum, M., Ma, R. H., Weeks, D., and Zhu, L. (2008). Controlling nanoparticle delivery in magnetic nanoparticle hyperthermia for cancer treatment: Experimental study in agarose gel. *International Journal of Hyperthermia*, **24**(4), pp. 337–345.

58. Bergey, E. J., Levy, L., Wang, X., Krebs, L. J., Lal, M., Kim, K. S., Pakatchi, S., Liebow, C., and Prasad, P. N. (2002). DC magnetic field induced magnetocytolysis of cancer cells targeted by LH-RH magnetic nanoparticles in vitro. *Biomedical Microdevices*, **4**(4), pp. 293–299.

59. Sefiane, K., Skilling, J., and MacGillivray, J. (2008). Contact line motion and dynamic wetting of nanofluid solutions. *Advances in Colloid and Interface Science*, **138**(2), pp. 101–120.

60. Wasan, D. T. and Nikolov, A. D. (2003). Spreading of nanofluids on solids. *Nature*, **423**(6936), pp. 156–159.

61. Routbort, J. (2009). Argonne National Lab, Michellin North America, St. Gobain Corp.

62. Wei, X., Zhu, H., Kong, T., and Wang, L. (2009). Synthesis and thermal conductivity of Cu$_2$O nanofluids. *International Journal of Heat and Mass Transfer*, **52**(19–20), pp. 4371–4374.

63. Karthikeyan, N. R., Philip, J., and Raj, B. (2008). Effect of clustering on the thermal conductivity of nanofluids. *Materials Chemistry and Physics*, **109**(1), pp. 50–55.

64. Zhu, H., Zhang, C., Tang, Y., Wang, J., and Ren, B. (2007). Preparation and thermal conductivity of suspensions of graphite nanoparticles. *Carbon (New York, NY)*, **45**(1), pp. 226–228.

65. Kwak, K. and Kim, C. (2005). Viscosity and thermal conductivity of copper oxide nanofluid dispersed in ethylene glycol. *Korea-Australia Rheology Journal*, **17**(2), pp. 35–40.

66. Li, X. F., Zhu, D. S., Wang, X. J., Wang, N., Gao, J. W., and Li, H. (2008). Thermal conductivity enhancement dependent pH and chemical surfactant for Cu-H$_2$O nanofluids. *Thermochimica Acta*, **469**(1–2), pp. 98–103.

67. Lee, D., Kim, J. W., and Kim, B. G. (2006). A new parameter to control heat transport in nanofluids: Surface charge state of the particle in suspension. *The Journal of Physical Chemistry B*, **110**(9), pp. 4323–4328.

68. Xian-Ju, W. and Xin-Fang, L. (2009). Influence of pH on nanofluids' viscosity and thermal conductivity. *Chinese Physics Letters*, **26**(5), 056601.

69. Sundar, L. S., Singh, M. K., Ramana, E. V., Singh, B., Grácio, J., and Sousa, A. C. (2014). Enhanced thermal conductivity and viscosity of nanodiamond-nickel nanocomposite nanofluids. *Scientific Reports*, **4**, 4039.

70. Van Trinh, P., Anh, N. N., Thang, B. H., Hong, N. T., Hong, N. M., Khoi, P. H., Minh, P. N., and Hong, P. N. (2017). Enhanced thermal conductivity of nanofluid-based ethylene glycol containing Cu nanoparticles decorated on a Gr–MWCNT hybrid material. *RSC Advances*, **7**(1), pp. 318–326.

71. Selvam, C., Lal, D. M., and Harish, S. (2016). Thermal conductivity enhancement of ethylene glycol and water with graphene nanoplatelets. *Thermochimica Acta*, **642**, pp. 32–38.

72. Buongiorno, J. and Hu, L. W. (2009). *8. Innovative Technologies: Two-Phase Heat Transfer in Water-Based Nanofluids for Nuclear Applications Final Report* (No. DOE/ID/14765-8). Massachusetts Institute of Technology Cambridge, MA 02139-4307.

73. Tao, F., Grass, M. E., Zhang, Y., Butcher, D. R., Renzas, J. R., Liu, Z., Chung, J. Y., Mun, B. S., Salmeron, M., and Somorjai, G. A. (2008). Reaction-driven restructuring of Rh-Pd and Pt-Pd core-shell nanoparticles. *Science*, **322**(5903), pp. 932–934.

74. Sınağ, A., Yumak, T., Balci, V., and Kruse, A. (2011). Catalytic hydrothermal conversion of cellulose over SnO_2 and ZnO nanoparticle catalysts. *The Journal of Supercritical Fluids*, **56**(2), pp. 179–185.

75. Trépanier, M., Dalai, A. K., and Abatzoglou, N. (2010). Synthesis of CNT-supported cobalt nanoparticle catalysts using a microemulsion technique: Role of nanoparticle size on reducibility, activity and selectivity in Fischer–Tropsch reactions. *Applied Catalysis A: General*, **374**(1–2), pp. 79–86.

76. Nutt, M. O., Hughes, J. B., and Wong, M. S. (2005). Designing Pd-on-Au bimetallic nanoparticle catalysts for trichloroethene hydrodechlorination. *Environmental Science & Technology*, **39**(5), pp. 1346–1353.

77. Hou, W., Dehm, N. A., and Scott, R. W. (2008). Alcohol oxidations in aqueous solutions using Au, Pd, and bimetallic AuPd nanoparticle catalysts. *Journal of Catalysis*, **253**(1), pp. 22–27.

78. Shi, X., Jiang, X., Lu, L., Yang, X., and Wang, X. (2008). Structure and catalytic activity of nanodiamond/Cu nanocomposites. *Materials Letters*, **62**(8–9), pp. 1238–1241.

79. Hunt, A. J. (1986). Solar radiant heating of gas-particle mixtures.

80. Haussener, S., Hirsch, D., Perkins, C., Weimer, A., Lewandowski, A., and Steinfeld, A. (2009). Modeling of a multitube high-temperature solar thermochemical reactor for hydrogen production. *Journal of Solar Energy Engineering*, **131**(2).

81. Hussein, A. K., Li, D., Kolsi, L., Kata, S., and Sahoo, B. (2017). A review of nano fluid role to improve the performance of the heat pipe solar collectors. *Energy Procedia*, **109**, pp. 417–424.

82. Goel, N., Taylor, R. A., and Otanicar, T. (2020). A review of nanofluid-based direct absorption solar collectors: Design considerations and experiments with hybrid PV/Thermal and direct steam generation collectors. *Renewable Energy*, **145**, pp. 903–913.

83. Taylor, R. A., Phelan, P. E., Otanicar, T. P., Adrian, R., and Prasher, R. (2011). Nanofluid optical property characterization: Towards efficient direct absorption solar collectors. *Nanoscale Research Letters*, **6**(1), p. 225.

84. Khan, M. S., Abid, M., Ali, H. M., Amber, K. P., Bashir, M. A., and Javed, S. (2019). Comparative performance assessment of solar dish assisted s-CO$_2$ Brayton cycle using nanofluids. *Applied Thermal Engineering*, **148**, pp. 295–306.

85. Otanicar, T. P., Phelan, P. E., Prasher, R. S., Rosengarten, G., and Taylor, R. A. (2010). Nanofluid-based direct absorption solar collector. *Journal of Renewable Sustainable Energy*, **2**(3), 033102.

86. Sani, E., Vallejo, J. P., Mercatelli, L., Martina, M. R., Di Rosa, D., Dell' Oro, A., and Lugo, L. (2020). A comprehensive physical profile for aqueous dispersions of carbon derivatives as solar working fluids. *Applied Sciences*, **10**(2), 528.

87. Tyagi, H., Phelan, P., and Prasher, R. (2009). Predicted efficiency of a low temperature nanofluid-based direct absorption solar collector. *Journal of Solar Energy Engineering*, **131**(4), 041004.

88. Han, D., Meng, Z., Wu, D., Zhang, C., and Zhu, H. (2011). Thermal properties of carbon black aqueous nanofluids for solar absorption. *Nanoscale Research Letters*, **6**(1), 457.

89. Otanicar, T. P. and Golden, J. S. (2009). Comparative environmental and economic analysis of conventional and nanofluid solar hot water technologies. *Environmental Science & Technology*, **43**(15), pp. 6082–6087.

90. Lu, L., Liu, Z.-H., and Xiao, H.-S. (2011). Thermal performance of an open thermosyphon using nanofluids for high-temperature evacuated tubular solar collectors. *Solar Energy*, **85**(2), pp. 379–387.

91. Veeraragavan, A., Lenert, A., Yilbas, B., Al-Dini, S., and Wang, E. N. (2012). Analytical model for the design of volumetric solar flow receivers. *International Journal of Heat and Mass Transfer*, **55**(4), 556–564.

Multiple Choice Questions

1. The size of dispersion phase particles in nanofluids is
   - (a) >1 nm
   - (b) Between 1 and 100 nm
   - (c) <100 nm
   - (d) All of the above
2. What is dispersion phase?
   - (a) It is the main phase.
   - (b) It is the basic colloidal layer.

    (c) The phase whose particles are distributed in another continues phase.

    (d) It is the solvent of the solution.

3. Dispersion phase is

    (a) Discontinuous phase

    (b) Continuous phase

    (c) Sometimes continuous and sometimes discontinuous

    (d) None of above

4. Dispersion medium is

    (a) Discontinuous phase

    (b) Continuous phase

    (c) Sometimes continuous and sometimes discontinuous

    (d) None of above

5. The nature of a nanofluid solution is

    (a) Homogeneous

    (b) Heterogeneous

    (c) Homogeneous and heterogeneous

    (d) None of above

6. Nanoscience can be studied with the help of

    (a) Quantum mechanics   (b) Newtonian mechanics

    (c) Macro-dynamics   (d) Geophysics

7. The nature of surfactants could be

    (a) Cationic in nature   (b) Anionic in nature

    (c) Zwitterion in nature   (d) All of the above

8. The full form of STM is

    (a) Scanning tunneling microscope

    (b) Scientific technical microscope

    (c) Systematic technical microscope

    (d) Super tensile microscope

9. What is fouling effect?

    (a) The deposition of any unwanted material on heat transfer surface

    (b) Also called corrosion

    (c) Form of oil layer on metal

    (d) None of above

10. Which is the correct full form of XRD?

    (a) X-ray diffraction   (b) X-ray diffractor

    (c) X-ray difference   (d) X-ray diagram

11. What is dispersion medium?

    (a) It is the main phase.

    (b) The medium in which nanoparticles (dispersion phase) are dispersed.

    (c) It is the basic colloidal layer.

    (d) The phase whose particles are distributed in another continuous phase.

12. Which is the correct form of TEM?

    (a) Tuning electron microscopy

    (b) Tunneling electron microscopy

    (c) Transmission electron microscopy

    (d) Trans electron microscopy

13. Which is the correct full form of SEM?

    (a) Scanning electron microscopy

    (b) Surface electron microscopy

    (c) Supra electron microscopy

    (d) Super electron microscopy

14. Which of the following is a bottom-up approach?

    (a) Chemical precipitation method

    (b) Sol–gel method

    (c) Hydrothermal/solvothermal method

    (d) All of the above

15. Which of the following is a top-down approach?

    (a) Lithography   (b) Chemical etching technique

    (c) Attrition-milling process   (d) All of the above

Answer Key

1. (b)   2. (c)   3. (a)   4. (b)   5. (b)   6. (a)   7. (d)

8. (a)   9. (a)   10. (a)   11. (b)   12. (c)   13. (a)   14. (d)

15. (d)

Short Answer Questions

1. Define nanofluid.
2. What are the two approaches for the synthesis of nanofluids?
3. What are the important applications of nanofluids?
4. What is the role of nanofluids in solar energy harvesting?
5. What is electro-wetting?
6. What is the role of surfactant in the stability of nanofluids?
7. Define zeta potential.
8. Write any five merits of nanofluids as a cooling agent in various heat-producing devices.

Long Answer Questions

1. Explain two approaches with their various methods for the synthesis of nanofluids.
2. Explain the role of nanofluids in organic reactions.
3. Explain various biomedical applications of nanofluids.
4. Explain the uses of nanofluids in electronic devices.
5. How nanofluid is beneficial for society? Explain with suitable examples.
6. Explain the role of nanofluids as heat transfer agents in various fields.
7. Explain various motorized applications of nanofluids.

Chapter 15

Nanoelectronics: Basic Concepts, Approaches, and Applications

Balwinder Kaur,[a] **Radhika Marwaha,**[a] **Subhash Chand,**[b]
and Balraj Saini[c]

[a]*Department of Chemistry, Punjabi University, Patiala, Punjab, India*
[b]*Lajpat Rai DAV College, Jagraon, Punjab, India*
[c]*Chitkara College of Pharmacy, Chitkara University, Punjab, India*
balwindertaheem1987@gmail.com

15.1　Introduction

In the 21st century, nanotechnology has emerged as a cutting-edge technology and has implausible applications in physics, chemistry, biology, material science, and medicine. The main research and development are toward the fabrication of novel materials by altering size, shape, and distribution of particles. Nanoelectronics is an offshoot of nanotechnology, i.e., engineering of tiny machines. It is a term that refers to the use of nanotechnology in electronic components. It is an art and science of miniaturization of matter at the nanoscale. The tinier the electronic component, the harder it is to manufacture.

Nanotechnology: Principles and Applications
Edited by Rakesh K. Sindhu, Mansi Chitkara, and Inderjeet Singh Sandhu
Copyright © 2021 Jenny Stanford Publishing Pte. Ltd.
ISBN 978-981-4877-43-5 (Hardcover), 978-1-003-12026-1 (eBook)
www.jennystanford.com

There has been a steady growth in nanoelectronics all through the past decades. Nanoelectronics include a diverse set of devices and nanomaterials with the common characteristic that they are so small that their physical effects alter material properties on the nanoscale. Interatomic interactions and quantum mechanical properties play a vital role in the functionality of these devices. Although much of this work in the field of nanoelectronics has long-term explanation, yet it has been considered most in III–V semiconductors and mainly in Ga(Al)As alloy systems where the technologies are used. In addition, the transport property of GaAs and high-quality heterojunction have made them accessible to dimensional range used in high-performance lithography [1].

In 1965, Gordon Moore, co-founder of Intel Corporation hypothesize a doubling every year in the number of components per integrated circuit. This set up standards and objectives in the field of growth of semiconductor technology, yet we have not touched the limit. The scientists had predicted the end of Moore's law by 2018 with 16 nm feature size of miniature transistor. Advanced chip technology works on the development of nanoelectronic devices, which overcome this limit of miniaturization [2]. Among the widely used materials in nanoelectronics, zero-dimensional quantum dots, one-dimensional materials such as nanotubes and nanowires, and C-based materials such as carbon nanotubes, fullerene, and graphene are included. In the next 10–20 years, enhancement of technology under molecular or quantum operations is expected [3, 4]. This chapter focuses on the approach of nanotechnology toward nanoelectronics, various nanoelectronic devices, and their emerging applications in the various fields of life.

15.2   Need of Nanoelectronics

Before we proceed further toward various nanoelectronic devices, it is essential to understand the reason behind the fascination of consumers toward these supermini toys. For the last few decades, the electronic industry has been keen to emphasize upon the simultaneous reduction in size and cost along with improvement in the performance of electronic products. This has resulted in the fascination of consumers toward various electronic products such as mobile phones, computers, and display sensors. The increase in the demand for miniatured electronic products has led to the desire of companies to be competitive and has opened the gateway for research in the field of nanoelectronic devices. The product size is reduced by the miniaturization of the individual electronic devices such as transistors. Small-size devices have direct economic benefits as the cost of integrated circuit chips is dependent on the number of chips produced per silicon wafer. An increase in the density of device decreases the cost per bit.

Several more reasons for shrinking the size of electronic devices include low consumption of energy, improved functionality along with improved data storage capacity. Researches in the field of nanoelectronic devices have led to the possibility of supermini low-power communication and computing devices along with embedded sensors. Incorporation of electronics with biological systems will be the outcome of further shrinking of devices. Furthermore, as the process of miniaturization continues, it would be possible to put the power of today's computers in the fingers of a person's hand.

15.3 Basic Underlying Principle

The behavior of nanoelectronic devices is entirely quantum mechanical, which cannot be realized without the knowledge of the following two effects: quantum tunneling and Coulomb blockade.

15.3.1 Quantum Tunneling

Quantum tunneling refers to the quantum mechanical phenomenon where a particle tunnels through a barrier that it classically could not surmount. Classical mechanics could not explain this phenomenon. The difference arises because classical mechanics treated matter as solely particles, whereas quantum mechanics emphasizes upon wave-like character of material particles, i.e., dual character of matter. Heisenberg's uncertainty principle imposes a limit on the simultaneous measurement of position and momentum of particles accurately and forms the basis. It implies that there are no regions of zero probability; hence, the probability of the given particle to exist on the opposite side of the barrier is nonzero, which leads to its tunneling on another side of the barrier.

15.3.2 Coulomb Blockade

Coulomb blockade is the term used for increased resistance at the small bias voltage of electronic devices having at least one low-capacitor tunnel junction. The resistance of the device as a result of this Coulomb blockade no longer remains constant but increases to infinity for zero biasing.

Classical electrodynamics imposes a restriction to the flow of current through an insulating barrier, but quantum mechanical laws state that there is a nonzero probability of electrons to be on the other side of the barrier upon zero biasing, so there would be a current. We can say tunnel junction behaves like a resistor with a constant current, termed ohmic resistor, with exponential dependence of resistance on the barrier thickness. The two conductors with an insulating layer behave like a capacitor along with the resistor, so termed dielectric. Due to the discreteness of electrical charge, one electron current is passed through the tunnel barrier, so the junction has one electron charge causing voltage $V = e/C$ where V is voltage, e is electronic charge, and C is capacitance. A large voltage build prevents another tunneling. Power then decreases under low biased voltage, and resistance no longer becomes constant. This increase in the resistance under zero biasing is called Coulomb blockade [5].

In the past few years, due to the shrinkage in the size of transistors, computers have grown more powerful [6]. At the same time, it is true that quantum mechanical laws and limitation of fabrication techniques, power dissipation, etc. will soon prevent further miniaturization of conventional field-effect transistors (FETs). Researchers have explored that in the next 15 years, mass-produced transistors will shrink from the present 250 nm dimension to 100 nm and below, which would become difficult and costly to fabricate. Also such ultra-dense circuits will face the challenge of proper functioning [7–16].

Attempts have been made to find an alternative to these conventional transistors for ultra-dense circuits and to continue this process of shrinking to nano and even to the molecular scale. New electronic devices on the nanoscale serve both as an amplifier and a switch like FETs. But unlike FETs, they do not operate on the basis of movement of bulk electrons, rather upon discreteness of electrons. Alternative nanoelectronic switches and amplifiers can be broadly divided into two categories: solid-state quantum effect and single electron devices and molecular electronic devices.

Before we go to alternatives for conventional FETs, it is important to examine conventional microelectronic transistors along with the possible limitations of miniaturization. In this chapter, we confine our discussion to solid-state quantum effect and single electron devices.

15.4   Microelectronic Transistors

Transistors can be used to set voltage and for amplification. The primary types of transistors are FETs and bipolar junction transistors (BJTs). Metal oxide semiconductor field-effect transistor, known as MOSFET, is a commonly used transistor. Three terminals are source, drain, and gate. The shrinking of device dimensions to nanoscale changes the mode of operation, whereas basic components (source, drain, and gate) remain in the same operational mode. The miniaturization to nanoscale thereby changes the channel through which current flows from source and drain.

The basis of the semiconductor is doped silicon, boron, and arsenic, which are used as dopant to produce charges for conduction. The negatively charged semiconductor is n-doped, and electrons are responsible for the conduction of current, whereas in p-doped semiconductors, holes are responsible for conduction.

The gate of MOSFET is a metal electrode separated from the semiconductor using an insulating oxide barrier. The voltage and the electric field of MOSFET's gate control the current flow from the source to the drain, and for this reason, it is called FET [17]. The miniaturization of electronic circuits to the nanoscale can be achieved by shrinking the dimensions of all the circuit components by a constant factor known as scaling; it preceded at a very high rate, thereby doubling the number of transistors on the chip. Platinum chips contain more than 3.2 million transistors with a minimum feature of 350 nm [18]. MOSFETs reach a minimum feature size of 100 nm, and lower than this may not persist [16, 17].

Transistors having a gate length of 40 nm are already in operation [19, 20]. Further reduction in gate length to 25 nm has been achieved by using GaAs [21]. The increase in the density of such transistors decreases the flow of current. Along with this, shrinking of electronic devices in FETs has the following limitations [22, 23]:

- Avalanche breakdown due to high-energy electrons created by a high electric field applied to a small region
- Heat dissipation due to increase in density
- Nonuniform device operation
- Shrinkage of depletion region [8]
- Shrinkage and nonuniformity of oxide layer

The thermodynamic limitation to the scaling of FETs, along with the various limitations to miniaturization, suggests that it would be desirable to find a replacement of FETs. This permits the construction of electrical circuits requiring few switching devices for performing the same function. Decrease in the effectiveness of doping and increase in the significance of quantum mechanical effects are some other limitations in this area. On approach to the nanoscale or even the molecular scale, bulk properties of a solid are replaced by quantum mechanical properties. Further, it would be advantageous if it depends on the quantum properties rather than the doped material [24].

15.4.1 Solid-State Quantum Effect and Single Electron Nanoelectronic Devices

Many nanoscale solid-state replacements have been suggested for semiconductor transistors to overcome the limitations to miniaturization. The operation of all these nanoscale devices is based on quantum mechanics.

The common feature of all these devices is the island composed of semiconductor or metal-containing electrons. It is the replacement of channel in the conventional FETs. The composition, shape, and size of the island decide the properties of solid-state nanoelectronic devices. Microelectronic devices make use of the silicon element belonging to group 14 of the order table, whereas most of the solid-state nanoelectronic devices make use of elements from groups 13 and 15, for example GaAs and AlAs [25, 26]. The devices made from 13 and 15 groups are cost effective and easy to fabricate. Solid-state quantum effect devices can be further divided as follows: quantum dots (QDs), resonant tunneling diodes (RTDs), and single electron transistors (SETs).

15.4.1.1 Quantum dots

A tiny speck of a matter concentrated into a single point is given the name quantum dot. Quantum dots are one of the widespread nanoparticles having diameter in the range of 2–10 nm [27]. On inorganic basis, semiconductors quantum dots pose crystalline arrangement of atoms or ions. The label quantum is due to the confinement of electrons. Quantum dots have combined properties of bulk semiconductor and discrete molecules. The high surface-to-volume ratio of these particles is responsible for this. Although they are crystals, they behave more like atoms and hence also termed artificial atom. Quantum dots can be single elements such as Si, Ge or compounds such as CdSe, CdS, PbSe, PbS, InAs, and InP.

15.4.1.1.1 *Working of quantum dots*

When some of the electrons from the dot material absorb a photon, they get excited to the conduction band from the valence band, as a result of which a hole is formulated in the valence band. The electron and hole pair together form an exciton.

The average distance of separation between electron and hole is termed Bohr's radius. When the size of the semiconductor is reduced in comparison to Bohr's radius, it is termed quantum dot. Energy is released when the excited electron returns to the hole in the valence band. The amount of energy release depends on the band gap. The color of light emitted depends on the size of the dot. Relatively small size of dots leads to consistent energy released from electron to electron. Large size dots of dimensions 5–6 nm yield lower energy (red and orange color), whereas small dots of dimensions 2–3 nm yield high energy in the blue and violet regions of the spectrum [28]. It is called the quantum confinement effect, indicating the predominance of constraints at the atomic level (Fig. 15.1).

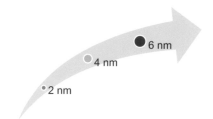

Figure 15.1 Size of nanoparticles (dots) varies with the color they emit.

15.4.1.1.2 *Synthesis of quantum dots*

In the quantum dots, electrons being confined in all the three dimensions, they are zero dimensional, but there exist some other quantum confined semiconductors such as quantum wires that contain the confined electrons and holes in two dimensions and leave the third dimension free for propagation. Quantum well confines electron and holes in one dimension, thereby leaving two dimensions free for propagation. Cadmium selenide, cadmium sulfide, or indium arsenide semiconductor materials are used as quantum dots.

In the process of synthesis of quantum dots, an appropriate metallic or organometal precursor such as zinc combines with a chalcogen precursor such as sulfur. The solvent used should be stable at high temperature so that particle aggregation can be controlled; another feature for a solvent is that it must act as a surfactant molecule so that the surface of quantum dot can be stabilized. There are numerous techniques for the synthesis of these tiny dots, and methods employed can be broadly divided into the following two categories:

15.4.1.1.2.1 Top-down synthesis

Quantum dots of large dimensions from semiconductor material are synthesized by the top-down approach. For the synthesis of quantum dots having specific size and shape, a series of experiments on the quantum confinement effect are needed. However, this process suffers from various defects and incorporate impurities, yet numerous methods employed under this technique are as follows:

- Electron beam lithography

- Focused ion beam technique
- Itching technique

15.4.1.1.2.2   Bottom-up synthesis

This type of synthesis can be broadly divided into two categories:

I. Wet chemical methods:

- Gel process
- Microemulsion process
- Hot solution decomposition process

II. Vapor phase methods:

- Molecular beam epitaxy
- Physical vapor deposition
- Chemical vapor deposition

15.4.1.1.3   *Application of quantum dots*

Quantum dots are novel innovations for helping the industry. The quantum dot technology has applications in numerous fields such as light-emitting diodes (LEDs), solar cells, biomedicine, instrumentation, quantum computing, and many more. This technology offers the biggest color gamut of latest inventions. The three key elements in the quantum dot structure are core, shell, and ligand. The core part absorbs and releases light. The shell is responsible for confining emission and passivating defect, whereas the ligand provides stability. Some of the applications are explained in detail.

15.4.1.1.3.1   Light-emitting diodes

People are experiencing a serious shortage of energy resources due to the increasing population and the consumption of a large amount of energy. Many researches have been performed to achieve efficient and low-energy light sources. Inorganic LEDs and organic light-emitting diodes (OLEDs) are the result of these continuous efforts to achieve solid-state light source [6]. The use of polymers and organic molecules as emissive layers to improve the characteristics of LEDs is the latest research in this area. The OLEDs display flexibility (tendency to deposit on plastic substrate), very thin displays, and transparency. An OLED is constructed by introducing a thin film of carbon-based material between an electron-emitting cathode and an electron-removing side anode, thereby considering one of electrodes transparent. The thin film of organic material called emitter is electroluminescent; it emits light upon excitation using electric current. The organic materials have conductivity between the conductors and insulators, so they behave like organic semiconductors. The highest occupied molecular orbital (HOMO) and the lowest unoccupied molecular orbital (LUMO) behave like valence band and conduction band. OLEDs have advantage over both LCDs and LEDs for being a thinner, lighter, more flexible, and brighter substrate.

Quantum-dots-based LED technology is the new type of LED technology based on nanoparticles although the structure of LEDs is similar to that of OLED technology. The difference is that a layer of quantum dots is placed between the electron and hole transporting layers. The application of an electric field generates electron and hole pairs in the quantum dot layer, combination of which emits a narrow spectrum of photon. For example, the full width half maxima (FWHM) for cadmium-based LED is 25–35 nm and for cadmium free LED is 40–50 nm.

The first structure of quantum-dot-based LEDs was studied in 1994. It consists of a layer of cadmium selenide quantum dots and a polymeric electron transport layer placed between two electrodes. The application of an organic cathode (CIM/LiF/Al) increases the quantum efficiency along with better device performance and brightness. This improvement is attributed due to balance electron/hole injection due to the presence of organic CIM. The current efficiency is low in n-type materials as compare to p-type materials. Hence, in p-type materials, injection of hole barriers of organic hole transport layer are required for improvement of QDLED. ZnCdSe core/multishell QD's employee brightness and efficient blue color. The efficiency and brightness can be improved by dropping polyvinyl (N-cabazole) PVK in the emissive layer [5].

It has low energy of HOMO, which balances the charge injection and reduces the barrier at the interface of quantum dots and the hole transport layer. The shell and their compounds determine the properties of quantum dots, and the diameter of the cluster plays a significant role in determining the band gap of quantum dots. With a small cluster diameter, band gap will be lower. Also thickness of the shell impacts the efficiency. The repulsive force between the polymer molecular chains prevents the accumulation of nanoparticles, so polymer quantum dots are good options.

As we know cadmium sulfide or cadmium selenide is used in quantum dots. These particles are great threats because during in vitro degradation, cadmium ions are discharged. The company Samsung uses cadmium-free quantum dot technology and, nowadays, makes use of indium as quantum dots [29, 30]. Approximately 1 billion color combinations can be produced for better light efficiency. The team at Samsung has improved the structure of quantum dots made of environment-friendly indium phosphide. The structure prevents oxidation of the core and builds a thick symmetrical shell around it to prevent leakage. The ligand on the shell surface has also been made shorter to allow it to absorb electron current faster. The quantum efficiency increased up to 21.4% and lifetime to million hours.

15.4.1.1.3.2 Photodetectors

One of the applications of quantum dots is their promising use in photodetectors for detecting infrared and visible light. In the infrared region, photodetectors can be used in biomedical imaging, quality control, night vision cameras, and product inspection. In the visible region, photodetectors find application in image sensors for the conversion of light to electronic signals. Quantum dots can also be used in various other fields such as surveillance, industry inspection, spectroscopy, and biomedical imaging. This is because quantum dots can be integrated with silicon electronics and flexible organic substrate. The tunability of optical absorption and emission spectra by adjusting the size of quantum dots stands behind this application. PbS dots are used in the infrared region, whereas CdSe and InP are used in the UV–visible region [31, 32]. Some examples of quantum dot photodetectors based on the latest researches are as follows:

(a) **In_2S_3 quantum dot photodetectors**

Low-dimensional materials, due to their unique photoelectric properties along with their quantum dots, gained so much interest in the latest researches. Graphene-like two-dimensional nanomaterials have large technological importance [33, 34]. In_2S_3 quantum dots belong to group 13 and 16 semiconductor materials and find application in photodetectors, photocatalytic degradation, and many more [35–40].

The method of preparing sulfide quantum dots can be studied under top-down and bottom-up approaches. The bottom-up methods such as thermal template and the microwave process have certain limitations. A novel method for the synthesis of indium sulfide at atmospheric conditions has been developed by using indium chloride and sodium sulfide as indium and sulfur sources, respectively. Investigations reveal that one of the methods of preparation of indium sulfide quantum dots is by using $Na_2S·4H_2O$ (sodium sulfide) and $InCl_3·4H_2O$ (indium chloride) [38, 42, 47].

A photodetector device based on indium sulfide quantum dots after fabrication shows stabilization at 10^{13} Jones under 365 nm ultraviolet irradiation at room temperature. This reflects the high potential application of indium sulfide in photodetectors. In comparison to others, this method is cost effective, fast, and environment friendly. It also opens the gateway to research in sulfide quantum dots in the field of photodetectors.

(b) $CH_3NH_3PbBr_3$ photodetectors

$CH_3NH_3PbBr_3$ perovskite halide quantum dots are used for visible region photodetectors. Perovskite quantum dots with the common formula $APbX_3$ (A = Cs, MA (methylammonium), FA (formamidium) and X = Cl, Br, I) are the latest development in quantum dot research. They have photoluminescence quantum efficiency up to 95% and its FWHM is 20–30 nm. They are the best alternative for cadmium-selenide-based quantum dots. The size and composition of halide can tune the emission wavelength to cover the entire visible region of the spectrum.

$CH_3NH_3PbBr_3$ quantum dots have been synthesized for application in the visible region photodetectors. From optical studies, it is confirm that the average size of fabricated quantum dot is 2–4.5 nm. Time-dependent photoresponse can be stabilized by several on and off cycles of lamp [43].

(c) Colloidal quantum dot multispectral photodetectors

Colloidal quantum dot (CQD) optical sensors work in the middle to large infrared region of the spectrum and are applied in a large number of fields such as thermal imaging, gas sensing, and investigation of hazards in the environment. These quantum dot sensors are the perfect replacement for costly and toxic mercury-based compounds containing quantum dot infrared photodetectors. Mercury-free CQDs can be used to detect light in the region of wavelength, which otherwise is difficult to access. The detector used here is lead sulfide based, which is compatible with CMOS manufacturing technology.

PbS CQD has emerges a promising detector in the size range of 1–2 micron, also suffers from the drawback that it has put a limit on the energy band gap. Lead sulfide has a lower limit of 0.3 eV; however, doping with iodine reduces this value to 0.1 eV [44]. Huge doping with iodine makes it possible to excite the electron with the photon using much lower energy than used before. New researches have now emerged that replace iodine with sulfur, using the ligand exchange process. This process of doping is done in large dots as they have exposed sulfur in comparison to smaller ones. The increase in the size of dots increases the wavelength of absorption of the infrared region. On iodine doping in PbS QD samples/matrix, a band like spectrum is obtained which pose utility in hyper spectral (multispectral) imaging which yield vital informations: compositional as well as visual information like sensing aspects.

15.4.1.1.3.3 Biomedical applications

Quantum dots reflect several advantages as luminous probes for biomedical imaging. It is attributed to their high photostability, broad absorption spectra, large extinction coefficient, and tunable emission wavelength. The surfaces of quantum dots can be developed in such a manner that they can attach surface groups such as COOH, NH$_2$ to allow conjugation with biomolecules such as antibodies, polysaccharides, and peptides. These bioconjugated quantum dots have been used as probes in the several areas such as DNA hybridization, receptor-mediated endocytosis, monitoring of parasite metabolism, and real-time visualization of tissue in cellular structure along with diagnostic applications [46–48].

Organic labeled dyes do not have near-infrared emission possibility. This accelerates the use of quantum dots as non-theranostic platform of sensing, imaging, and therapy. A number of researches in the use of quantum dots in the field of sensors, drug delivery, and biomedical imaging have been published. Some of reviews of latest developments available in the literature regarding the use of quantum dots for medical applications are as follows:

(a) **Water-soluble quantum dots** have short life due to colloidal instability, which is a major limitation to their exploitation. PbS nanoparticles capped with dihydrolipoic acid–polyethylene glycol (DHLA–PEG) ligands terminated with functional groups such as –COOH, –NH$_2$, and –N$_3$ find their applications in the in vivo imaging [49, 50]. The PEG-coated PbS QDs have high quantum yield and near IR based emission in optical studies but low absorption of biological tissues. Near-infrared imaging of in vivo biodistribution is noted at wavelength greater than 1000 nm with low absorption of tissues and light scattering.

(b) **Photothermal therapy (PTT)** has the potential for cancer treatment. The novel liquid exfoliation method has led to the development of photothermal agents based on the two-dimensional antimonene quantum dots (AMQDs). PEG is used for the surface modification of AMQDs, which enhances their biocompatibility and stability. PEG-coated AMQDs have a photothermal conversion efficiency of 45.5%. This platform of AMQD-based photothermal agents also exhibits the unique feature of near-infrared-induced rapid degradability [45]. The 2D antimonene (AM) has expanded its utility in biomedical applications through the development of photothermal agents [57].

15.4.1.1.3.4   Solar cells

A quantum dot solar cell (QDSC) utilizes quantum dots as the absorbing photovoltaic material. This new invention replaces the bulky materials such as silicon or copper indium gallium selenide used in conventional solar cell devices. By 2030, solar PV market will grow by a multiple of 10's. QDSCs can be a cost-effective and highly efficient technology.

As we know, quantum dots have tunable energy band gaps. The alteration in size results in a variety of band gaps without changing the underlying material. This property makes them desirable for solar cells. Lead sulfide colloidal quantum dots are capable of giving energy in the far-infrared region. Most of the solar energy reaching the earth's surface is in the infrared region. QDSCs have higher accessibility for the infrared energy than other photovoltaic solar cell devices.

Researchers at the US Department of Energy's National Renewable Energy Laboratory have shown that nanotechnology may greatly increase the amount of electricity produced by solar cells. A photovoltaic solar cell produces three free electrons upon the absorption of a single photon of sunlight. In this process, the maximum amount of energy is lost in the form of heat. However, QDSCs reduce the heat loss and convert sun's energy to electricity, which make them cost competitive in comparison to conventional power sources.

The conversion of solar energy to electricity by QDSCs can be explained with the help of the multiple excited generation process. In this process, the absorption of a single photon of light produces multiple pairs of electrons and holes. The light-absorbing layer present on the quantum dot surface contains transparent electrodes, which absorb the photon of the light; the electron will travel/migrate toward electrode to produce current.

The advantages of using QDSCs are as follows:

- Yield high efficiency with favorable power-to-weight ratio
- Compact and flexible
- Low power consumption
- Cost effective
- Versatile in nature as they can be used in windows other than rooftops.

QDSCs also suffer from some limitations. CdSe used in QDSCs is toxic. For this reason, they require a stable polymer shell. The accumulation of toxic cadmium and selenium materials in the human body along with quantum dot metabolism are still major challenges. QDSCs may also have some surface defects that trap the recombination of charge carriers. Research is still needed in this direction before the commercialization of these devices [52].

15.4.1.2   Resonant tunneling diodes

Introduction
A resonant tunneling diode (RTD) is a diode in which electrons tunnel through some states that are in resonance. These diodes are operated by quantum mechanical tunneling, i.e., when a potential barrier is applied, electrons tunnel through certain energy levels. The tunneling of electrons is measured graphically via current–voltage characteristics. It was first proposed by Tus et al. [53]. The study of RTD is highly advantageous as circuit diagram is simple and negative differential resistance to yield to lower power consumption [54]. RTDs utilize microwave source (high frequency) in terahertz range [55]. RTDs have found their application in oscillators, optoelectronics, and detection of photons.

Choice of semiconductor materials for RTDs
Groups II-V and II-VI combination semiconductors are highly accepted for the fabrication of RTD. The stated semiconductor groups have attractive physical properties as follows:

- Tunable band gap
- Quantum confinement effect
- High power capacity
- Holding capacity of high power terahertz emission at room temperature

Binary III–V semiconductor materials create a potential barrier (multiple or double) in the conduction and valence bands. One of the III-nitride double barrier structures has helped to expose polarization-coupled carrier dynamics, which is found complicated in quantum cascade lasers. Three parameters that tune RTD activity are as follows:

1. The height of potential barrier
2. The impurity concentration between two ohmic layers
3. Thickness of the active layer

The peak current falls at a much higher rate with an increase in barrier thickness. The height of barrier and the thickness of layer should be chosen properly to maintain peak current density. To attain high peak current density, the electron density near the barrier of current flow should be high. Introduction of a step heterojunction emitter space allows high current density [56]. Also the introduction of sub-quantum wells can alter the tunneling operation, i.e., 3D–2D conversion to 2D–2D leads to improved emitter carrier injection efficiency [57].

RTDs can also be fabricated by using Si/SiGe materials. But their use is limited due to discontinuities between the conduction and valence bands of Si and SiGe alloy. Here, only at a low temperature, negative differential resistance was observed. Rong et al. [58] introduced $Al_{0.69}In_{0.31}N$ as an electric barrier to study peak current measurements.

15.4.1.3 Single electron transistor

Introduction
A single electron transistor (SET) is a new type of switching device that uses controlled electron tunneling to amplify current. This transistor is constructed based on quantum mechanical principle. The electron tunnels between the electrical contact and the nanoscale object. The gate voltage controls the electron tunneling through islands, i.e., it is a critical parameter along with the path length between placed nanoparticles. The above-stated conditions and the parameters are achieved by the following three approaches: break junction, electron deposition, and electromigration (EM).

The widely operated approach is electromigration as it includes the dynamics of the materials to which the driven force is provided by high current density. The following are the merits of the electromagnetic approach:

- Its easy-to-control algorithm

- Performed in ultrahigh vacuum
- Compatibility with gate electrode

Further, the use of nanoparticles is important due to their fascinated physical and chemical properties as follows:

- Spatial atomic arrangements
- High surface-to-volume ratio
- Quantum confinement effect

Due to these remarkable properties, nanoparticles are used in memory devices in the nanoelectronic industry [59]. The general block diagram of SET is shown in Fig. 15.2.

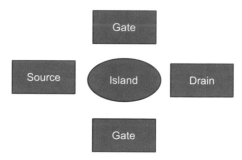

Figure 15.2   Illustration of SET with two gates.

Types of SET

Single electron transistors are of many types depending on the type of nanomaterials used in the matrix. Generally, semiconductor nanoparticles and conducting carbon nanotubes are used. Some of the important categories are as follows:

1. **Graphene-based materials**

   The interest in graphene nanostructures (quantum dots) has arisen due to their noise characteristics when used in electronic devices [60]. A decade ago, microscale electronic devices gained significant attention in important biological aspects like biosensors [61, 62] and specific interaction with biomolecules [63, 64]. Sophistication came into existence by significant observation in the current variations in these materials. Nowadays, interest has built up because of their noise characteristics due to nanoscale dimensions. The electronic property of noise characteristics is basically optimized by (a) surface defects and (b) specific substrate. Noise characteristics in the previous studies have been observed on cryogenic conditions, which need high precautions. Jasper et al. have studied the noise characteristics in the liquid-gated environment [60]. They have made an inference in terms of fluctuations in source, drain, and gate potentials. Graphene has established its existence due to electronic properties such as weak spin orbital coupling and forbidden hyperfine interaction. The binary quantum dots of GaAs are highly used in SETs due to the remarkable measurements observed by both high charge sensitivity and high speed for studying discrete electron transport [65]. Lin et al. reported a device in which graphene quantum dots are placed and the distances between nanostructures are measured by the etched area. The SET and graphene quantum dots produced a strong coupling due to their proximity to each other. Here, the substrate is Si element.

No doubt, semiconductor materials are very important for the electronic industry. But with the passage of time, miniaturization of devices is highly demandable. We have studied that SET is made with a chemical material that needs to possess fast electron transmission. Graphene possesses a honeycomb lattice with fast electron transfer. It is an island material. Figure 15.3 shows a block diagram with two quantum dots placed side by side, which has a big advantage that, more the path length (of graphene QD), less will be obstacle to the passage of current rather than shorter path length [66]. Also an increase in the gate voltage increases the current. Last but not the least, temperature also has an impact on the current passage. The highest current passage is recorded at a lower temperature (cryogenic condition) because the SET system is dependent on the Fermi probability function, which has a direct relation with the low-temperature condition.

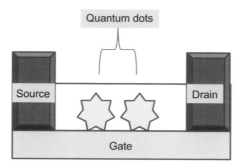

Figure 15.3 SET with double quantum dots.

2. **Carbon nanotubes**

The excellent electrical and mechanical properties [67] of carbon nanotubes have made them a widely used material in SETs. Since these materials are conducting in nature and also have well-defined dimensions, they are suitable for current passage at room temperature. Other parameters such as temperature and gate voltage have a direct impact on current passage. The current transport is governed by the length difference of two closely placed nanotubes. Mostly, carbon nanotube SET devices are fabricated by chemical vapor deposition [68]. Due to the mechanical properties, carbon nanotubes have outstanding performance at the smallest diameter of 1 nm. Single-walled carbon nanotubes exhibit Coulomb blockade in SETs as compared to multiwalled nanotubes and from this noise properties are investigated.

15.5 Conclusion and Future Aspects

In continuation of the exponential rate of shrinking of electronics into the next century, one must be aware of the obstacles of miniaturization. The fundamentals of thermodynamics and quantum mechanics along with the certain obstacles arising from the cost and fabrication challenges, heat dissipation, etc. impose a limit on miniaturization. This chapter explains the functions and applications of various solid-state devices, along with the evolution of nanoelectronic devices from the microelectronic devices and circuits. This worldwide enterprise of engineering nanoscale devices is still ongoing, and it will definitely transform our computing and technology infrastructure in the future.

Acknowledgment

The authors Dr. Balwinder Kaur and Dr. Subhash Chand are thankful to the UGC-SAP and the Chemistry Department, Punjabi University, Patiala, and Lajpat Rai DAV College, Jagraon, Ludhiana, for providing necessary facilities. Dr. Balwinder Kaur is highly obliged to UGC-New Delhi for providing MANF.

References

1. Beaumont, S. P. (1996). III–V nanoelectronics. *Microelec. Engine.*, **32,** pp. 283–295.

2. Edenfeld, D., Kahng, A. B., Rodgers, M., and Zorian, Y. (2004). Semiconductor technology roadmap for semiconductors. *Computer*, **37**, pp. 47–56.

3. Yu, B. and Meyyappan, M. (2006). Nanotechnology: Role in emerging nanoelectronics. *Solid-State Elec.*, **50**, pp. 536–544.

4. Sangwan, V. K. and Hersam, M. C. (2020). Neuromorphic nanoelectronic materials. *Nat. Nanotechnol.*, **15**, pp. 517–528.

5. Crippa, A., Tagliaferri, M. L. V., Rotta, D., De Michielis, M., Mazzeo, G., Fanciulli, M., Wacquez, R., Vinet, M., and Prati, E. (2015). Valley blockade and multielectron spin-valley Kondo effect in silicon. *Phys. Rev. B*, **92**, pp. 035424.

6. Muller, R. S., Kamins, T. I., Chan, M., and Ko, P. K. (1986). *Device Electronics for Integrated Circuits*. Wiley.

7. Bate, R. T., Frazier, G., Frensley, W., and Reed, M. (1989). An overview of nanoelectronics. *Texas Instr. Tech. J.*, **6**, pp. 13–20.

8. Frazier, G. (1988). An ideology for nanoelectronics. In *Concurrent Computations*, pp. 3–21. Springer, Boston, MA.

9. Keyes, R. W. (1988). Miniaturization of electronics and its limits. *IBM J. Res. Develop.*, **32**, pp. 84–88.

10. Keyes, R. W. (1992). The future of solid-state electronics. *Phys. Today*, **45**, pp. 42–48.

11. Keyes, R. W. (1993). The future of the transistor. *Sci. Am.*, **268**, pp. 70–78.

12. Seitz, C. L. and Matisoo, J. (1984). Engineering limits on computer performance. *Phys. Today*, **37**, pp. 38–45.

13. Meindl, J. D. (1995). Low power microelectronics: Retrospect and prospect. *Proc. IEEE*, **83**, pp. 619–635.

14. Stix, G. (1995). Toward point one. *Sci. Am.*, pp. 90–95.

15. Hamilton, D. and Takahashi, D. (1996). Silicon slowdown: Scientists are battling to surmount barriers in microchip advances. *Wall Street Journal*, December 10, A1.

16. Broad, W. (1997). Incredible shrinking transistor nears its ultimate limit: The laws of physics. *New York Times*, February 4. Available at https://www.nytimes.com/1997/02/04/science/incredible-shrinking-transistor-nears-its-ultimate-limit-the-laws-of-physics.html.

17. Rabaey Jan, M., Anantha, C., and Borivoje, N. (1996). *Digital Integrated Circuits: A Design Perspective.* Pearson Education India.

18. Levin, C. (1994). How far out is nanotechnology? *PC Magazine*, **13**, pp. 32.

19. Ono, M., Saito, M., Yoshitomi, T., Fiegna, C., Ohguro, T., and Iwai, H. (1993). Sub-50 nm gate length n-MOSFETs with 10 nm phosphorus source and drain junctions. In *Proceedings of IEEE International Electron Devices Meeting*, Washington, DC, USA, pp. 119–122.

20. Ono, M., Saito, M., Yoshitomi, T., Fiegna, C., Ohguro, T., and Iwai, H. (1995). A 40 nm gate length n-MOSFET. *IEEE Trans. Electron Devices*, **42**, pp. 1822–1830.

21. Han, J., Ferry, D. K., and Newman, P. (1990). Ultra-submicrometer-gate AlGaAs/GaAs HEMTs. *IEEE Electron Device Lett.*, **11**, pp. 209–211.

22. Taur, Y., Buchanan, D. A., Chen, W., Frank, D. J., Ismail, K. E., Lo, S. H., Sai-Halasz, G. A., Viswanathan, R. G., Wann, H. J., Wind, S. J., and Wong, H. S. (1997). CMOS scaling into the nanometer regime. *Proc. IEEE*, **85**, pp. 486–504.

23. Goldhaber-Gordon, D., Montemerlo, M. S., Love, J. C., Opiteck, G. J., and Ellenbogen, J. C. (1997). Overview of nanoelectronic devices. *Proc. IEEE*, **85**, pp. 521–540.

24. Bate, R. T. (1988). The quantum-effect device: Tomorrow's transistor? *Sci. Am.*, **258**, pp. 96–101.

25. Wang, C. T. (1990). *Introduction to Semiconductor Technology: GaAs and Related Compounds.* Wiley-Interscience.

26. Montemerlo, M. S., Love, J. C., Opiteck, G. J., Goldhaber-Gordon, D., and Ellenbogen, J. C. (1996). *Technologies and Designs for Electronic Nanocomputers*. McLean: MITRE.

27. Chakraborty, T. (1999). *Quantum Dots: A Survey of the Properties of Artificial Atoms*. Elsevier.

28. Ninan, A., Dolby Laboratories Licensing Corp. (2015). Quantum dot modulation for displays, US Patent 9,010,949.

29. Li, Y., Hou, X., Dai, X., Yao, Z., Lv, L., Jin, Y., and Peng, X. (2019). Stoichiometry-controlled InP-based quantum dots: Synthesis, photoluminescence, and electroluminescence. *J. Am. Chem. Soc.*, **141**, pp. 6448–6452.

30. Tessier, M. D., Baquero, E. A., Dupont, D., Grigel, V., Bladt, E., Bals, S., Coppel, Y., Hens, Z., Nayral, C., and Delpech, F. (2018). Interfacial oxidation and photoluminescence of InP-based core/shell quantum dots. *Chem. Mater.*, **30**, pp. 6877–6883.

31. Konstantatos, G. and Sargent, E. H. (Eds.) (2013). *Colloidal Quantum Dot Optoelectronics and Photovoltaics*. Cambridge University Press.

32. Atatüre, M., Englund, D., Vamivakas, N., Lee, S. Y., and Wrachtrup, J. (2018). Material platforms for spin-based photonic quantum technologies. *Nat. Rev. Mater.*, **3**, pp. 38–51.

33. Callicó, G. M. (2017). Image sensors go broadband. *Nat. Photon.*, **11**, pp. 332–333.

34. Souissi, R., Bouguila, N., and Labidi, A. (2018). Ethanol sensing properties of sprayed β-In_2S_3 thin films. *Sens. Actu. B: Chem.*, **261**, pp. 522–530.

35. Ho, C. H. (2012). The study of below and above band-edge imperfection states in In_2S_3 solar energy materials. *Phys. B: Cond. Matter*, **407**, pp. 3052–3055.

36. Butanovs, E., Butikova, J., Zolotarjovs, A., and Polyakov, B. (2018). Towards metal chalcogenide nanowire-based colour-sensitive photodetectors. *Optica. Mater.*, **75**, pp. 501–507.

37. Ho, C. H., Lin, M. H., Wang, Y. P., and Huang, Y. S. (2016). Synthesis of In_2S_3 and Ga_2S_3 crystals for oxygen sensing and UV photodetection. *Sens. Act. A: Phys.*, **245**, pp. 119–126.

38. Yu, K., Ng, P., Ouyang, J., Zaman, M. B., Abulrob, A., Baral, T. N., Fatehi, D., Jakubek, Z. J., Kingston, D., Wu, X., and Liu, X. (2013). Low-temperature approach to highly emissive copper indium sulfide colloidal nanocrystals and their bioimaging applications. *ACS Appl. Mater. Inter.*, **5**, pp. 2870–2880.

39. Zhang, X., Zhang, N., Gan, C., Liu, Y., Chen, L., Zhang, C., and Fang, Y. (2019). Synthesis of In_2S_3/UiO-66 hybrid with enhanced photocatalytic activity towards methyl orange and tetracycline hydrochloride degradation under visible-light irradiation. *Mater. Sci. Semicond. Process.*, **91**, pp. 212–221.

40. Bera, A., Mandal, D., Goswami, P. N., Rath, A. K., and Prasad, B. L. (2018). Generic and scalable method for the preparation of monodispersed metal sulfide nanocrystals with tunable optical properties. *Langmuir*, **34**, pp. 5788–5797.

41. Buchmaier, C., Rath, T., Pirolt, F., Knall, A. C., Kaschnitz, P., Glatter, O., Wewerka, K., Hofer, F., Kunert, B., Krenn, K., and Trimmel, G. (2016). Room temperature synthesis of $CuInS_2$ nanocrystals. *RSC Adv.*, **6**, pp. 106120–106129.

42. Wang, P., Xie, J., Xiao, K., Hu, H., Cui, C., Qiang, Y., Lin, P., Arivazhagan, V., Xu, L., Yang, Z., and Yao, Y. (2018). $CH_3NH_3PbBr_3$ quantum dot-induced nucleation for high performance perovskite light-emitting solar cells. *ACS Appl. Mater. Interface.*, **10**, pp. 22320–22328.

43. Li, Yanbo, et al. (2008). High-performance UV detector made of ultra-long ZnO bridging nanowires. *Nanotechnology*, **20**(4), p. 045501.

44. Konstantatos, G. and Sargent, E. H. (2007). PbS colloidal quantum dot photoconductive photodetectors: Transport, traps, and gain. *Applied Physics Letters*, **91**(17), p. 173505.

45. Michalet, X., Pinaud, F. F., Bentolila, L. A., Tsay, J. M., Doose, S. J. J. L., Li, J. J., Sundaresan, G., Wu, A. M., Gambhir, S. S., and Weiss, S. (2005). Quantum dots for live cells, in vivo imaging, and diagnostics. *Science*, **307**, pp. 538–544.

46. Jamieson, T., Bakhshi, R., Petrova, D., Pocock, R., Imani, M., and Seifalian, A. M. (2007). Biological applications of quantum dots. *Biomaterials*, **28**, pp. 4717–4732.

47. William, W. Y., Chang, E., Drezek, R., and Colvin, V. L. (2006). Water-soluble quantum dots for biomedical applications. *Biochem. Biophys. Res. Commun.*, **348**, pp. 781–786.

48. Susumu, K., Mei, B. C., and Mattoussi, H. (2009). Multifunctional ligands based on dihydrolipoic acid and polyethylene glycol to promote biocompatibility of quantum dots. *Nat. Protocols*, **4**, p. 424.

49. Mei, B. C., Susumu, K., Medintz, I. L., and Mattoussi, H. (2009). Polyethylene glycol-based bidentate ligands to enhance quantum dot and gold nanoparticle stability in biological media. *Nat. Protocols*, **4**, p. 412.

50. Tao, W., Ji, X., Xu, X., Islam, M. A., Li, Z., Chen, S., Saw, P. E., Zhang, H., Bharwani, Z., Guo, Z., and Shi, J. (2017). Antimonene quantum dots: Synthesis and application as near-infrared photothermal agents for effective cancer therapy. *Angew. Chemie Inter. Edition*, **56**, pp. 11896–11900.

51. Ouellette, O., Lesage-Landry, A., Scheffel, B., Hoogland, S., García de Arquer, F. P., and Sargent, E. H. (2020). Spatial collection in colloidal quantum dot solar cells. *Ad. Funct. Mater.*, **30**, p. 1908200.

52. Collins, V., Fungura, F., and Zasada, Z. Quantum Dot-Sensitized Solar Cells.

53. Tsu, R. and Esaki, L. (1973). Tunneling in a finite superlattice. *Appl. Phys. Lett.*, **22**, pp. 562–564.

54. Kim, T., Jeong, Y., and Yang, K. (2006). Low-power static frequency divider using an InP-based monolithic RTD/HBT technology. *Electr. Lett.*, **42**, pp. 27–29.

55. Davies, A. G., Burnett, A. D., Fan, W., Linfield, E. H., and Cunningham, J. E. (2008). Terahertz spectroscopy of explosives and drugs. *Mater. Today*, **11**, pp. 18–26.

56. Gao, R., Manut, A. B., Ji, Z., Ma, J., Duan, M., Zhang, J. F., Franco, J., Hatta, S. W. M., Zhang, W. D., Kaczer, B., and Vigar, D. (2017). Reliable time exponents for long term prediction of negative bias temperature instability by extrapolation. *IEEE Trans. Electron Devices*, **64**, pp. 1467–1473.

57. Lin, S., et al. (2020). III-nitrides based resonant tunneling diodes. *J. Phys. D Appl. Phys.*, **53**(25), p. 253002.

58. Rong, T., Yang, L. A., Yang, L., and Hao, Y. (2018). Theoretical investigation into negative differential resistance characteristics of resonant tunneling diodes based on lattice-matched and polarization-matched AlInN/GaN heterostructures. *J. App. Phys.*, **123**, p. 045702.

59. Sattar, A., Fostner, S., and Brown, S. A. (2013). Quantized conductance and switching in percolating nanoparticle films. *Phys. Rev. Lett.*, **111**, p. 136808.

60. Fried, J. P., Bian, X., Swett, J. L., Kravchenko, I. I., Briggs, G. A. D., and Mol, J. A. (2020). Large amplitude charge noise and random telegraph fluctuations in room-temperature graphene single-electron transistors. *Nanoscale*, **12**, pp. 871–876.

61. Arjmandi-Tash, H., Belyaeva, L. A., and Schneider, G. F. (2016). Single molecule detection with graphene and other two-dimensional materials: Nanopores and beyond. *Chem. Soc. Rev.*, **45**, pp. 476–493.

62. Puczkarski, P., Swett, J. L., and Mol, J. A. (2017). Graphene nanoelectrodes for biomolecular sensing. *J. Mater. Res.*, **32**, pp. 3002–3010.

63. Chen, Q., Sun, J., Li, P., Hod, I., Moghadam, P. Z., Kean, Z. S., Snurr, R. Q., Hupp, J. T., Farha, O. K., and Stoddart, J. F. (2016). A redox-active bistable molecular switch mounted inside a metal–organic framework. *J. Am. Chem. Soc.*, **138**, pp. 14242–14245.

64. Zhou, C., Li, X., Gong, Z., Jia, C., Lin, Y., Gu, C., He, G., Zhong, Y., Yang, J., and Guo, X. (2018). Direct observation of single-molecule hydrogen-bond dynamics with single-bond resolution. *Nat. Commun.*, **9**, pp. 1–9.

65. Gotz, G., Steele, G. A., Vos, W. J., and Kouwenhoven, L. P. (2008). Real time electron tunneling and pulse spectroscopy in carbon nanotube quantum dots. *Nano Lett.*, **8**, pp. 4039–4042.

66. Khademhosseini, V., Dideban, D., Ahmadi, M., and Ismail, R. (2020). Current analysis of single electron transistor based on graphene double quantum dots. *ECS J. Solid State Sci. Technol.*, **9**, p. 021003.

67. Dresselhaus, G. and Riichiro, S. (1998). *Physical Properties of Carbon Nanotubes*. World Scientific.

68. Soh, H. T., Quate, C. F., Morpurgo, A. F., Marcus, C. M., Kong, J., and Dai, H. (1999). Integrated nanotube circuits: Controlled growth and ohmic contacting of single-walled carbon nanotubes. *Appl. Phys. Lett.*, **75**, pp. 627–629.

Multiple Choice Questions

1. The prefix nano comes from
   (a) French word meaning billion
   (b) Greek word meaning dwarf
   (c) Spanish word meaning particle
   (d) Latin word meaning invisible

2. What exactly is a quantum dot?
   (a) A semiconductor nanostructure that confines the motion of a conduction band electrons, valence band holes, or excitons in all three spatial directions.
   (b) Fiction for endpoints of worm holes.
   (c) Explain the sports that appear in the electron microscopy smaller than 1 nm image of a nanostructure.
   (d) The sharpest possible tip of an atomic force microscope.

3. Nanoparticle targets the rare _____causing cells and remove them from blood.

   (a) Tumor (b) Fever

   (c) Infections (d) Cold

4. In which field nanoparticles are used with silica-coated iron oxide?

   (a) Magnetic application

   (b) Electronics

   (c) Medical diagnosis

   (d) Structure and mechanical materials

5. What is the purpose of increasing the wafer diameter from 200 to 300 mm?

   (a) The price of the 300 mm wafer is low

   (b) Easier to fabricate

   (c) To produce more silicon devices from a single wafer

   (d) To increase the size of a die

6. What is n-type semiconductor (SC)?

   (a) Silicon semiconductor without impurities

   (b) Silicon semiconductor with impurities from column V of Mendeleev's periodic table

   (c) Silicon semiconductor with impurities from columns III and V of Mendeleev's periodic table

   (d) Silicon semiconductor with impurities from group III of Mendeleev's periodic table

7. A semiconductor diode (SC) is made of

   (a) The junction of two n-type SCs

   (b) The junction of n- and p-type SCs

   (c) The junction of two non-doped SCs

   (d) Junction of two p-type SCs

8. Which concept makes a fundamental difference between the dynamics of electromagnetic field (Maxwell equation) and Schrodinger equation?

   (a) Wave nature (b) Momentum

   (c) Potential (d) Mass

9. Moore's law is due to

   (a) Price of the silicon reduces over time

(b) Large wafers

(c) Transistor size scaling

(d) Modern circuits architecture

10. The size of atom is nearly

(a) 0.01 nm

(b) 0.1 nm

(c) 1 nm

(d) 10 nm

Answer Key

1. (b)   2. (a)   3. (a)   4. (c)   5. (c)   6. (b)   7. (b)
8. (d)   9. (c)   10. (b)

Short Answer Questions

1. What is nanoelectronics?
2. What is Moore's law?
3. What is MOSFET?
4. What is island?
5. What are 13–14 and 14–15 semiconductor devices?
6. What is exciton?
7. Quantum dots are considered artificial atoms. Justify.
8. What are the dimensions of quantum dots?

Long Answer Questions

1. Explain the Coulomb blockade effect?
2. What are quantum dots? Give the band theory to explain the working of quantum dots.
3. What are the obstacles to the miniaturization of electronic devices?
4. Give the advantages of using quantum dots in solar cells.
5. What are the advantages and disadvantages of QDLED over OLED?

Index

acoustic wave 254, 255
active ingredient 134, 135, 163, 344
active site 83, 85, 86, 90, 91, 93, 94
additives 340, 351, 353, 406, 414, 423, 425
adsorption 18, 75, 100, 252, 378, 391
  fatty acid 165
  oligonucleotide 264
aerospace 206, 208, 268
AFM *see* atomic force microscopy
agent 15, 16, 25, 50, 60, 71, 79, 128, 274
  active 59, 134, 139
  antibacterial 337
  anti-browning 343
  anticaking 159
  antimicrobial 159, 274, 335, 337, 339, 341, 344, 348, 407, 416
  capping 314, 381, 382
  carcinogenic 351
  chemotherapeutic 48
  conducting 83
  dispersing 11
  drying 337
  energy-distributing 206
  gelating 341
  infectious 275
  nano-priming 131, 303
  oxidizing 78, 479
  particle-active 59
  pest-control 134
  photocatalytic disinfecting 337
  photosensitizing 60
  screening 7
  staining 215

  therapeutic 8, 158
  viscosifying 420
alcohol 12, 222, 273, 479
alumina-toughened zirconia (ATZ) 224, 225
amalgamation 217, 347, 369, 444, 447
amplification 138, 275, 278, 501
amplifier 252, 259, 500, 501
analytes 157, 250, 251, 254, 255, 258, 259, 267–271, 275–277, 349, 378
angiography 443, 444
anode 72, 81–83, 86, 88, 93, 95, 98, 506
antibody 21, 22, 44, 48, 138, 162, 185, 187, 249, 251, 266, 269, 270, 272, 277
antimicrobial packaging 407, 415, 416
antioxidant 151, 154, 156, 163, 266, 335, 343, 345, 348, 414, 424
application
  aerospace 220
  analytical 256
  automotive 220, 473
  bioimaging 4
  biological 266
  catalyst support 89
  channelized scientific 392
  clinical 2
  commercial 277
  cutting-edge 211
  dental implant 223
  diagnostic 59, 184, 268, 510
  electronic 73, 227, 228, 462, 471